KU-186-398

MOLECULES OF EMOTION

Why You Feel the Way You Feel

CANDACE B. PERT, PH.D.

WITH A FOREWORD BY
DEEPAK CHOPRA, M.D.

POCKET
BOOKS

LONDON · SYDNEY · NEW YORK · TOKYO · SINGAPORE · TORONTO

First published in the USA by Scribner, 1997
First published in Great Britain by Simon & Schuster UK Ltd, 1998
This edition first published by Pocket Books, 1999
An imprint of Simon & Schuster UK Ltd
A Viacom Company

The names of some of the individuals in this book have been changed

1 3 5 7 9 10 8 6 4 2

Simon & Schuster UK Ltd
Africa House
64-78 Kingsway
London WC2B 6AH

Simon & Schuster Australia
Sydney

A CIP catalogue record for this book is available from the British Library

ISBN 0-671-03397-2

Printed and bound in Great Britain by Caledonian International Book
Manufacturing, Glasgow

THIS BOOK IS DEDICATED TO THE CORE OF
MY EMOTIONAL LIFE, TO THOSE CLOSEST
WHO NEVER CEASED THEIR GENEROUS SUPPORT OF MY EFFORT:

my mother,
Mildred Ruth Rosenberg Beebe;
my husband,
Michael Roland Ruff;
our children,
Evan Taaved Pert, Vanessa Carey Pert,
and Brandon Mulford Pert;
and my sisters,
Wynne Ilene Beebe and Deane Robin Beebe Fitzgerald.

ACKNOWLEDGMENTS

It seems as though my book project has been in development for an eternity. Its completion had to await (or did it cause?) certain powerful transformations in the universe, or at least in my own bodymind. For all this, I am most grateful to God, who manifested Herself in the form of devoted professional angels whose participation was essential for bringing my book into the world.

I will be thankful forever for the help of Susan Moldow, whose enthusiasm and brilliant vision provided constant inspiration and energized my effort; Nancy Griffith Marriott, old friend, consciousness writer and consultant, who helped me set my thoughts down in an understandable way; Beth Rashbaum, who pushed me to greater heights of clarity and cohesiveness; Muriel Nellis, my fairy godmother, the best, most marvelous agent in the world; Jane Roberts, for insightful, constant aid of all kinds; and Bernardo Issel, for seminal library assistance and his angelic manifestations during times of need.

CONTENTS

FOREWORD

I have admired Candace Pert and her work for many years. In fact, I can remember the first time I heard her speak and my delight at realizing: Finally, here is a Western scientist who has done the work to explain the unity of matter and spirit, body and soul!

In exploring how the mind, spirit, and emotions are unified with the physical body in one intelligent system, what I call "the field of intelligence," Candace has taken a giant step toward shattering some cherished beliefs held sacred by Western scientists for more than two centuries. Her pioneering research has demonstrated how our internal chemicals, the neuropeptides and their receptors, are the actual biological underpinnings of our awareness, manifesting themselves as our emotions, beliefs, and expectations, and profoundly influencing how we respond to and experience our world.

Her research has provided evidence of the biochemical basis for awareness and consciousness, validating what Eastern philosophers, shamans, rishis, and alternative practitioners have known and practiced for centuries. The body is not a mindless machine; the body and mind are one.

I have lectured and written about the important role of perception and awareness in health and longevity—how awareness can actually transform matter, create an entirely new body. I also have said that the mind is nonlocal. Now Candace provides us with a vivid scientific picture of these truths. She shows us that our biochemical messengers act with intelligence by communicating information, orchestrating a vast complex of conscious and unconscious activities at any one moment. This information transfer takes place over a network linking all of our systems and organs, engaging all of our molecules of emotion, as the means of communication. What we see is an image of a "mobile brain"—one that moves throughout our entire body, located in all places at once and not just in the head. This bodywide information network is ever changing and dynamic, infinitely flexible. It is one gigantic loop, directing and admitting information simultaneously, intelligently guiding what we call life.

There is a revolution taking form that is significantly influencing how the Western medical community views health and disease. Candace Pert's contribution to this revolution is undeniable; and her professional integrity in the pursuit of scientific truth, wherever it had to take her, regardless of its personal or professional cost, underscores the feminine, intuitive potential of science at its best.

—Deepak Chopra, M.D.
La Jolla, California

1

THE RECEPTOR REVOLUTION: AN INTRODUCTORY LECTURE

SCIENTISTS, by nature, are not creatures who commonly seek out or enjoy the public spotlight. Our training predisposes us to avoid any kind of overt behavior that might encourage two-way communication with the masses. Instead, we are content to pursue our truth in windowless laboratories, accountable only to members of our highly exclusive club. And although presenting papers at professional meetings is encouraged, in fact required, it's rare to find one of us holding sway to standing-room-only crowds, laughing, telling jokes, and giving away trade secrets.

Even though I am a long-standing club member and bona fide insider myself, I cannot say that it has been my trademark to follow the rules. Acting as if programmed by some errant gene, I do what most scientists abhor: I seek to inform, to educate, and inspire all manner of people, from lay to professional. I try to make available and interpret the latest and most up-to-date knowledge that I and my fellow scientists are discovering, information that is practical, that can change people's lives. In the process, I virtually cross over into another dimension, where the leading edge of biomolecular medicine becomes accessible to anyone who wants to hear about it.

This mission places me in the public spotlight quite often. A dozen times a year, I am invited to address groups at various institutions, and so, when not engaged in my work at Georgetown University School of Medicine, where I am a research professor in the Department of Biophysics and Physiology, I go shuttling from coast to coast, sometimes even crossing the great blue waters. It was never my plan to become a scientific performer, to act as a mouthpiece for educating the public as well as practitioners in the alternative health movement, so wed was I for most of my career to the mainstream world of the lab and my research. But it's been a natural evolution, and I am now at home in my new role. The result of translating my scientific ideas into the vernacular seems to have been that my life in science and my personal life have transformed each other, so that I have become expanded and enriched in myriad unexpected ways by the discoveries I've made, the science I've done, and the meaning I continue to uncover.

Writing this book was an attempt to put down on paper, in a much more detailed and usable form, the material I've been presenting in lectures. My goal in writing, as in speaking, was twofold: to explain the science underlying the new bodymind medicine, and to give enough practical information about the implications of that science, and about the therapies and practitioners embodying it, to enable my readers to make the best possible choices about their personal health and well-being. Perhaps my journey, intellectual as well as spiritual, can help other people on their paths. And now—on with the "lecture"!

ARRIVAL

Whenever possible I try to arrive at the lecture hall early, before the members of the audience take their seats. I get a thrill out of sitting in the empty room, when all is quiet and there exists a state of pure potentiality in which anything can happen. The sound of the doors swinging open, the muffled voices of the crowd as they file slowly into the room, the clinking of water glasses and screeching of chairs—all of this creates a delightful cacophony, music to my ears, the overture for what is to come.

I watch the people as they move toward their seats, finding their places, chatting with a neighbor, and getting comfortable, preparing themselves to be informed, hopefully entertained, unaware that my goal is to do more: to reveal, to inspire, to uplift, perhaps even to change lives.

"Who's this Candace Pert?" I may ask, retaining my anonymity as I playfully engage the person now seated next to me. "Is she supposed to be any good?" The response is sometimes informative and always amusing, allowing me a brief entry into the thoughts and expectations of those I am about to address. I nod knowingly in response and pretend to arrange myself more comfortably, more attentively.

I often find myself addressing very mixed audiences. Depending on the nature of my host's organization, the crowd is either weighted toward mainstream professionals—doctors, nurses, and scientific researchers— or toward alternative practitioners—chiropractors, energy healers, massage therapists, and other curious participants—but frequently includes members from both camps in a blend that can best be described as the Establishment meets the New Paradigm. This sort of composition is very different from the more homogeneous audiences present at the hundreds of talks I've given over the past twenty-four years to my fellow scientists, colleagues, and peers. For them, I deliver my more technical remarks in the language of the club, not needing to translate the code we all understand. I still address such groups, making the yearly round of scientific meetings, but now I also venture into a foreign land, where few of my fellow scientists dare—or wish—to go.

Breathing deeply for a moment or two, I relax into my seat and close my eyes. My mind clears as I offer a brief prayer to enter a more receptive state. Calling on an intuitive sense of my audience's expectations and mood, I can feel the wall coming down, the imaginary wall that separates us, scientist from lay person; the expert, the authority, from those who do not know—a wall I personally stopped believing in some time ago.

THE AUDIENCE

As the room fills, I can feel the excitement building. When I open my eyes and glance around at one of these mixed crowds, I notice first that, in marked contrast to the more scientific gatherings, there are usually large numbers of women present. It still surprises me to see so many of them, dressed beautifully in their flowing California-style robes of many colors. I am always stunned by the many shades of purple in their dress, more shades than I ever knew existed! Then, looking beyond the surface, I try to assess the various components of my audience and what might have motivated them to come today.

My attention goes first to the doctors and other medical professionals, whose contingent is almost always dominated by males. The men sit erect in their well-tailored dark suits and crisp white shirts, while nearby their female counterparts look officiously around, checking the room for the faces of their colleagues.

Scattered more sparsely throughout the room are the neophytes, earnest young men and women with packs on their backs and dreams in their eyes. Their posture is perky and eager, revealing their sincerity and also their uncertainty about what they want or where they are going.

As the room settles and voices are hushed to a low din, I wonder: What do all these people expect me to tell them? What do they want to know, what are they hoping for?

Some are here because they saw me on Bill Moyers's PBS special *Healing and the Mind*, a program that also included segments with Dean Ornish, Jon Kabat-Zinn, Naomi Remen, and a number of the other doctors, scientists, and therapists who are trying to make the same mind-body connections that have become my life's work. Being interviewed by such a well-informed, receptive journalist made it possible for me to speak of the molecules of mind and emotion with a passion and humor not ordinarily associated with medical research scientists. I tried to make it easy for a television audience to understand the exciting world of biomedicine, molecular theory, and psychoneuroimmunology, revealing information usually shrouded by an impenetrable language, letting them know that they have a stake in understanding this body of knowledge, because it could give them the power to make a difference in the state of their own health.

The physicians, nurses, health care professionals—what brings them out? Have they touched on some new situation that their current knowledge cannot explain? Many of them know me as a former chief of brain biochemistry who toiled at the National Institutes of Health for thirteen years, demonstrating and mapping biochemicals I later came to call the physiological correlates of emotion. Some may know that I left the National Institutes of Health when I developed a powerful new drug for the treatment of AIDS and couldn't get the government interested. All of them seem to be aware that science marches on, and that much of what they were taught in medical school twenty years ago, even ten years ago, is no longer current, even applicable. They know that my work is in a breaking field—no less a chronicler of contemporary culture than Tom Wolfe himself has pronounced neuroscience the "hottest field in the academic world" in a recent issue of *Forbes*—and that it's just now finding its way into medical schools around the world.

Then there are the many massage therapists, acupuncturists, chiropractors—the so-called alternative medicine practitioners who offer their patients approaches that are not part of the mainstream. I'm aware that these people have been marginalized for years, rarely taken seriously by the powers that be—the medical schools, insurance companies, the American Medical Association, the Food and Drug Administration—

although it is well documented that the public spends billions yearly on their services. Later, in the Q&A sessions that follow the talks, they tell me they believe I have done the research that will lead to the validation of their theories, their beliefs. They have read about my theory of emotions, about how I have postulated a biochemical link between the mind and body, a new concept of the human organism as a communication network that redefines health and disease, empowering individuals with new responsibility, more control in their lives.

The philosophers, the seekers, they're here too. Some are very silent—listeners, not talkers—these pale, earnest young men and women who tell me after the lecture that they've been traveling in India or living in Asia. They see my work as proof of what their gurus and masters have long been saying, and they want more answers, perhaps about the meaning of it all. Maybe they've heard me quoted as the scientist who said "God is a neuropeptide." They know I'm not afraid to use what most scientists consider a four-letter word—*soul*—in my talks, and they want me to address their spiritual questions today.

Many come simply because they are curious. Perhaps they've heard of my reputation as a young graduate student who laid the foundation for the discovery of endorphins, the body's own pain suppressors and ecstasy inducers. Or they may know me as the young woman who was passed over for a prestigious pre–Nobel Prize and dared to challenge her mentor for the recognition she felt she deserved. They may recall how the resulting front-page controversy exposed a system that was sexist and unjust at its core, and caused a shake-up that embarrassed a medical dynasty.

Others are here because they need to have hope. The sick, the wheelchair-bound, I see them positioned on the aisles, near the doors. They know I've been on the cutting edge with my research, crossing disciplines and researching for breakthroughs in cancer, AIDS, mental illness. I always feel a little nervous when I see them sitting in my audience. Are they expecting me to deliver their miracle cure like a preacher at a revival meeting? *Hope* is a dirty, rarely uttered word in the circles I frequent, and it still tugs uncomfortably at my self-image as a scientist. To think I'm being viewed as a healer—God forbid, a faith healer! Yet I can't ignore the expressions of desperation and suffering that I see on their faces. Information. Yes, at least I can give them that, something they can use in seeking alternatives, these people for whom mainstream medicine offers no further answers, no treatment, no hope.

Regardless of their profession, orientation, or expectations emotional or intellectual, I've come to believe that most of the lay people who find their way to my lectures are hoping to hear science demystified, de-jargonized, described in terms they can understand. They want to be more in control of their own health and to learn more about what is going on in their own bodies, and they have been deeply disappointed, disillusioned by the failure of science to deliver on its promises to provide cures for the major diseases. Now they want to take back some power into their own hands, and they need to know about what the latest scientific discoveries mean for obtaining optimal health.

Perhaps you, my reader, see yourself in one or more of the groups described above. If so, I hope for your sake, as I always hope for the members of my audiences, that some part of the information presented in this book will make a difference in your life.

TAKING THE STAGE

A sudden hush descends on the room, catching me off guard, and my head turns as I glimpse a figure walking slowly across the stage toward the spotlit podium. What follows is generally a lavish detailing of my list of accomplishments. I feel genuinely moved by the appreciation expressed by my host or hostess, but always a bit embarrassed and undeserving of such flattering words.

Over the years, I've learned to keep my ego reigned in by saying a quiet blessing during these introductory remarks. I ask that I not be cowed by my mission, nor swept up in it. I remind myself that, in spite of the spotlight I am about to step into, first and always I am a scientist, a seeker of the truth—not a rock star! I silently vow that I won't let any of this go to my head—although that could easily happen, and did happen occasionally at one time.

At last I hear my name and rise from my chair to begin the long walk onto the stage. I remember to breathe deeply as I pass the front row and feel all eyes in the room turn to focus on me. A few whispered words reach my ears as I move along: "There she is! Is that her? She doesn't look like a scientist!"

What did they expect? I wonder with an inward chuckle. I am still a woman, a wife, and a mother. Don't I fit their pictures of the scientist? Of course, they have their own ideas, and many of them fit the standard cliché of the conservatively dressed, intense-looking, usually male scientist. Not too long ago, I wore those serious little boxy suits, the dress-for-success uniform, conforming to the more buttoned-down image people

expect. But now, my own transformation is boldly reflected ɪ̄
present myself, an image that better matches my message theʂ
keeping with the evolution of my scientific ideas, my dress has eᵥ
that I now look more like the ladies in the flowing robes, my ɪ   ʂ
looser and more colorful, more comfortable, even more purple! These
days I dare to be more outrageous, although those who know me insist
that outrageousness has always been the hallmark of my personality,
however submerged I've tried to keep it at times to survive.

Taking my place at the podium, I wait while the technicians fumble
with my mike and make last-minute adjustments to the projection
screen at my side. As I look out on the sea of upturned faces, I am struck
by how perfectly still people sit. I know they won't move until I crack a
joke, giving them permission to enjoy themselves and explode in laughter, animating the room and filling it with energy.

My audience is ready and so am I—hundreds, sometimes thousands
of people are seated before me waiting for my words. I take one last
minute to focus inwardly on my mission: to tell the truth about the facts
that were discovered by my colleagues and myself. First and foremost, I
am a truth-seeker. My intention is to provide an understanding of the
metaphors that express a new paradigm, metaphors that capture how
inextricably united the body and the mind really are, and the role the
emotions play in health and disease.

The house lights dim as I clear my throat and my first slide comes up
on the screen.

SETTING THE TONE

There is something incredibly intoxicating about standing in front of a
huge room full of people who are all laughing uproariously. I have
become quite addicted to this experience, ever since 1977 when I gave a
lecture to the National Endocrine Society and accidentally brought
down the house with a joke that was intended to cover a mistake I'd
made. Now I don't waste any time. I start right off with a cartoon that
never fails to elicit hearty, if sometimes nervous, laughter.

My first slide looks like this:

I use this joke to make the point that as a culture we are all in denial about the importance of psychosomatic causes of illness. Break the word *psychosomatic* down into its parts, and it becomes *psyche*, meaning mind or soul, and *soma*, meaning body. Though the fact that they are fused into one word suggests some kind of connection between the two, that connection is anathema in much of our culture. For many of us, and certainly for most of the medical establishment, bringing the mind too close to the body threatens the legitimacy of any particular illness, suggesting it may be imaginary, unreal, *unscientific*.

If psychological contributions to physical health and disease are viewed with suspicion, the suggestion that the soul—the literal translation of *psyche*—might matter is considered downright absurd. For now we are getting into the mystical realm, where scientists have been officially forbidden to tread ever since the seventeenth century. It was then that René Descartes, the philosopher and founding father of modern medicine, was forced to make a turf deal with the Pope in order to get the human bodies he needed for dissection. Descartes agreed he wouldn't have anything to do with the soul, the mind, or the emotions— those aspects of human experience under the virtually exclusive jurisdiction of the church at the time—if he could claim the physical realm as his own. Alas, this bargain set the tone and direction for Western science over the next two centuries, dividing human experience into two distinct and separate spheres that could never overlap, creating the unbalanced situation that is mainstream science as we know it today.

But much of that is now changing. A growing number of scientists recognize that we are in the midst of a scientific revolution, a major paradigm shift with tremendous implications for how we deal with health and disease. The Cartesian era, as Western philosophical thought since Descartes has been known, has been dominated by reductionist methodology, which attempts to understand life by examining the tiniest pieces of it, and then extrapolating from those pieces to overarching surmises about the whole. Reductionist Cartesian thought is now in the process of adding something very new and exciting—and holistic.

As I've watched as well as participated in this process, I've come to believe that virtually all illness, if not psychosomatic in foundation, has a definite psychosomatic component. Recent technological innovations have allowed us to examine the molecular basis of the emotions, and to begin to understand how the molecules of our emotions share intimate connections with, and are indeed inseparable from, our physiology. It is the emotions, I have come to see, that link mind and body. This more holistic approach complements the reductionist view, expanding it rather than replacing it, and offers a new way to think about health and disease—not just for us scientists, but for the lay person also.

In my talks, I show how the molecules of emotion run every system in our body, and how this communication system is in effect a demonstration of the bodymind's intelligence, an intelligence wise enough to seek wellness, and one that can potentially keep us healthy and disease-free without the modern high-tech medical intervention we now rely on. In this book I've tried to give pointers about how to tap into that intelligence, and, in the Appendix, I've provided a listing of organizations that practice various aspects of bodymind medicine, so that those of you who are interested can get some guidance on getting the most out of that intelligence, allowing it to do its job without interference. The Appendix also contains some basic tips for healthful living, distilled from my own experience.

SHIFT HAPPENS! The Ptolemaic earth at the center of the universe can give way to the Copernican sun-centered theory—but not without considerable resistance. Witness Galileo, who was brought before the Inquisition for his role in promulgating that theory over a century after it was first proposed! Or ask Jesse Roth, who in the 1980s found insulin not just in the brain but in tiny one-celled animals outside the human body. This gave the reigning medical paradigm a good shake, because everyone "knew" that you needed a pancreas to make insulin! In spite of his eminence as clinical director for the National Institutes of Health, Dr. Roth couldn't get his papers published in a single reputable scientific journal for quite a while. The reviewers sent them back with comments such as: "This is preposterous, you must not be washing your test tubes well enough." Jesse retaliated by using new test tubes and repeating his results often enough so that other researchers, intrigued by his findings, began doing similar experiments and reporting similar results.

Jesse's story illustrates one of the paradoxes of scientific progress: Truly original, boundary-breaking ideas are rarely welcomed at first, no matter who proposes them. Protecting the prevailing paradigm, science moves slowly, because it doesn't want to make mistakes. Consequently, genuinely new and important ideas are often subjected to nitpickingly intense scrutiny, if not outright rejection and revulsion, and getting them published becomes a Sisyphean labor. But if the ideas are correct, eventually they will prevail. It may take, as in the case of the new discipline of psychoneuroimmunology, a good decade, or it may take much longer. But, eventually, the new view becomes the status quo, and ideas that were rejected as madness will appear in the popular press, often touted by the very critics who did so much to impede their acceptance. Which is what is happening today as a new paradigm comes into being.

And not a moment too soon as far as the holistic/alternative health crowd is concerned. They've been disgusted with the reigning medical model for years and have, in fact, been working actively to overturn it. It's largely through their efforts that such formerly dismissed techniques as acupuncture and hypnosis have gained the credibility they now have. But even when I talk with the average health-conscious consumer, people who have no ideological animus one way or the other, I'm always astonished at how deep their anger at our present health system is. It's obvious the public is catching on to the fact that they're the ones paying monstrous health care bills for often worthless procedures to remedy conditions that could have been prevented in the first place.

IN ORDER TO grasp the enormity of this revolution, you have to first understand some of the fundamentals of biomolecular medicine, which is what I like to explain at the beginning of my talks. How many of us can close our eyes and picture or define a receptor, or a protein, or a peptide? These are the basic components that make up our bodies and minds, yet to the average person, they are as exotic and remote from everyday experience as the Abominable Snowman. If we're to understand what role our emotions may play in our health, then understanding the molecular-cellular domain is a crucial first step. I also like to provide some historical context to help people understand the impact of the recent discoveries. It's a version of one of those lectures I'm putting on the page here to provide a broad overview of my work, the basic science that makes it all decipherable, and fun.

But I also have a story to tell, one that is more personal than scientific, even though parts of it do make their way into some of my more informal public lectures. The narrative of how I was transformed by the science I did, and how the science I did was inspired and influenced by my growth as a human being, especially by my experience as a woman, is as informative, I believe, as the facts of my scientific adventures, and equally as important. For this reason, I have included my personal narrative in this book, sandwiched in between sections of my lecture, where

I hope it provides a perspective that enlightens as it reveals the human story behind the molecules of emotion. As befitting my own evolution, the personal and the scientific do eventually intertwine as my story progresses, underscoring the fact that science is a very human pursuit and cannot be truly appreciated if it appears as a cold and emotionless abstraction. Emotions affect how we do science as well as how we stay healthy or become ill.

THE BASICS

And now on with the science!

The first component of the molecules of emotion is a molecule found on the surface of cells in body and brain called the *opiate receptor.* It was my discovery of the opiate receptor that launched my career as a bench scientist in the early 1970s, when I found a way to measure it and thereby prove its existence.

Measurement! It is the very foundation of the modern scientific method, the means by which the material world is admitted into existence. Unless we can measure something, science won't concede it exists, which is why science refuses to deal with such "nonthings" as the emotions, the mind, the soul, or the spirit.

But what is this former nonthing known as a receptor? At the time I was getting started, a receptor was mostly an idea, a hypothetical site believed to be located somewhere in the cells of all living things. The scientists who most needed to believe in it were the pharmacologists (those who study and invent drugs) because it was the only way they knew to explain the action of drugs in the organism. Dating back to the early twentieth century, pharmacologists believed that for drugs to act in the body they must first attach themselves to something in it. The term *receptor* was used to refer to this hypothetical body component, which allowed the drug to attach itself and thereby in some mysterious way to initiate a cascade of physiological changes. "No drug acts unless it is fixed," said Paul Ehrlich, the first modern pharmacologist, summarizing what he believed to be true, even though he had no real evidence. (Only he said it in Latin to emphasize the profundity of the concept.)

Now we know that that component, the receptor, is a single molecule, perhaps the most elegant, rare, and complicated kind of molecule there is. A *molecule* is the tiniest possible piece of a substance that can still be identified as that substance. Each and every molecule of any given substance is composed of the smallest units of matter—atoms such as carbon and hydrogen and nitrogen—which are bonded together in a configuration specific to that substance, which can be expressed as a chemical formula, or, more informatively, drawn as a diagram.

Invisible forces attract one molecule to another, so that the molecules cohere into an identifiable substance. These invisible forces of attraction can be overcome if enough energy is applied to the substance. For example, heat energy will melt ice crystals, turning them into water, which will then vaporize into steam as its molecules move so fast, with so much energy, that they break loose of each other and fly apart. But the chemical formula remains the same for each state—in this case H_2O, two hydrogen atoms bonded to one oxygen atom—whether that state is an icy solid, a watery liquid, or a colorless vapor.

In contrast to the small, rigid water molecule, which weighs only 18 units in molecular weight, the larger receptor molecule weighs upwards of 50,000 units. Unlike the frozen water molecules that melt or turn into a gas when energy is applied, the more flexible receptor molecules respond to energy and chemical cues by vibrating. They wiggle, shimmy, and even hum as they bend and change from one shape to another, often moving back and forth between two or three favored shapes, or conformations. In the organism they are always found attached to a cell, floating on the cell surface's oily outer boundary, or membrane. Think of them as lily pads floating on the surface of a pond, and, like lilies, receptors have roots enmeshed in the fluid membrane snaking back and forth across it several times and reaching deep into the interior of the cell.

The receptors are molecules, as I have said, and are made up of proteins, tiny amino acids strung together in crumpled chains, looking something like beaded necklaces that have folded in on themselves. If you were to assign a different color to each of the receptors that scientists have identified, the average cell surface would appear as a multicolored mosaic of at least seventy different hues—50,000 of one type of receptor, 10,000 of another, 100,000 of a third, and so forth. A typical neuron (nerve cell) may have millions of receptors on its surface. Molecular biologists can isolate these receptors, determine their molecular weight, and eventually crack their chemical structure, which means identifying the exact sequence of amino acids that makes up the receptor molecule. Using the biomolecular techniques available today, scientists are able to isolate and sequence scores of new receptors, meaning that their complete chemical structure can now be diagrammed.

Basically, receptors function as sensing molecules—scanners. Just as our eyes, ears, nose, tongue, fingers, and skin act as sense organs, so, too, do the receptors, only on a cellular level. They hover in the membranes of your cells, dancing and vibrating, waiting to pick up messages carried by other vibrating little creatures, also made out of amino acids, which come cruising along—*diffusing* is the technical word—through the fluids surrounding each cell. We like to describe these receptors as "keyholes," although that is not an altogether precise term for something that is constantly moving, dancing in a rhythmic, vibratory way.

All receptors are proteins, as I have said. And they cluster in the cellular membrane waiting for the right chemical keys to swim up to them through the extracellular fluid and to mount them by fitting into their keyholes—a process known as *binding*.

Binding. It's sex on a molecular level!

And what is this chemical key that docks onto the receptor and causes it to dance and sway? The responsible element is called a *ligand*. This is the chemical key that binds to the receptor, entering it like a key in a keyhole, creating a disturbance to tickle the molecule into rearranging itself, changing its shape until—click!—information enters the cell.

THE TIES THAT BIND

If receptors are the first components of the molecules of emotion, then ligands are the second. The word *ligand* comes from the Latin *ligare*, "that which binds," sharing its origin with the word re*lig*ion.

Ligand is the term used for any natural or manmade substance that binds selectively to its own specific receptor on the surface of a cell. The ligand bumps onto the receptor and slips off, bumps back on, slips back off again. The ligand bumping *on* is what we call the binding, and in the process, the ligand transfers a message via its molecular properties to the receptor.

Though a key fitting into a lock is the standard image, a more dynamic description of this process might be two voices—ligand and receptor—striking the same note and producing a vibration that rings a doorbell to open the doorway to the cell. What happens next is quite amazing. The receptor, having received a message, transmits it from the surface of the cell deep into the cell's interior, where the message can change the state of the cell dramatically. A chain reaction of biochemical events is initiated as tiny machines roar into action and, directed by the message of the ligand, begin any number of activities—manufacturing new proteins, making decisions about cell division, opening or closing ion channels, adding or subtracting energetic chemical groups like the phosphates—to name just a few. In short, the life of the cell, what it is up to at any moment, is determined by which receptors are on its surface, and whether those receptors are occupied by ligands or not. On a more global scale, these minute physiological phenomena at the cellular level can translate to large changes in behavior, physical activity, even mood.

And how is all this activity organized, considering it is going on in all parts of the body and brain simultaneously? As the ligands drift by in the stream of fluid surrounding every cell, only those ligands that have molecules in exactly the right shape can bind to a particular kind of receptor. The process of binding is very selective, very specific! In fact, we can say that binding occurs as a result of *receptor specificity*, meaning the receptor ignores all but the particular ligand that's made to fit it. The opiate receptor, for instance, can "receive" only those ligands that are members of the opiate group, like endorphins, morphine, or heroin. The Valium receptor can attach only to Valium and Valium-like peptides. It is this specificity of the receptors that allows for a complex system of organization and insures that everything gets to where it's supposed to be going.

Ligands are generally much smaller molecules than the receptors they bind to, and they are divided into three chemical types. The first type of ligand comprises the classical *neurotransmitters*, which are small molecules with such unwieldy names as acetylcholine, norepinephrine, dopamine, histamine, glycine, GABA, and serotonin. These are the smallest, simplest of molecules, generally made in the brain to carry information across the gap, or synapse, between one neuron and the next. Many start out as simple amino acids, the building blocks of protein, and then get a few atoms added here and there. A few neurotransmitters are unmodified amino acids.

A second category of ligands is made up of *steroids*, which include the sex hormones testosterone, progesterone, and estrogen. All steroids start out as cholesterol, which gets transformed by a series of biochemical steps into a specific kind of hormone. For example, enzymes in the gonads—the testes for males, the ovaries for females—change the cholesterol into the sex hormones, while other enzymes convert cholesterol into other kinds of steroid hormones, such as cortisol, which are secreted by the outer layer of the adrenal glands under stress.

I've saved the best for last! My favorite category of ligands by far, and the largest, constituting perhaps 95 percent of them all, are the *peptides*. As we shall see, these chemicals play a wide role in regulating practically all life processes, and are indeed the other half of the equation of what I call the molecules of emotion. Like receptors, peptides are made up of strings of amino acids, but I'm going to save the details about peptides until a later point in my lecture. Meanwhile, one way to keep all this in your mind is to visualize the following: If the cell is the engine that drives all life, then the receptors are the buttons on the control panel of that engine, and a specific peptide (or other kind of ligand) is the finger that pushes that button and gets things started.

DIAGRAM OF A TETRAPEPTIDE
(CONTAINING FOUR AMINO ACIDS)

= AMINO ACID #1

AMINO ACID #2

#3

#4

THE CHEMICAL BRAIN

At this point, I'd like to move away from the purely molecular level, and, with our new knowledge of the receptor and its ligands, focus for a moment on how scientists now view the brain, and how that view differs from our earlier, more limited understanding.

For decades, most people thought of the brain and its extension the central nervous system primarily as an electrical communication system. It was common knowledge that the neurons, or nerve cells, which consist of a cell body with a tail-like axon and treelike dendrites, form something resembling a telephone system with trillions of miles of intricately crisscrossing wiring.

The dominance of this image in the public mind was due to the fact that we scientists had tools that allowed us to see and study the electrical brain. Only recently did we develop tools that allowed us to observe what we may now call the chemical brain.

But, yet-to-be-named *neuroscience* was so focused, for so long, on the concept of the nervous system as an electrical network based on neuron-axon-dendrite-neurotransmitter connections, that even when we had the evidence, it was hard to grasp the idea that the ligand-receptor system represented a second nervous system, one that operated on a much longer time scale, over much greater distances. The nerves were the classical subject of neuroscience, the route science had taken in its first explorations of the brain and central nervous system, so it was only with some disgruntlement that people could contemplate the idea of a second nervous system. Especially difficult to accept was that this chemical-based system was one indisputably more ancient and far more basic to the organism. There were peptides such as endorphins, for instance, being made inside cells long before there were dendrites, axons, or even neurons—in fact, before there were brains.

Until the brain peptides were brought into focus by the discoveries of the 1970s, most of our attention had been directed toward neurotransmitters and the jump they made from one neuron to another, across the little moat known as the *synaptic cleft*. The neurotransmitters seemed to carry very basic messages, either "on" or "off," referring to whether the receiving cell discharges electricity or not. The peptides, on the other hand, while they sometimes act like neurotransmitters, swimming across the synaptic cleft, are much more likely to move through extracellular space, swept along in the blood and cerebrospinal fluid, traveling long distances and causing complex and fundamental changes in the cells whose receptors they lock onto.

This, then, was as much as we understood about the receptor and its ligands by 1972, before researchers had actually found a drug receptor, and well before the breakthrough involving the immune system in 1984, which used receptor theory to define a bodywide network of information and to provide a biochemical basis for the emotions. In the wake of discoveries in the 1980s, these receptors and their ligands have come to be seen as "information molecules"—the basic units of a language used by cells throughout the organism to communicate across systems such as the endocrine, neurological, gastrointestinal, and even the immune system. Overall, the musical hum of the receptors as they bind to their many ligands, often in the far-flung parts of the organism, creates an integration of structure and function that allows the organism to run smoothly, intelligently. But I'm getting way ahead of my story. Let's take a break from the science and look at how some of these ideas developed historically.

A BRIEF HISTORY OF RECEPTORS

While the idea of the receptor mechanism had originated with pharmacologists in the early twentieth century, many university physiology departments took it up as well because they found it a useful concept to explain the new chemical substances being found in the nervous system—the neurotransmitters. These chemical communicators, which were secreted across the synapse, or gap between neurons, also functioned in a way that could be understood by the receptor-ligand model, even though biochemistry had yet to develop a way to measure what was happening.

The chemical formula of acetylcholine, the first neurotransmitter to be discovered, was still decades away from being diagrammed when physiologist Otto Loewi did his early neurotransmitter experiments following a dream he had one night! These first experiments, performed in 1921, involved the action of a neurotransmitter on a frog heart. Removed from the frog and placed still beating in a large beaker, the heart slowed down dramatically when Loewi applied juice extracted from the vagal nerve to it. The mysterious "vagusstuff" turned out to be the neurotransmitter acetylcholine. Made by the nerves, acetylcholine causes a slowing of the heartbeat and a rhythmic stimulation of the digestive muscles after eating, which together contribute to the feeling of relaxation. For both of these processes, scientists theorized that there were acetylcholine "receptor sites," some on the heart muscles, others on the digestive tract muscles, and still others on voluntary skeletal muscles, but they couldn't actually demonstrate their existence.

Early-twentieth-century theory became reality in 1972, when Jean-Pierre Changeux addressed a pharmacology conference in England. With a dramatic flourish, the biochemist pulled from his breast pocket a tiny glass tube with a single narrow blue band across its middle. The tube contained pure acetylcholine receptors taken from the body of an electric eel and separated from all the other eel molecules and stained blue. This was the first time a receptor had been isolated in the lab.

Changeux explained how the feat had been made possible by an unholy alliance between a cobra and an electric eel, with the former supplying the venom to isolate the receptors from the latter. In higher animals, the cobra's venom acts by entering a victim's body and diffusing to the acetylcholine receptors, including those on the diaphragm muscles, which regulate breathing. The venom blocks the access of natural acetylcholine to its receptors. Since acetylcholine is the neurotransmitter that's responsible for muscle contraction, the resulting paralysis of the diaphragm muscles causes death by suffocation.

Now, it just so happens that the densest concentration of acetylcholine receptors to be found anywhere is in the electric organ of the electric eel. Scientists had found that snake venom contained a large polypeptide, called alpha-bungero toxin, that bound specifically and irreversibly to the acetylcholine receptors in this organ that supplies the eel's jolt. It literally stuck like glue. By introducing radioactive atoms to the toxin in the snake's venom, Changeux could follow it to where it stuck to the acetylcholine receptors of the eel's electric organ, and thereby isolate those receptors. That is how he had obtained the blue-stained substance in his test tube. The process of making a ligand hot, or radioactive, by introducing radioactive atoms into it was a brilliant innovation, but it was—and still is—a very tricky procedure, because the radioactive substance can destroy the ligand's ability to bind, thereby defeating the whole point of the process.

Another major stream that had contributed to "receptorology," as we jokingly dubbed the emerging field, was the discipline of endocrinology, the study of ductless glands and their secretions. Endocrinologists, like the pharmacologists and physiologists before them, needed a way to explain how the chemical substances known as hormones acted at a distance from their sites of release on their targeted organs. But in those days—we're talking the 1950s and 1960s—it wasn't very likely that an endocrinologist would be found talking to a pharmacologist. Each field of study occupied its own little niche and was separated from the others by strictly drawn boundaries that defined the disciplines. Those working within a given discipline were generally unaware of and uninterested in what their fellow scientists were doing elsewhere. So people in each field kept making parallel discoveries without understanding what these discoveries had in common.

In the 1960s, endocrinologist Robert Jensen had been able to use a microscope to see estrogen receptors that had bound with radioactive estrogen he'd injected into female animals. As predicted, the radioactive estrogen went to receptors in breast, uterine, and ovarian tissue—all the known target organs for this female hormone. Later, estrogen receptors, as well as receptors for testosterone and progesterone, were unexpectedly found in another organ, the brain, with amazing consequences for sexual identity. But that's a later part of our story.

In 1970, endocrinologists Jesse Roth and Pedro Cuatrecasas, working on separate teams at the National Institutes of Health, were able to measure the insulin receptor by following Changeux's approach of rendering their ligand—insulin—radioactive. Before, Cuatrecasas had been able to get close enough to show that insulin receptors were located on the outside surface of cells. But the new techniques for labeling substances with radioactive atoms were among the key advances that allowed for the actual measurement of the receptor, a tremendous breakthrough in this field.

A NEW IDEA

My own work in "receptorology" began in 1970, in the halls of the pharmacology department of Johns Hopkins University, where I was able to earn my doctoral degree studying under two of the world's experts on insulin receptors and brain biochemistry. At that time, the insulin receptor was the only receptor being studied with the new methods that had been developed for trapping the more slippery ligands, that is, those that, unlike the snake toxin when it bound to the acetylcholine receptor, did not stay irreversibly stuck to its receptors. No one had tried the new methods on any other drugs. But there was clearly a need to study other receptors to try to trap other kinds of ligands.

In my own field, for example, the prevailing dogma was, as I mentioned earlier, that no drug could act unless fixed. This presented an interesting challenge to neuropharmacology, the particular area of pharmacology in which I had become interested, because, theoretically, it meant that if a drug worked, there had to be a receptor, and our job should be to find it. The drugs we were studying at the time were drugs that obviously changed behavior—I almost said consciousness, but back then nobody used the C-word, except the hippies. Yet everyone recognized that these drugs, which included heroin, marijuana, Librium, and PCP ("angel dust"), precipitated a radical change in the emotional state, that is, altered the state of consciousness of those who used them. That's why, when I began my career in the early 1970s, such drugs were our main tool for studying the chemistry of the brain.

The problem was that our drugs were all from plants, and it was well known that once in the body these plant-derived ligands bound to receptors so briefly before exiting the body in the urine that they were difficult, if not impossible, to catch and measure on their receptors.

The challenge I would eventually make my own was to use the new methodology to trap the small morphine molecule on its receptor in a test tube—a receptor that many people didn't even believe existed. The proof that it did would have ramifications beyond my wildest dreams. In completely unexpected ways, the discovery of the opiate receptor would extend into every field of medicine, uniting endocrinology, neurophysiology, and immunology, and fueling a synthesis of behavior, psychology, and biology. It was a discovery that touched off a revolution, a revolution that had been quietly under way for some time—about which more will be revealed in the future lecture sections in this book. But now my own story must begin.

One warm summer afternoon, shortly after I had been accepted into graduate school at Johns Hopkins University, I was packing for the move to Edgewood, Maryland, where I would live with my husband, Agu Pert, and our small son, Evan. As the material objects of domestic life—the dishes, the clothes, the iron I'd used to iron Agu's white shirts—began to disappear into boxes, I became aware of a growing sense of panic. By the time Agu came home, I was immobilized, slumped in a chair and fighting back tears.

"What's with you?" he asked, not taking much notice of my disturbed state. Always the calm and steady one, he said nonchalantly, "It looks like you got a lot done."

"I know," I responded, trying to rally myself. "But graduate school . . . graduate school . . . it's an hour away. How will I ever . . ." I trailed off, overwhelmed by the thought of the challenges that lay ahead of me. How would I balance the chores of my role as wife and mother with the demands of earning a Ph.D. degree, commuting daily to Baltimore, and working full-time in a laboratory? I gestured pathetically at the boxes on the floor.

"Don't worry," Agu declared. "I'll do it all! I'll do the cooking, the cleaning, I'll make sure Evan gets to day care. Your job is to concentrate on going to school and learning psychopharmacology."

And that's exactly what I did.

2

ROMANCE OF
THE OPIATE RECEPTOR

DESTINY

Looking back over twenty-five years, it seems that destiny played an important role in the unfolding of events that led to the discovery of the elusive opiate receptor. Although it was my fierce belief and passionate devotion that drove me in the final stages, I had only my curiosity and a series of seemingly serendipitous occurrences to put me on the track of proving that there did indeed exist within the brain a chemical mechanism that enabled drugs to act.

My first encounter with the opiate receptor was in the summer of 1970, after I'd graduated with a degree in biology from Bryn Mawr College and before I entered medical graduate school at Johns Hopkins University in the fall. That encounter was personal, not professional. In June I had accompanied my husband and small son to San Antonio, Texas, where we were to live for eight weeks while Agu completed his required medical corps basic training for the army. Agu had completed his Ph.D. in psychology at Bryn Mawr, and now it was time to fulfill his deferred military obligation. I was looking forward to a summer off, maybe even a vacation, after four years of grueling, married-with-child college life. I also intended to bone up on some basics before entering the doctoral program in the fall, so I brought with me a copy of *Principles of Drug Action* by Avram Goldstein. Since the program I was entering at Hopkins would focus on neuropharmacology, the study of the action of drugs in the brain, I wanted to prepare myself and figured Goldstein's book was the best place to start.

But real-life experience preempted the academic learning, and instead of reading about the opiate receptor, I got to experience its effects firsthand. A horseback-riding accident put me flat on my back in a hospital bed, where, doped to the gills on Talwin, a morphine derivative I was given to ease the pain of a compressed lumbar vertebra, I remained for most of the summer. My body immobilized by the injury and my attention span shanghaied by the drug, I was unable to concentrate enough to read the selected text or any other book, and instead spent my days lying around in a blissful altered state while my back healed.

Later, when I was off the drug and able to sit up, I read part of Goldstein's book, which included a thorough introduction to the concept of the opiate receptor. I remember marveling at how there were tiny molecules on my cells that allowed for that wonderful feeling I'd experienced every time the nurse had injected me with an intramuscular dose of morphine. There was no doubt that the drug's action in my body produced a distinctly euphoric effect, one that filled me with a bliss bordering on ecstasy, in addition to relieving all pain. The marvelous part was that the drug also seemed to completely obliterate any anxiety or emotional discomfort I had as a result of being confined to a hospital bed and separated from my husband and young child. Under its influence, I'd felt deeply nourished and satisfied, as if there weren't a thing in the world I wanted. In fact, I liked the drug so much that, as I was ending my stay at the hospital, I very briefly toyed with the idea of stealing some to take with me. I can see how people become addicts!

This intense overlap of physical and emotional experience, both originating from a single drug, fascinated me and sparked anew my interest in the connection between brain and behavior, mind and body— a connection that had originally come to my attention during my freshman year in college. On my own for the first time in my life, I had subsisted for an entire semester on a diet of peach pie, and thereby had thrown myself into both a thyroid blowout and a major depression. So it happened that I received my official introduction to the idea that something happening in the body could affect the emotions. Now, as I began graduate school, I was about to explore the connection scientifically, and begin the work to which I would eventually devote my life. And it all had to do with these strange little things called opiate receptors.

THAT FALL, at the age of twenty-four, I officially entered Johns Hopkins University Medical School as a doctoral candidate in the department of pharmacology. Unofficially, it was the beginning of my apprenticeship in neuroscience, a discipline that did not yet exist, and would not for almost a year. I didn't know it then, but I had walked right into the center of a revolution that was brewing, one in which the boundaries of distinct disciplines such as biochemistry, pharmacology, neuroanatomy, and psychology would dissolve to make way for the new interdisciplinary field of neuroscience.

I remember the first morning I arrived and parked behind the old Johns Hopkins medical laboratory building. I was literally trembling as I got out of my car, painfully aware that with the exception of a couple of high school science-fair projects and a borderline senior science project in college, I'd never performed a real experiment. As an undergraduate biology student, I had been incapable of bringing myself to kill and dissect an animal. But that was strictly my own shortcoming, not that of my education, which was excellent.

At Bryn Mawr, my early science training had been in the classroom of a Miss Oppenheimer, a fine teacher who almost threw me out of the department because of my stubborn, albeit principled refusal to kill a frog for dissection. There was some emotion in me that would not allow me to kill an animal. The thought of pulling apart a creature that I myself had just killed, no matter how marvelous its structure or incredible its fluids, made me sick to my stomach.

"Don't be squeamish!" Miss Oppenheimer exclaimed. "How can you ever expect to study the brain if you don't get over this? You've got to put this nonsense behind you if you ever want to do great work."

Miss Oppenheimer had become my role model, my heroine, and I would have done almost anything to please her, because she had actually taken me seriously when I told her of my interest in the crossover between physiology and psychology, but this I couldn't do. Only much later, after I became sensitive to the complex sexual politics of science, did I understand her vehemence on the subject. Miss Oppenheimer had been trained in another era, when the belief that women couldn't do good science prevailed. Women who survived did so by becoming hard and cold on the surface, adopting a persona I later came to refer to as the "science nun." I'd see them at meetings, these severe and often brilliant women, wearing all-black clothing, their hair pulled back and tightly knotted. They were rarely married and had no children, as if their female natures had been obliterated by their need to prove they were just as strong, just as exacting, and just as relentless as the guys.

Already married and a mother at twenty, I had two strikes against me as a novitiate to this particular order. In addition, my display of female squeamishness over spilling blood was almost too much for my teacher to stand. I know that Miss Oppenheimer spent more than a few distraught moments weighing my obvious devotion and creative gifts against her better instincts, which told her I would have no future in science. Somehow, she let me slip through, and while I was grateful for the leeway, I knew I'd never be able to fall back on this bit of female maneuvering once I got to the big leagues—especially if I wanted the guys to take me seriously, which I did.

All this flashed through my mind as I stood at the entrance of the Hopkins medical building, trembling and literally unable to move, feeling like a complete fraud, although ecstatically anxious at the same time. A fraud, yes, but a sincere and eager fraud, one who was willing to do whatever needed to be done, to learn whatever was required! What kept me from running back to the car and driving away that morning, I don't know. The only thing I knew for certain was that, in spite of my near-total lack of experience, I was there because I wanted to be. And the tug of fate was undeniable—everything had unfolded magically to bring me to this point where I now stood.

Geography had limited my choice of graduate schools to two, Johns Hopkins and the University of Delaware, both within commuting distance from Edgewood Arsenal in Edgewood, Maryland, where Agu would be stationed. Agu was completing his military requirement in the experimental psychology laboratory, planting tiny tubes in monkey brains to locate the centers of pleasure and pain. We lived on the base, a long but not impossible drive from downtown Baltimore, where Hopkins, my first choice, was located. Even though I had a small child at the time, there had never been any question but that I would continue my education uninterrupted while Agu finished his stint in the army. We were a scientific team, Agu and I, a blend of his knowledge of the behavioral sciences with my blossoming expertise in biology. Together, we expected to do great science.

At the entrance interview I'd had the winter before at Hopkins, the man who interviewed me was obviously amazed that a wife and mother, especially the wife of a soldier who might be called at any moment to the jungles of Vietnam, was seriously seeking entrance to one of the country's finest graduate schools to study biomedicine. I wasn't too surprised a month later when I received a letter from the medical school's biology department rejecting my application. Delaware's biology department had accepted me, and so I decided that was the end of it, that's where I would go. But fate intervened, and before my first fees came due for admission to Delaware, something happened that changed the course of my life forever, putting me on a trajectory headed straight into the center of the neuroscientific revolution.

Fate's opening gambit occurred in the spring, when I attended my first scientific meeting, the annual Federation of American Societies of Experimental Biology Conference in Atlantic City, New Jersey, a gathering attended by some 20,000 biologists from all over the world. During a break in the presentations, I found myself at the edge of a small circle formed around a scientific publisher who was gossiping about a new researcher at Johns Hopkins, a neuropharmacologist and psychiatrist by the name of Dr. Sol Snyder. This unusual combination of specialties caught my attention. Here was someone, this Dr. Snyder, who was studying the chemistry of the brain while at the same time bringing to it a knowledge and understanding of human behavior. I remember thinking, "That's exactly what I want to do!" But, unfortunately, I'd been accepted by Delaware, not Hopkins.

Fate was not going to abandon my cause, however. When I returned home to Bryn Mawr after the conference, Agu and I went to a lecture sponsored by the psychology department featuring Dr. Joe Brady, a psychologist from the medical school of Johns Hopkins who'd done pioneering mind-body studies on monkeys, linking the stress of having no control over a situation with the development of severe ulcers. After the lecture, we attended the departmental party, where Brady showed himself to be a real Gene Kelly on the dance floor. At one point he called out, "Anybody here know the Peabody?"

As it happened, I had learned the fast-moving 1930s dance from my Estonian in-laws during the many Brooklyn basement parties I'd attended while engaged to Agu, so I took up the challenge. We Peabodied like champs for the next hour before collapsing in a sweaty heap, kicking a lamp over on our way down. Later, over drinks, we made small talk, and he asked me what my goals were for after college.

"I want to study the brain," I told him, "because I'm interested in understanding behavior from the angle of biology." Joe Brady nodded attentively and then said, "Well, then, the guy for you is Sol Snyder, someone new at Hopkins medical school, a real wild man doing just that. Send me something about yourself, and I'll see that Sol gets it."

Ignoring the fact that one department at Hopkins had already turned me down, I wrote a long letter telling Joe my fondest dreams and desires, and I included a transcript of my courses and grades. Within a short time, I got a phone call from the wild man himself.

"You're accepted," Dr. Snyder said in a friendly, crisp tone. "Now apply."

And that's how it all began, a series of events that unfolded almost according to some predestined script, bringing me to Sol and to the tiny lab in the west wing of the Hopkins medical building, which was nestled in the slums of downtown Baltimore.

ENCHANTMENT

Hopkins offered a Ph.D. program through its pharmacology department that was research-oriented rather than academic. There were plenty of lectures, readings, and coursework, but the heart of the program was the labwork. Each student was expected to apprentice under four different scientists, rotating in and out of their labs every two months. It was made very clear that the successful completion of the program depended on performance in the lab.

Home base for me was Dr. Snyder's lab, where I began my training in lab technique and performed my first experiment. Consisting of only three benches crammed into a single room, the lab was but a dim foreshadowing of its future self. But to me it was heavenly. The centrifuges hummed and the radioactive counter clicked, while handsome postdocs scurried hither and yon, cracking sophisticated jokes and performing highly technical maneuvers at their benches. The realization that I had landed in a lab that was pushing the frontier of brain research, exploring the biological basis for mental illness, was almost too exciting for me to bear.

Solomon H. Snyder, I quickly came to see, more than lived up to his wild-man reputation. At thirty-four, he was already at the top of the profession, an acknowledged prodigy who was routinely described by his colleagues as brilliant and ambitious. He had been trained as a psychiatrist, but had apprenticed in neuropharmacology at the National Institutes of Health, studying the effects of drugs on the brain. There he had acquired an appetite and a skill for experimentation that led him to persuade Hopkins to give him both a private practice and a lab. Sol became the youngest full professor in the history of Hopkins when he was only thirty-one. His dual appointment in pharmacology and psychiatry positioned him on both sides of the mental health front, giving him a uniquely well-balanced, comprehensive perspective. He treated patients with the current mind drugs and monitored their effects, while a few paces down the hall, he directed research in the lab for the next generation of medicine in mental health.

At first, I wondered why Dr. Snyder rarely appeared in the lab. He preferred, I found out, to do his science from the "throne room," which was how his students referred to his office. The room was huge and immaculate, with an oversized desk at one end and a leather couch at the other. A genuine Kandinsky dominated one wall, while Sol's awards and prizes were prominently displayed on another: the Outstanding Young Scientist award from the Maryland Academy of Sciences, the John Jacob Abel Award, and many others. Sol's desk was always in perfect order, belying the volume of paperwork he did, enough to keep three full-time secretaries busy. They sat in the outer office, cranking out grant proposals and handling the constantly ringing phones.

At the time of my arrival, the focus in the lab was on identifying new chemical neurotransmitters—those "juices" secreted in the brain that were thought to carry information and direct the activities of the organism. Jumping the synapse between brain cells, the neurotransmitters bind to receptors on other brain cells, or neurons, causing an electrical charge that redirects the neural pathways. The effect on the organism is to change the physical activity, including behavior and even *mood*—the closest word to *emotion* in the lexicon of hard science.

Sol had developed a method for determining which substances were neurotransmitters and which were not. This involved measuring the "re-uptake" mechanism, a cellular operation that insured that the excess juices left over after binding would be sucked back into the neuron and destroyed. If the substance under investigation was found in the brain, and its re-uptake could be measured, then we had a neurotransmitter. Before Sol's method had been developed, only two neurotransmitters were well studied and understood—acetylcholine and norepinephrine (which is also called noradrenaline). But by the time I joined the lab, Sol and other neuroscientists were in the process of adding five more: dopamine, histamine, glycine, GABA, and serotonin.

Sol knew that the work being done in his lab was at the center of a revolution, and communicating this to his students was part of his charisma. He had a way of letting us know we were on the cutting edge, caught up in a grand and glorious gamble, which, if we won, would make us all stars. Yet at the same time we knew we were in about the most secure spot any apprentice scientist could ever ask for.

Sol, I discovered, was one of the Golden Boys of the medical establishment—well connected and well funded. While we were gathering data down in the scientific trenches, Sol was out in front, jetting around the country and the world to explore the furthest frontier. He'd fly in from Zurich or Palm Springs and the next day round us all up for a report on the latest and hottest news from labs around the globe: who was working on what, and what was breaking next where. We loved it and hung on his every word.

Unlike most scientists, who move forward slowly with tiny baby steps, afraid of taking any real risks, Sol liked to think big and bold. He had a profound disregard bordering on arrogance for the tedious side of science, directing only those experiments that were both very simple and focused on the really big questions. He displayed absolutely no respect for boundaries, tramping his way onto the highly protected turfs of other researchers to satisfy his huge interdisciplinary intellect. His specialty was spotting projects that showed promise for an imminent breakthrough, research where maybe nine-tenths of the work had been done, and all that was missing was a bold formulation, a risky adjustment.

"Let's take advantage of this situation," he would cry. "Bang! Let's get on it! Let's beat them to the prize!"

Sol saw science as a game, and took every advantage to win. A master at motivating us, he awed us with the way he commanded his resources and people. I was so inspired by his tactics, his charisma, and his brilliance that I was eager to do anything, including work all night or arrive in the lab at some ungodly hour in the early morning to take a time point on an experiment. I lived to please him and bring him good data.

If we saw Sol as just short of God, he in turn worshiped his mentor, Dr. Julius Axelrod, one of the founders of the field of neuropharmacology and an omnipresent force behind the scenes. At the National Institutes of Health, Sol had come up the ranks through Julie's lab, and he was one of "Julie's boys," a group of scientists who had learned a research style from their mentor that would lay the foundation of modern neuropharmacology. JULIE'S BOYS SKIM THE CREAM had been written on the wall of the lab years ago, to describe their hugely effective approach to research. And Julie's boys formed a scientific dynasty, sharing information and using their influence to support each other at funding time, often rotating favored students and postdocs through each other's labs in a giant game of chess. When Julie won the Nobel Prize in medicine shortly after I arrived at Hopkins for his work with noradrenaline, one of the nervous system's two principal neurotransmitters, the news electrified our lab. We all felt forever blessed, securely positioned in a line of succession that was part of Julie's patrimony.

The blessing extended far beyond giving us access to information and funding. At the heart of this chain of brilliant and aggressive minds was a philosophy, one that I came to understand in terms of the following dictums: Do not accept the conventional wisdom. Do not accept the idea that something can't be accomplished because the scientific literature says it can't. Trust your instincts. Allow yourself a wide latitude in your speculations. Don't depend on the literature—it could be right or it could be completely wrong. Spread all your hunches out before you, and go with the ones that you think are most probable. Select the one that you can test easily and quickly. Don't assume it has to be overly complicated to be of value, since often the simplest experiment yields the most unequivocal result. Just do the experiment! And if you can keep it to a one-day experiment, so much the better.

This was our inheritance, handed down from Julie Axelrod to his disciples, including Sol, and then from Sol on to me. Eventually, I would pass it on to my students, and they to theirs, in an unbroken chain of methodology and philosophy that I'm sure will continue holding sway long after we're all gone.

INITIATION

With a generous wave of his hand, Sol showed me where I would be working, assigning me to a lab bench, a chest-high slab of marble with drawers below it and shelves above. "Now go find Ken Taylor," Sol ordered me in his best fatherly tone. "He'll show you how to do the histamine assay."

The assay is a procedure that is at the foundation of experimental research. It provides us scientists with a method for measuring the quantity of a chemical substance, such as a neurotransmitter, in a series of samples such as tissue or blood. The point is measurement! Before you can ask any serious questions, you have to be able to give a numerical value to the chemicals in each of your samples, a number that corresponds to high or low concentrations of the substance you are studying.

Ken, it turns out, was incredibly handsome, a New Zealander who regularly organized the department's pub parties on Friday afternoons. He was also a very focused researcher and rigorous in his instruction. His presence at my lab bench was absolutely erotic, a fact that increased my desire to be the best at whatever he instructed me to do. But I kept my feelings hidden and carefully avoided any female maneuvering. I'd been well indoctrinated with an almost religious approach to science by Miss Oppenheimer at Bryn Mawr, and even though I'd balked at her approach, I wasn't going to take any chances and risk being dismissed as a less than serious student here at Hopkins. Instead, my attitude was one of a novitiate in the Church of Science being taught her first catechism by a virile young priest. Later, when I had my own lab, I would see the potential for combined male-female energy as a positive force to do great science.

Over the course of the next few weeks, Ken filled me in on the basics. Histamine is a chemical normally secreted by cells of the immune system, causing allergic reactions such as sneezing and itching (which is why we take antihistamines to relieve allergy symptoms). Contrary to conventional wisdom, Ken and Sol had recently found histamine in the brain, a find that had led them to speculate on the possibility that histamine could be a neurotransmitter, one more of the brain's information-carrying messenger chemicals that they'd been identifying. Even at this point, I knew that research on neurotransmitters was the hottest thing going, and I was ecstatic to be included in any part of it.

Soon I fell into a daily routine. My first job every morning was to number fifty test tubes and place them on a rack. After I'd done that, I would get the day's brain tissue samples from Ken and distribute them evenly among the numbered test tubes. Then the fun began. Using a thin, delicate, hand-blown glass straw called a pipette, I carefully transferred tiny amounts of various substances into the test tubes, the first of about ten steps that would turn each test tube into a numerical result by the end of the day.

I learned later that the histamine assay I was doing was based on Sol's own early work. Devising a method of carrying out the measurement, which he had done while working with Julie Axelrod, was Sol's first major accomplishment in the lab, and this method now served as a link in the process to determine if histamine was a neurotransmitter.

What we were doing with the histamine assay is an example of the way that most biomedical research unfolds. First, a technique is discovered that provides answers to questions that have previously been unanswerable. Then we deploy that new technique to the nth degree, putting to it every possible question that might apply or be of further research value, until we've exhausted the possibilities—or until some newer technique comes along to make the previous one obsolete.

I loved sitting at my lab bench, day in and day out, pipetting my chemicals and wearing my crisp, white lab coat. (It's only in Hollywood movies that scientists wear white lab coats. In real life, the novices, not the real scientists, are the ones who wear them.) I loved it so much that it wasn't unusual for me to spend ten hours or more in the lab at a stretch. The atmosphere was charged, often very intense, which made for a special feeling of aliveness. I thrived on the incessant conversations about everything from science to art to politics.

I soon became aware of an unspoken but formal hierarchy in the lab. Rank, it seemed, was a matter of longevity. Those who were there the longest generally had the most power—unless you were a woman. (In that case—and a rare case it was, since there weren't many women in important labs like Sol's—you were seen not as a wise senior but as an old shoe, comfortable, nonthreatening, reliable.) Advancement was the result of higher-ups leaving to embark on careers of their own, allowing the now-seasoned novitiates to occupy their slots. But this wasn't always the case. Doing "hot science" brings its rewards. Any kind of major discovery can zoom an underling straight to the top of the heap—something I was to find out firsthand in the not-too-distant future.

After a few months of practice at the histamine assay, when my data was crystalline and my technique well honed, I was summoned before a panel of senior scientists in the program. Their goal was to grill me mercilessly on every aspect of the histamine assay in order to determine if I was worthy to pass on to the next stage of the program. Even though I had prepared thoroughly, I was so nervous under their cold and unfriendly scrutiny that suddenly I couldn't remember anything.

In short order, my inquisitors demolished what little knowledge I had, while seeming to relish every minute of my ordeal. As I realized later, it was a ritual to them, like a fraternity hazing, putting young scientists in their place, reminding them they really didn't know much yet.

Sometime after enduring this unsettling little game, I was told, to my great relief, that I had passed the review and was now permitted to enter the next stage, selecting an original research project for my Ph.D. dissertation. I was well aware that there was no hope of my ever becoming a real scientist without a Ph.D. Those who stopped short and stayed at the M.A. level were forever relegated to the bench, seldom acknowledged on scientific papers, regardless of how much they contributed. But once I had a Ph.D., which had been my plan from the start, I'd be in the club, and all I needed for admission was to turn out an original piece of research, one good enough for publication in a reputable scientific journal.

After conferring with Sol, it was decided that I should work on the choline re-uptake mechanism for my Ph.D. dissertation. One of Sol's postdocs, Hank Yamamura, had already used Sol's formula to measure choline re-uptake in the brain. Now Sol assigned me to follow up on his findings by measuring it in the ileum of a guinea pig (research that was related to work being done at a lab in Scotland, where they were investigating the role of various neurochemicals binding to as-yet-unidentified receptors on cell surfaces in the guinea pig ileum, causing muscle contraction).

The ileum, the upper part of the small intestine, contains the cholinergic nerve that releases the neurotransmitter acetylcholine. Here was another opportunity to be part of the neurotransmitter research I'd been so excited about when I'd first arrived, but my enthusiasm was noticeably flat. I couldn't help but think it was dull and derivative, a kind of hand-me-down project with a fairly predictable outcome, one that did nothing to inspire or excite my imagination.

Putting my distaste aside, I threw myself into the work. This was my first time preparing a scientific experiment from scratch, depending only on a few earlier reports, and after several tries, I had it up and running. I remember thinking about Dr. Frankenstein as I organized the procedure, first removing a section of the guinea pig's gut, then squirting buffer through it to act as an enema. After that, I dissected it to the muscle, which was heavily lined with nerves. I then minced the muscle into neuron-containing fragments and placed them in beakers, adding a radioactive form of choline. The "hot" choline acted as a trace to give a signal that could be measured easily, showing that it was quickly taken up by the neurons in the nerve-containing muscle and converted into the neurotransmitter acetylcholine.

The choline assay was a meat-and-potatoes kind of project, a sure bet to an easy Ph.D., and one for which any sensible doctoral candidate would have been grateful. For weeks I toiled away at it, but remained profoundly unmoved in the process.

• • •

WHILE I WAS still in the set-up stages of the hated choline project, I came across a flyer posted on the department bulletin board announcing a lecture by a Dr. Pedro Cuatrecasas, an endocrinologist and newly appointed professor of pharmacology at Hopkins. The lecture was being offered as part of a department series and served as a way for the highly renowned researcher from the National Institutes of Health to introduce himself and his work to the Hopkins faculty and student body.

"If anyone around here ever wins the Nobel Prize," Sol had told me, "it's going to be Pedro!"

I marked my calendar and began to make some inquiries about the upcoming lecture.

Dr. Cuatrecasas, I found out, had been part of one of the NIH teams that was the first to isolate and measure a receptor on the surface of a cell wall—the receptor for the hormone insulin. As explained earlier, the ability to actually measure a receptor meant that one of the biggest mysteries of modern medicine had been solved. Central to his method was the Multiple Manifold Machine, a device that had been built for Marshall Nirenberg to use in the experiments he had done a few years before at the NIH when he was trying to crack the DNA amino acid code. The Triple M, as I came to call it, had revolutionized filtration, a process that allowed for the rapid separation of bound from unbound ligands, making it possible to measure receptor-specific binding. In his lecture it was expected that Dr. Cuatrecasas would be showing how he and his team had used this device to discover the insulin receptor.

The day of the lecture, I arrived early and stood in the lobby of the lecture hall with the waiting crowd. Sol was unable to attend himself, but had urged me to go and insisted I give him a full report when I returned. The excitement in the air was almost palpable, not the usual prelude to a department lecture, and I remember feeling a sense of anticipation that what we'd be hearing would be far from the usual stuff. This was about science that was newly breaking, the leading edge! When the doors opened, I entered and hurried to a seat in the front row.

Inside, the room was buzzing. At the podium stood the lecturer, not a particularly commanding figure but nonetheless an attractive and intense man, his Latin heritage visible in the sparkle of dark eyes and, as he talked, in his expressive enthusiasm for his subject. All eyes in the room followed him closely as he paced back and forth, showing graph after graph, curve after curve, proving beyond a doubt that he had indeed discovered a method to measure insulin as it bound to specific receptors on fat cells as well as to receptors on cells in the liver, mediating the entry and storage of sugar in these cells. As his remarks came to a close, he looked directly at me, pausing for a moment to flash a stunning smile as our eyes met. Did he know I was Sol's graduate student and that Sol had sent me to check him out? I couldn't tell for sure, but I decided there on the spot that he was someone I wanted to work with. As soon as I got back to Sol's lab, I put in a request to be rotated into Dr. Cuatrecasas's lab as part of the requirements for my course of studies.

A few days after the lecture, Sol and his wife invited Agu and me to a small dinner party at their home, where the only other guests were Pedro Cuatrecasas and his wife. I was thrilled and felt very privileged that Sol had included me in such an intimate gathering—me, a lowly graduate student!

It was at that dinner that I first experienced the powerful and romantic allure of doing great science. As I listened to their conversation, I felt that nothing could compare to learning and working in the company of men such as these. I was thrilled by all of it, the gossip, the politics, the breakthroughs, even the sense of competition with other labs. I loved it all, though in my naivete I really knew very little about it.

It was also at this dinner that I was nudged farther along the path that would lead me to my life's work. This happened during the course of a conversation that began when Sol politely asked Agu and me about life in the army. Agu had started military life as a captain, having been a reservist during college and graduate school, and because he had more seniority than his colleagues, he was at the time running the psychology lab at Edgewood Arsenal. The talk turned to our experience at boot camp in Texas, where, I explained, I'd spent three weeks flat on my back in the base hospital, pumped full of opiate-derived morphine and feeling no pain. I mentioned that I'd brought with me a copy of Avram Goldstein's *Principles of Drug Action*, which I'd repeatedly tried to read while in the hospital, never getting any further than the section on opiate receptors.

It was fortuitous that I had mentioned Goldstein's book and my experience of being medicated with morphine in the same breath, for it reminded Sol of his own interest in the subject of the opiate receptor. In fact, during the same summer I spent lying on my back, Sol had been a participant at the Gordon Conference, an exclusive, prestigious scientific gathering. Avram Goldstein himself was one of the presenters, and at the conference, he outlined a plan for how he intended to discover the opiate receptor in the brain. Sol was skeptical about the technique Goldstein proposed to use, which seemed crude and probably not up to the job of isolating a reliable signal out of the thousands of chemicals that occupy any single drop of neurojuice. But he was interested in the project itself. After the conference, Sol had looked up some of Goldstein's papers and brought them back to his office, intending to study them further.

"The opiate receptor?" I asked Sol the next day in his office, as he handed me one of Goldstein's papers on the subject.

"Yes, it's just like the insulin receptor—only it's for morphine," Sol replied.

An intriguing discussion followed in which I learned that finding receptors in the brain was expected to be many times more difficult than locating them in the rest of the body. At the time of our discussion, only one receptor for a known neurochemical had ever been found. In the experiment I described earlier in my lecture, the Frenchman Jean-Pierre Changeux had just recently isolated and measured the receptor for acetylcholine in the electric organ of an electric eel. But this was considered a special case. Fully 20 percent of the electric organ of this species of eel consisted of acetylcholine receptors, a huge target compared to the one-millionth of a percent of the brain that the opiate receptor was estimated to be. What made the search for an opiate receptor even more problematic, as Sol had said, was that while receptors had been located for chemicals that originated inside the body (endogenous), like insulin and acetylcholine, no one had ever found a receptor that fit drugs that originated outside of the body (exogenous), opiates like morphine, heroin, or marijuana.

As I listened, I was remembering my ordeal at the army hospital and the blissful state of consciousness I'd experienced every time I got an injection of painkilling morphine. The opiate receptor! Now here was a goal I could easily imagine pursuing, a project worthy of my dreams, my ambitions, my aspirations. To take part in unraveling the mystery of how the opiates worked to produce their magical, otherworldly effects—what could be more exciting? To find a receptor for morphine, the drug over which wars had been fought, kingdoms lost, the mystical substance that suffused the writings of Coleridge and DeQuincey and was part of the revolution in sensibility wrought by the great nineteenth-century romantics Byron, Shelly, Keats, and Wordsworth. Morphine was named in honor of Morpheus, the god of dreams in Greek mythology, and it was a drug that I knew firsthand, a drug that had fascinated me by its effects on mind and body.

I decided that night I would ask Sol to let me switch projects and leave the choline re-uptake assay behind and begin a search for the opiate receptor. I knew he believed that finding the opiate receptor was a near-impossibility, and I figured it wasn't likely to be something he'd even think of assigning to a Ph.D. candidate. But I didn't care. His skepticism only added to the glamour, enhanced the appeal. And, besides, it would be an original project—not a hand-me-down based on someone else's research, something he knew I found tedious. I remember feeling inspired and thinking, from my somewhat limited perspective at the time, that if I succeeded, then I just might become famous—as a graduate student who had actually done an exciting and original piece of work for her doctoral dissertation.

When I approached Sol, he was ambivalent.

"It's a long shot," he warned me. "The choline study is a bread-and-butter project, easily achievable," Sol insisted. "Are you sure you want to take the chance when you've got a sure thing?"

"Absolutely," I replied.

"Well, read over Goldstein's paper and think about how you would approach the project," he told me.

As I read over the paper and reflected on its contents, a single thought stood out in my mind: What could Goldstein have done if he'd had Pedro's Triple M?

THRESHOLD

It was 1972, and the Nixon administration had just begun an all-out war on drugs. Heroin and heroin addicts were to be a focus of Nixon's campaign, and the administration announced it was allocating over six million dollars for addiction research. There was a lot of talk in the media about the possibility of creating a "magic bullet," a new drug to cure heroin addiction, but in the lab, we knew that all this talk was just that—we didn't have a clue how opiates and heroin worked on the brain. Sol had already set his focus on amphetamine research, and he'd written a grant and submitted it to the newly established National Institute of Drug Abuse. With Sol's history of successfully obtaining money for his projects, there was no doubt he'd get funded for this one.

As researchers on the breaking edge of neuroscience, we knew that all the talk about coming up with new drugs in the struggle against addiction didn't make sense. How could we come up with a drug to fight drug addiction when the mechanism for how opiates like heroin and morphine operated in people was still so far from being understood? For any real breakthroughs to occur in the battle against addiction, a logical first step would be to find the opiate receptor. And so Sol, part of whose brilliance lay in combining great leaps of the imagination with equal doses of conservative science, added a very brief appendix to his amphetamine proposal, requesting funds to support research aimed at finding the opiate receptor.

While Sol waited for the bureaucracy to give us the green light, I headed off to Pedro's lab for a two-month rotation. My mission was to get some firsthand knowledge of the Triple M and how Pedro had used it to understand and measure the insulin receptor.

Pedro's laboratory pulsated around the clock. I loved the graceful but exhilarating rhythm, a samba compared to the rock-and-roll pace I was used to in Sol's lab. But in the back of my mind, I stayed focused on my goal, which was to practice and hone my skills in preparation for my soon-to-begin pursuit of the opiate receptor.

Pedro himself kept fairly regular hours in order to be with his family at dinnertime, but his students and postdocs stayed late into the night, often hovering over their work until the first signs of light. In the lab, Pedro would make his appearance with the flourish of a showman. Unlike Sol, who was usually busy with equally important matters elsewhere, Pedro enjoyed appearing right when the most critical step of an experiment was about to be performed, giving it his special spin, and demonstrating his techniques to us in the process.

The bulk of the lab's activity centered on Pedro's efforts to expand his earlier work with the insulin receptor. With the team that had made the discovery now disbanded, there was a wild race going on for what in science are called the "follow-up discoveries." Since he was one of the original group, Pedro was able to command generous amounts of funding as well as personnel in the high-stakes race against the competition. As can be expected with most scientific achievements, a dispute arose over who was actually responsible for the successful discovery of the insulin receptor, and who would earn the prizes that would ultimately follow. At Hopkins, the rumor was that Pedro was responsible for the bulk of the work, and that it was his ability to make pure, active, radioactive insulin that enabled the group to eventually prove the existence of the insulin receptor.

Before getting started on the Triple M, Pedro wanted me to tie up loose ends on an experiment he'd begun shortly before leaving the NIH to come to Hopkins. With Nobel laureate Chris Anfinson, he'd created something called an "affinity column," a piece of state-of-the-art technology for purifying molecules, which used the natural affinity of one molecule for another to purify enzymes from solution. He'd put me to work using this technique to try to isolate an important enzyme. Living things use hundreds of enzymes, which are protein catalysts that efficiently perform in seconds chemical reactions that would take weeks in a test tube—if they could happen at all.

For weeks I was the darling of Pedro's lab, loading the cellular soup into the glass column and successfully building up the supply of enzymes in a pure form. Midway through the process, however, Pedro got news that a rival lab had beaten us to the punch. They'd completed the purification while we were still toiling away. One morning he came into the lab, slapped a paper down on my bench, and exited without saying a word. I read the document and, in my naivete, thought—*Great!* Our work is confirmed. But as far as Pedro was concerned, all that mattered was that a rival lab had beaten us to the punch. The experiment was over. We had been scooped, and since Pedro was not interested in follow-up work, I was taken off the project, pronto. Looking back, I see that this incident was my first close-up glimpse of the gamesmanship that goes on in the big leagues, where winning is everything.

I consoled myself by learning how to measure insulin receptors with the help of the Triple M. Every day I'd pour my cellular concoction, a mixture that contained radioactive insulin and liver cell membranes, through the filtering mechanism. The radioactive insulin acted like a ligand, first binding to its receptor, then slipping off, then rebinding and rereleasing. As long as the receptor stayed in a wet state, this activity would continue without stopping. We were looking for a way to rapidly dry the organic material and trap the ligand in a bound state, while the unbound substances were washed away.

Pedro's Triple M provided state-of-the-art rapid filtration. With the Triple M, numerous test tubes could be dumped onto the filtering mechanism one by one. The cellular soup would be sucked away with a strange *whhhhoooosssshhhhkkk* sound, leaving only the bound material stuck onto the receptor. The Triple M had worked perfectly on the insulin receptor, but would it help us find the opiate receptor?

As a result of Sol's connections, as well as his sterling reputation, the grants committee gave us the money for the opiate research, but they were skeptical of our chances. Along with their approval for the funding, the committee sent a letter covering themselves should we fall flat on our faces, as they obviously expected we would. As far as they were concerned, our chances were slim to none. Indeed, if anyone other than Sol Snyder had had the audacity to submit the proposal, it surely would have been rejected, never to have seen the light of day.

EXPERIMENT

An experiment is divided into two parts: first the design, then the implementation. But rarely do they follow one another smoothly.

My experimental design was planned with Pedro's insulin receptor work in mind, borrowing especially from the rapid filtration technique I'd learned on his Triple M. At the same time, I was mindful of the work Avram Goldstein was doing. In the paper Sol had given me, Goldstein described his attempts to isolate the opiate receptor with a method that involved pulverizing mouse brains into a solution and adding radioactive opiates. The batter was then spun very fast in a centrifuge, causing cell nuclei and nerve endings to separate out at different speeds.

It was obvious to both Sol and me that the technique Goldstein used could not work for a number of reasons. As indicated by the handwritten notes he had scrawled all over Goldstein's paper, Sol believed that one of Goldstein's problems was that the radioactive tag on his binding substance hadn't been hot enough. He deduced this because although a signal had been found, it wasn't very clear or distinct. So in our experiment, we decided to zero in on only one opiate, morphine, and to make sure that we got it to the point where it was as hot as the existing technology allowed—a big challenge. Today there are catalogs from which you can order any number of already-purified radioactive opiates, but back then we had to ship the "cold" morphine to a special lab to get it tagged with radioactive isotopes. When we got it back, we had to put it through a purification process to separate out any possible contaminants. This was a crucial step, since if our hot morphine was contaminated, it wouldn't give off a signal clear enough to be picked up by the counter.

While Goldstein hadn't been able to get his experiment to work, he did have a clever idea that we would find useful in implementing our own experiment. Since the morphine in his soup could bind indiscriminately to anything, he needed a way to show it was binding to an opiate receptor and not to some artifact, that is, to something that had been created in the artificial environment of the test tube. To do this, he used a stereoisomer, a synthetic opiate designed specifically in a laboratory and having two mirror-image forms. Both forms have the same chemical structure, but the left-handed version, called levorphanol, is an extremely potent opiate, while the right-handed version, called dextrophan, is almost inert.

Goldstein tagged the two forms of the opiate with a radioactive trace and mixed each one into a test tube containing his cellular soup. He then predicted that only the potent levorphanol opiate would bind to the receptor, because it alone had the correct configuration, or fit, to do so. The other one, the dextrophan opiate, had the wrong shape to fit the receptor, similar to the way that a right hand has the wrong shape to fit into a left-handed glove. Goldstein's prediction that test tubes incubated with radioactive levorphanol would show higher numbers on the counter, reflecting a higher degree of binding in comparison to the radioactive dextrophan, was borne out, but the difference was so small—less than 2 percent—that no one believed it. Nor could anyone repeat it. Still, Goldstein was on to something, and Sol and I intended to make full use of it. What I didn't know was that Sol had also recognized the possibility of scoring a coup, slipping in front of Avram and waltzing away with the prize.

With the benefit of hindsight, I can see how willingly I embraced this macho ethos—the intense rivalry, the competition for credit, and the overriding drive to make the big score, regardless of who got burned in the process. With the lack of any female role models, I thought that to accomplish great breakthroughs in science, you had to be tough and aggressive. Most of the women I saw were stuck on the lower rungs of the hierarchy, rarely rising above their assigned stations, always stepping in to do the menial labor and then becoming invisible when it came time to hand out the credit.

This wasn't going to be the case for me, I vowed. The raw ambition that permeated the atmosphere in both Sol's and Pedro's labs was having an effect on me. I was beginning to dream of what it would be like to command the money and resources to run my own lab. From what I had observed, the only way to accomplish this was to make a really great discovery, and it had occurred to me, although fleetingly at this point, that the opiate receptor could be my ride to the top.

Every morning I made my concoction, the soup du jour consisting of either homogenized rat brain or minced guinea pig intestine, with the addition of the new, hotter radioactive morphine. In order to make my soup, however, I had to overcome one obstacle, and that was my lingering squeamishness about killing a rat in order to obtain the fresh brains I needed. I hadn't yet killed an animal, having avoided the issue ever since Miss Oppenheimer had let me slip by back at Bryn Mawr. But now I was going to have to do the deed. And it did not come naturally.

I knew I had to desensitize myself if I was to succeed, and so I began the gradual process of rewiring my nervous system a good week in advance of my first day on the opiate-receptor project. Each day I forced myself to stand a little closer to the door of the room where they did the killing. After a few days, I was able to stand in the doorway and watch as the animals were decapitated, a procedure done with a slick little guillotine, allowing the brain to be quickly scooped out and immersed in a cold liquid buffer that kept the neurons alive and nourished while freezing the internal chemistry. Soon, I was able to stand right next to the bench and watch. Then I killed one myself. My hands trembled and my heart pounded, but I forced myself to do it. It was so traumatic, I had to sit down after that first time to regain my composure.

Eventually, the ordeal became easier, but never did it become something I could do with a total lack of feeling, coldly or cruelly. There was always a ritual to it, an awareness that this was a sacrifice of life for life. Taking the lives of these animals so that research could be done for potential cures to save human lives seemed like a fair trade-off, especially if done with respect and without inflicting suffering.

Some will argue that human life is no more valuable than animal life, and while I can well understand that viewpoint, when it came down to it, I made the choice I believed was the right one. These white rats had been bred for research, and scientists use them in ways I consider appropriate. In my career, I have never seen animals mistreated or killed in ways that promoted suffering. And if that has gone on in the past, as animal-rights activists claim, it can no longer be the case. Today there are strict regulations protecting animals that are used for experimentation, and researchers must apply to committees for approval of their testing methods.

But back to my experiment. After a period of incubating the animal organs with the hot morphine, I dumped my soup into the Triple M rapid filtration machine that Pedro had loaned me, and I rinsed away the unbound material, leaving the bound ligand to be dried right on the receptor. But how to tell if the receptor was really the opiate receptor and not something else the ligand had bound to? Taking Goldstein's lead, I introduced the stereoisomer opiate dextrophan into my soup, along with the radioactive morphine. In another test tube, I placed the levorphanol. Since the dextrophan was an incorrect fit, I knew it wouldn't compete with the morphine at the receptor site, so I could expect to see relatively high counts showing that the hot morphine had bound. The levorphanol would compete, bumping the hot morphine right off the receptor, and thus reduce my counts. I figured the difference between the two counts, a high one for the dextrophan mixture and a low one for the levorphanol mixture, would give a measure of the opiate receptor.

When developing a new technique such as we were doing, the rule of thumb is to first try the least complex combination of conditions and ingredients, and hope they work. If they don't, and you're faced with a busted experiment, you can go back and try to tell which one of the variable conditions—time, temperature, concentrations, number of washings, and so forth—caused it to go wrong. Every experiment is like a long chain, one that is only as strong as its weakest link. If this experiment was to work, I knew that I had to find that weak link, and to do it I'd need to be relentless, keeping a serious focus on mixing the ingredients of each day's cocktail, as well as attending to the other variables.

The doors of my world slammed shut as I put all my energy into the pursuit of the opiate receptor. Each day I altered the recipe, hoping to create the perfect bathing solution for my morphine to show a specific binding to the opiate receptor. I'd usually work long hours, transferring the contents of the last test tube around 7 or 8 P.M. Then I'd have to put them all in the counter, a device that kept track of the radioactivity in each test tube, somewhat like a Geiger counter. The counter sounded just like a casino slot machine as it ticked off the radioactivity in numbers that were then printed out on paper.

I loved the counting room—a feeling not shared by many of my fellow students. It was a place where the truth couldn't be fudged: Either you were successful or you weren't, because in the counting room there was no middle ground. Within its four walls, it wasn't unusual to hear wailing, groaning, or the occasional joyful whoop.

I remember how I'd place my samples into the counter every night, and, like a mother hen, I'd hover over it, listening, waiting for the initial clicking sounds, and saying a little prayer for the successful outcome of the experiment. Then I'd go home, hoping that when I returned the next morning, there'd be something, some numerical results that would have made our work worthwhile.

I was also in love with the raw data, which issued from the machine on a narrow slip of paper like that of an adding machine. In the morning, I would gingerly gather the data on these slips of paper in order to enter it in my lab notebook. It was almost like a formal ceremony to me, and it heightened my desire to get there in the morning to learn the results from the night before.

My devotion to the data was part of the romance. After the gathering comes the crunching, scanning the data for the particular pattern we hoped would be there. Then I would do the massaging, or organizing the data, to bring the pattern into sharper focus. I loved it all.

Regrettably, my opiate data provided nothing to massage or crunch. Day in and day out, there was only hazy noise, nothing but nonsense. I couldn't even repeat Goldstein's weak signal. Weeks went by, and there still wasn't any good news for me to plunk down on Sol's desk. I began to have pangs of despair. My work was my child, a child conceived by my imagination, but now I feared it would be aborted. There were some mornings when I had the urge to cry as I filled my notebook with senseless numbers, and often I had to fight the temptation to throw the notebook into the trash bin.

I was paying my dues, as they say, learning that to get an experiment right was often a case of figuring out exactly what was going wrong. Every day I'd make the forty-five-minute drive to the lab from the army base where we lived at Edgewood and go over my work again and again, looking for where I might have made a mistake, or what I could do to make the numbers mean something. But despite my exhaustive attempts to manipulate the conditions and materials, I kept coming up with gibberish.

My intuition told me that the opiate receptor was there. When I got deep into the literature, I noted that several opiate "experts" claimed they couldn't find it, so in their minds it couldn't exist. They were ignoring that chemists had been developing new synthetic opiates for years, all based on theories and hypotheses about an opiate receptor. Even at this early stage of my career, I employed one of the main manifestos of my mentor and refused to believe the experts. I'd adopted Question Authority as my motto, and I knew I had to live by it, especially now when the "experts" seemed to be winning.

Failure again, and again no hint of a signal. Certainly, Sol had been right about needing a hotter trace, but evidently this hadn't been enough. There was still another link in the chain that had to be strengthened before I would see a clear enough signal on the counter. I went back to my soup and continued to fiddle endlessly with the other variables in the hope that something would work.

Two things kept me going during this time. One was my fascination with the romance of opium and the pivotal part it had played in the Romantic movement of the nineteenth century. I was fascinated by the possibility of understanding the workings of a drug that had inspired a whole generation of artists and literati, thus sparking a revolution in thought and sentiment that swept through Europe at the time. The idea that there was a mechanism in the human brain that allowed all this to happen, and that I might discover it, was absolutely thrilling to me. When I wasn't in the lab, I spent hours researching everything there was to read on the subject, delving deep into the stacks and poring over articles that went back to the turn of the century, when heroin, the first synthetic version of opium, began its infamous public career. Touted as a nonaddictive cough medicine by the Bayer Company (of later aspirin fame), it was eventually discovered to be addictive and was later criminalized.

Another reason I persevered was purely scientific. In the medical literature I'd found a review that mentioned the recent work of a Dr. Hans Kosterlitz, a German pharmacologist who'd fled to the British Isles from Hitler's Germany and was doing research at the University of Aberdeen in Scotland. He'd proven that he'd been able to manipulate a guinea pig's ileum by using opiates such as codeine, morphine, heroin, and Demerol to induce constipation in vitro. He'd also emphasized that these same opiates were involved in human analgesia, or pain relief. How could this be, I wondered, unless there were identical opiate receptors in the human brain and the guinea pig gut?

I knew there was an opiate receptor, and if I was failing to find it, it could only mean that I was doing the experiment wrong. I continued the painstaking job of altering variables, trying to find a way to get a signal from the hot morphine and then measure it against the stereoisomers. If I could do this, I believed, I could show for certain that the opiate receptor did exist.

EUREKA!

In the midst of this anguish, Sol sent me and another student, Anne Young, off to Nashville, Tennessee, for an intensive tutorial funded by the American College of Neuropsychopharmacology. Fifty outstanding American graduate students had been selected for participation in a program that was intended to thoroughly indoctrinate us in the reigning biomedical paradigm and give us an opportunity to rub up against the scientific superstars of pharmacology.

On the last day of the program, the big guns jetted in to attend an extravagant banquet held in their honor, and then jetted out again. They were all handsome, happy, sparkling, and male, and I was entranced. Some of them I knew by name from conversations I'd had with Sol. Among them was Dr. Julius Axelrod, who refrained from giving a presentation, as befitted his still-fresh Nobel stature.

I had met Julie earlier that year, just a few months after he'd won the Nobel Prize. We were in Chicago to attend a *festschrift* that Sol had organized in Julie's honor. Sol introduced me at the dinner, dragging me, thrilled and nervous, over to Julie's table, proudly referring to me as "my little girl." It was a term of endearment he'd used back at the lab that I'm embarrassed to admit I didn't mind at all. The dinner lecture was given by Floyd Bloom, who'd come out of the NIH with a reputation for being a prodigy and was at the time fulfilling that reputation at the Salk Institute. After the applause, Floyd stepped down from the podium and headed straight for our table, where he chose a seat right next to mine. The experience of being in his presence was near-orgasmic, so powerful was his allure, and my heart pounded wildly as I listened to the conversation, too in awe of this bastion of scientific superstars to say a single word. After the dinner, a group of us drove with Julie to a nearby bluegrass joint, and I thought I'd die from excitement and joy. In an attempt to put us all at ease, Julie joked, "What kind of a drug is bluegrass?" But I was so overwhelmed by the experience of being in the same car with a full-blown Nobel laureate that I misunderstood and thought he was trying to have a serious discussion about something I just wasn't getting.

IT'S OFTEN a good idea, when faced with a busted experiment, to put the mind on vacation and wait for the unconscious to come up with the right answer. I tried to do this all during the sweltering eight weeks in

Nashville, but the opiate-receptor assay was never far from my awareness. I returned to Hopkins in late August with renewed vigor. Having had the experience of being in the company of world-class scientists at the seminar, I wanted more than ever to be part of the group, which required making a fabulous discovery. But first I had to figure out how to get my experiment to work.

I was beginning to get the idea that the hot morphine wasn't a good choice of trace. This insight came to me while in Nashville, when I had come across a rather complex article authored by a British scientist named W.D.M. Paton. He presented his notion of how two nearly identical drugs could bind to the same receptor. One, the *agonist*, could enter the receptor and create changes in the cell, while the other, the *antagonist*, could block the receptor by occupying it, which would have no observable effect on cell activity. Paton called this the Ping-Pong Theory, according to which the magnitude of drug action is proportional to how many times the drug hits, or "pings," the receptor, which in turn affects how long the drug remains on the receptor. Since the antagonist doesn't ping over and over again, it can stay longer on the receptor, and thus block the agonist's access.

If Paton was right, I needed a radioactive antagonist as a trace. Only a hot antagonist would remain on the receptor long enough for rapid filtration to separate bound and free drugs from one another. By the time I returned home from Nashville, I'd convinced myself that the hot antagonist was the missing piece separating me from my much-coveted Eureka!

It was in the midst of this latest insight that Sol called me into his office and told me he was going to shut down the opiate-receptor assay. I was crushed. There were too many other things to do, he told me, to waste precious money, time, and resources on what was looking more and more like a dead end. In addition, Sol explained, he was responsible for seeing that I got my Ph.D., and no one was going to give me an advanced degree for a busted experiment. In most cases, he assured me, the Ph.D. was awarded for dull and insignificant projects that had been wrung dry by endless explorations of the variables.

"But Sol," I pleaded, "you've got to let me continue. I know I'm close! All we need is a new hot trace!"

But my words fell on deaf ears. His only response was a disapproving grimace that spoke volumes. I headed for the office door, knowing that my days of looking for the opiate receptor were numbered, and if I was going to try anything new, I'd better get started right away. But I stopped as he issued his parting shot: "No! You can't spend any more money on wild-goose chases for new hot traces!" I looked away, bit my lip, and made my exit.

That afternoon, I brooded through the weekly lab meeting, making my unhappiness obvious to everyone in the room. But Sol was unflinching. From his perspective, a huge amount of time and resources had already gone down the drain, and what originally was intended as a quick raid to steal the prize from Goldstein was turning into a major campaign with no end in sight. He no doubt felt he'd been excessively generous: He'd given me a wide berth and plenty of resources to find the opiate receptor, and I hadn't done it. Sol knew it had been a long shot from the start, and now he'd decided it was time to stop trying. He was washing his hands of it.

The next day, I begged him one more time to let me give it another chance, but he was adamant and refused to listen to any of my ideas. The opiate project was over, he told me in no uncertain terms, and now I was expected to gracefully dive back into the original choline project.

But that was not my plan.

Naloxone, I had figured out, would be the best choice of antagonist to use. I knew that if a heroin addict was injected with a few milligrams of this potent drug, it would totally reverse the effects of overdosing, even in the case of a coma. It was assumed that this was due to the ability of the naloxone to bump the heroin from its receptor, displacing it and then occupying the receptor site itself—in other words, acting as an antagonist. Naloxone had the right chemical configuration to bind to the opiate receptor, but because it was an antagonist, it could not trigger any activity in the cell, could not create the euphoria or analgesia that the opiates did.

Secretly, I decided to get some naloxone and repeat the experiment with it when no one was around. The only glitch in my plan was I didn't know how to get my hands on some hot naloxone. Then I remembered that Agu had some cold, or nonradioactivated, naloxone at his lab in Edgewood, which he'd been using to reverse analgesia in his test monkeys. All I needed to do was to borrow some and send it to a lab in Boston where it would be labeled with a radioactive isotope. It would take a few weeks, and I'd have to hold my breath, hoping that no one in Boston would call Sol to confirm the order. But I decided it was worth the gamble. When the bill arrived from Boston, the jig would be up, and by then I'd be either a heroine or a donkey. But I didn't let myself think about the latter possibility.

I sent off the naloxone. Several weeks went by, and then one day in October, a call came that a package had arrived and was waiting for me in Radiation Control. I quickly retrieved it and stealthily put the hot naloxone through what I hoped was an adequate purification process. This done, I returned to Sol's lab and stashed the paperwork inside my workbench.

I decided to do the actual binding experiment that same Friday afternoon after everyone had left early for the regular TGIF pub party or had gone home to get an early start on the weekend. It wasn't unusual for me to work long hours into the night, so no one would be surprised if I waved them all good night and kept on working in the lab. But around four o'clock, I got a phone call from Agu with the news that the baby-sitter had gotten sick and couldn't pick up five-year-old Evan from preschool. My heart raced as I tried to think of what to do.

An hour later, I was flying down the freeway toward the day-care center where Evan was waiting. Instead of taking him home, I had decided to bring Evan back to the lab with me. It was risky, because there were strict rules regarding children being present, especially around radioactivity, but I needed at least one more hour to get everything set up. The actual data wouldn't be in until Monday, but I had to be there to transfer the brain membrane filters and put the test tubes in the counter. Somehow, I sneaked him past the guards, and we were suddenly safe and alone inside the lab.

"Mommy, what's that?" Evan asked, pointing to the Triple M.

"That's a big suitcase," I told him, because in fact this magical device did look like one of those oversized metallic suitcases. "Now, don't touch anything," I added in a hushed tone. I realized that I'd have to give him something to do to keep him busy while I did the transfers. I looked around the lab quickly and seized on the thirty-six empty, brand-new vials in which I would have to transfer the brain membrane filters.

"Honey, how about helping Mommy take the caps off these?" Evan lit up. I lifted him onto one of the stools and spread out the small plastic vials on the large, pristine workbench. He became totally involved, and with one eye on Evan and the other on my membrane filters, I completed the task.

When the last of the thirty-six test tubes was on the counter, I turned the machine on. I had included three variables, one for each group of twelve test tubes. In the first group of tubes was pure radioactive nalox-one; in the second, levorphanol in combination with the hot naloxone; and in the third, dextrophan in combination with the hot naloxone. Nor-mally, I would have waited around and stood by the counter, getting quick counts of the numbers as they ticked out, a rough but gratifying peek at the results within a half hour. But with Evan in the lab, I couldn't do this. And anyway, I told myself, if the results weren't any good, I didn't want to ruin my weekend. I set the counter to run through the weekend, gathered Evan up, and left the lab.

I arrived extra early on Monday morning and went straight to the counting room. I hastily ripped the ticker tape containing the numbers from the counter and returned to my bench. Opening my notebook, I turned to the protocol page and proceeded to slowly copy down the numbers, one by one. My hope was that the tubes containing levor-phanol plus hot naloxone would have a low count, since the very potent opiate levorphanol should block the naloxone from binding to the opiate receptor. Low numbers would indicate that a receptor binding race— not unlike sperm competing for entrance to the ovum—had taken place, and the naloxone had lost. Conversely, I hoped that the test tubes con-taining dextrophan, the inactive opiate, would have high counts, since the dextrophan was unable to bind to the opiate receptor, and gave the naloxone no competition. Of course, the tubes containing hot naloxone alone, with no competition from either isomer, should have the highest counts of all.

I had arranged the test tubes in a staggered order: lev, dex, nal only, lev, dex, nal only, etc. The careful protocol was the key that kept it straight. Playing a game with myself, I kept the left side of my notebook page covered—the side where I'd written the contents of each test tube—as I meticulously copied a single number onto each correspond-ing line. Low, High, High, Low, High, High. I could feel the growing excitement buzzing in my stomach as the first six numbers matched up to what I'd predicted.

I forced myself to keep my eyes from running down the counter tape and seeing the remaining numbers, continuing my careful recording. Low, High, High, Low, High, High. My heart was really pounding now. I copied the last numbers and rearranged the data into three columns. The results were staggeringly clean, and right on target. I had gone from no signal at all to a signal so loud it practically shrieked in my face, and all because of a single variable, the hot naloxone instead of morphine!

This was the killer experiment of my dreams, and I'd done it. I'd found the opiate receptor.

Anne Young was working at the bench next to mine when I entered the final numbers into my notebook. When I finished, I closed the book and turned to her.

"Anne," I said, my voice cracking from a dry throat. "I think we should go to a bar and have a drink."

Anne was always ready to have a good time, but now she glanced up from her work with an expression of concern. "Why? Are your results so bad that you need to get drunk?" she asked.

"No," I said, my voice getting louder. "No, it's just the opposite!" I was shouting. "They're that *good!* Let's go get some champagne and celebrate!"

The date was October 25, 1972.

THE NEXT morning, Sol returned from a conference. He was usually a little cranky on his first day back from a trip, so I wasn't surprised to be greeted with a scowl when I burst into his office.

"Sol, you're not going to believe it!" I exclaimed, laying my open notebook down on his desk. "It worked! It worked! We've found the opiate receptor!"

Intent and silent, he studied the numbers I had written in the notebook, lingering over them for a full minute while I stood barely breathing at his side.

"Fuck," he said in a low voice, continuing to look at the numbers. I began to feel apprehensive. Was he getting mad because I'd gone ahead and done the experiment against orders?

"Fuck, fuck, fuck!" he began to sputter, and then looked up at me, his face lighting up in a wild grin. He jumped up from his chair and began to pace excitedly around the office.

"The ball is in your court," he turned and announced to me suddenly. "You can have whatever you need. You can have Adele as your technician. Get her to repeat the experiment, and if it works, you've got her for good!"

I was relieved and thrilled. A second wave of pleasure—almost as intense as the one I'd felt the day before, when I saw the data for the first time—washed over me. I had made Sol happy! And now, as a reward, he was doing the unthinkable, plucking me out of my lowly status as a graduate student and thrusting me into a league light-years beyond, for only important senior scientists are privileged to have their own technicians to perform experiments.

Despite my excitement, however, I couldn't help but notice a strange glint in Sol's eye, one I'd never seen before. He seemed ecstatic yet oddly preoccupied, as if some grand plan were unfolding in the inner workings of his brilliant mind. What I didn't realize was that, in the world of big-league science, Sol had just seen how he could win the game. And, very soon, I would be the one he'd send out on the playing field, to score the points and claim the victory for my team, an ascending star at the very center of the action.

3

PEPTIDE GENERATION:
A CONTINUED LECTURE

I LOOK UP from my lectern out into the darkened auditorium, where my audience, barely visible, awaits my further words. A quick glance assures me they are still with me, and I click confidently to the next slide. As carried away as I get telling the personal side of my story, I remind myself I'm here to focus on the science, to explain the molecules of emotion and give my listeners some context for appreciating why the discoveries my colleagues and I have made may have profound implications for their lives. One of my favorite slides comes up: three rats, rolled over on their backs, limbs floppy, eyes closed, obviously in a deep swoon. "These, ladies and gentlemen, are rats in bliss," I usually say, and then pause for the laughter. "You can tell by their body language that they are totally satisfied and don't have a care in the world—the result of injecting our furry friends with a substance called endorphin, the body's own natural morphine : . . which your bodies make, too."

FINDING THE KEY

A shocking but exciting fact revealed by the opiate-receptor findings was that it didn't matter if you were a lab rat, a First Lady, or a dope addict—everyone had the exact same mechanism in the brain for creating bliss and expanded consciousness.

The discovery of the opiate receptor touched off a mad scramble among scientific researchers to find the natural substance in the body that used the receptor—the key that would fit the lock. We knew that the brain receptor didn't exist to serve as a binding mechanism for external plant extracts, such as morphine and opium. No, the only reason that made any sense for an opiate receptor to be in the brain in the first place was if the body itself produced some kind of substance, an organic chemical that fit the tiny keyhole itself—a natural opiate.

Less than three years after the discovery of the opiate receptor, the natural opiate was indeed found. A Scottish research team at the University of Aberdeen directed by John Hughes and Hans Kosterlitz (the same man whose work on the action of opiates in guinea pigs had confirmed my own hunch about the existence of the opiate receptor) demonstrated that a substance they'd isolated from pig brains was the brain's own morphine, an endogenous ligand that fit the opiate receptor and created the same effects that exogenous opiates such as morphine did. They called this substance enkephalin (Greek for "from the head"). Later, in a much-contested bout of one-upmanship, American researchers had named their version of the substance "endorphin," meaning endogenous morphine. (Scientists would rather use each other's toothbrushes than each other's terminology.) The American version is the name that has stuck.

BASIC BUILDING BLOCKS

Let's take a closer look at what a peptide is and how chemists came to work with them. Peptides are tiny pieces of protein, and proteins—from *proteios*, meaning primary—have long been recognized as the first materials of life. While it took over a century for chemists to be able to determine the chemical structure of a protein and write a formula that clearly described its elemental content and organization, we now know, as I covered earlier in my lecture, that a peptide consists of a string of amino acids, each joined together like beads in a necklace. The bond that holds the amino acids together is made up of carbon and nitrogen, and is extremely tough, able to be severed only after hours, or in some cases days, of boiling in strong acid. When there are approximately 100 amino acids in the chain, the peptide is considered to be a *poly*peptide; after 200 amino acids, it's called a protein.

In order to identify a new peptide, a chemist must first extract the substance and then isolate it from all other biochemical impurities. Then the challenge is to characterize it, a process that involves naming each of the amino acids that makes it up—there are twenty known major amino acids—and writing these names in the precise sequence of their arrangement. The result is the chemical structure of the peptide.

Now, some of you may be starting to doze, but you need to know the alphabet before you can learn to read. Amino acids are the letters. Peptides, including polypeptides and proteins, are the words made from these letters. And they all come together to make up a language that composes and directs every cell, organ, and system in your body.

Amino acids were the first substances extracted from living things to have their structures written by organic chemists—a process that began in 1806. The isolation and identification of the body's twenty amino acids was the result of a series of discoveries that took place between then and 1936, starting with L-asparagine, which was first isolated after the evaporation of a water extract of asparagus shoots. (You may have noticed a strong smell in your urine a few hours after you eat a large portion of asparagus—that's asparagine!) Threonine, the most recently discovered amino acid, and one that would be crucial to work I myself would be doing in the future, was isolated from a human blood clot, a substance that contains the protein fibrin, which must be boiled in acid for days to break its chemical bonds. In order of their discovery, from the first to the last, the twenty common amino acids are L-asparagine, cystine, L-leucine, glycine, DL-tyrosine, L-aspartic acid, DL-alanine, L-valine, L-serine, L-glutamic acid, L-phenylalanine, L-arginine, L-lysine, L-histidine, L-proline, L-tryptophan, L-hydroxyproline, L-isoleucine, methionine, and threonine.

It took over a century of work to discover the chemical structures of each of these amino acids, with chemists performing extraction after extraction of mysterious organic sources of protein such as silk, pancreas glands, wheat gluten, or casein from milk, until all they were left with was white crystals, indicating that what they had was the pure stuff.

NAMING THE BABY

But I want to return to our discussion of the substances the amino acids add up to—the peptides. In order to determine the chemical structure of any substance, peptide or otherwise, to write its name in terms of the atoms that make it up, the substance must first be purified out of the organic source that contains it, be it pig brains, guinea pig ileum, or the human brain. Once a sample has been extracted and purified of everything but the molecule in question, there are techniques to figure out how many atoms of hydrogen, of carbon, of this or that, it contains.

Finally, there are physical methods to determine how these atoms are arranged in space, eventually yielding the formula for the complete structure of the peptide, that is, its chemical name.

However, it took decades of ingenious chemical discovery before these methods were developed. Learning how to break the peptides apart, amino acid by amino acid, then atom by atom, was an immensely complicated endeavor. Thus, for many substances, their biologically active components were identified and measured years before their actual chemical structure could be written, because the biological explorations were based on a body of knowledge that considerably predated the more sophisticated kind of analysis involved in peptide chemistry.

And indeed those early biological explorations, which took place when peptides were still an obscure interest of a few far-flung scientists, looked so crude that they seemed downright primitive. Down in the dark basements of a dozen or so laboratories, men (alas only men) labored over huge simmering vats from which wafted the putrid odors of many pounds of pig pituitary glands, intestines, frog skins, sheep hypothalami, and so forth. Solvents such as acidified acetone were added to the vile soup in the vats in order to extract, or purify, the substance that was believed to be the source of a given biological activity. The resulting yellow brew, isolated from its confounding sludge of impurities, was evaporated of all solvents, until there was nothing left of it but a dirty powder. Then the powder was carefully doled out into glass dishes, each containing a particular animal tissue, and observed for signs of activity. Could the powder make an intestinal or uterine muscle contract, a blood vessel relax? Some chemists injected a solution of the powder into the whole animal, and watched to see if its ears turned red, or its blood pressure rose, or its sex hormones surged. If these bioassays showed clear signs of activity, then the powder would be further processed until there remained only pure white crystals. Again, the purified substance—which had now been reduced to what the researchers believed to be a single peptide molecule—would be put through the bioassays, and if its ability to tickle receptors on the tissue was still there, then it could be said that the peptide had been isolated. At that point the attempt to do the chemical analysis that would enable its structural formula to be written down could begin.

By 1975, scientists had worked out the chemical formulas for only thirty or so peptides, including the polypeptide insulin. Formulas were written as a sequence of three-letter abbreviations, each representing a specific amino acid. In late December of that same year, writing in the highly prestigious science journal *Nature*, the Scottish team triumphantly published their chemical analysis of the brain's own morphine, which consisted of a pair of peptides, each five amino acids in length. And so it happened that two more peptides were added to the slowly growing peptide family. But these were very special peptides, as I'll be explaining shortly.

Enkephalin's chemical structure was summed up by the following formula: Tyr-Gly-Gly-Phe-Met and Tyr-Gly-Gly-Phe-Leu. With this brief shorthand, a peptide chemist had all the information needed to whip up a batch of enkephalin from amino acid starting materials in just a few days.

A PEPTIDE HISTORY

Every peptide has its story, and as far as we know today, there are eighty-eight stories in the naked city called the body. I say "as far as we know today" because we still can't say that all the peptides have been discovered and their stories told. Research brings new peptides to light every year. There will probably be over three hundred by the time we are finished finding them.

The first peptide was discovered in the gut around the turn of the century and was classified a hormone when it was shown to act on a dog's small intestine to stimulate the secretion of pancreatic juices. This astounded physiologists who up until that time had assumed that all physiological functions were controlled by electrical impulses from the nerves. They named the substance secretin, but it took another sixty years to isolate it in its pure form and to determine its chemical structure. Another gastrointestinal hormone called gastrin, which turned out to be a small piece of a longer peptide called cholecystokinin (CCK), was discovered a few years later and found to carry messages from the pancreas to the gall bladder.

Another peptide, cryptically named "Substance P," was partially isolated from horse brain and intestines in 1931 by Ulf von Euler. He won a Nobel Prize for his feat, even though Substance P remained a chemically undefined "powder" for forty years, until Susan Leeman determined its eleven-amino-acid structure in 1971. Susan Leeman, at this writing, has yet to win a Nobel Prize; in fact, she was denied tenure at Harvard, where, after she defined Substance P's structure, she discovered that the peptide's activities went beyond those we knew about—

lowering blood pressure and contracting smooth muscle—to the transmission of pain through certain nerve fibers.

The first peptide ever to be replicated outside the body was oxytocin. This is the substance that is released from the pituitary gland during childbirth to bind with receptors in the uterus, where it causes the uterine contractions that will eventually do the work of expelling the baby. As early as 1902, people knew there was something in crude extracts of farm animal pituitary glands that could be used by obstetricians to aid women who'd been in labor a long time.

Contemporary pharmacologists, neurobiologists, and physiologists like Sue Carter, Tom Insel, and Jaak Panksepp have shown that oxytocin not only contracts the uterus in labor, but also produces the uterine contractions of sexual orgasm in females. In the brain, it acts to produce maternal behavior, stops infanticide, and seems to help some male rodents find long-term, monogamous relationships. This unifying function of peptides, coordinating physiology, behavior, and emotion toward what seems to be a coherent, meaningful end, is very characteristic of humans and animals, and is something we will explore in more detail later.

It took many months to make a synthetic form of oxytocin, but the feat was finally accomplished in 1953 by Victor du Vigneaud in his New York laboratory. As befitting the difficulty of his accomplishment, the extremely dedicated du Vigneaud spent his nights on a cot in his office instead of leaving his experiments and returning to his family on Long Island. He wanted to be available around the clock to add key ingredients with perfect timing to his difficult synthesis. For his trouble, he received the Nobel Prize the following year. Although he produced only a tiny amount of synthetic oxytocin for all his labors, it exhibited the chemical traits and biological activity that proved to the world that he had indeed replicated the simple nine-amino-acid-long peptide that was the active ingredient in the pituitary gland. Today, a synthetic analog known as Pitocin is used routinely to induce and speed up labor when the doctor has decided it's time for the baby to be born, even if the mother's body and the fetus don't agree.

Oxytocin is the darling of the peptide revolution. Its importance in peptide history cannot be overstated, because once it was demonstrated with oxytocin synthesis that chemists could make something identical to what the body could make, they also realized they could attempt to improve on nature. Now scientists began to make a series of analogs, or substances with almost the same structure, by substituting this amino acid or that amino acid for the amino acids in the authentic sequence, and testing these various analogs for their activity. The resulting therapeutic analogs, or drugs, could be made more potent, longer-lasting, and more resistant to decay than the body's own self-healing substances.

A few years after du Vigneaud's nightly vigil in the laboratory, Bruce Merrifield of Rockefeller University speeded up the process of synthetic peptide production by inventing solid-phase peptide synthesis. This was accomplished by attaching one end of a growing peptide to a tiny plastic bead and gradually adding the amino acids one by one, in a cleverly controlled cycle of chemical reactions. Now easy-to-harvest peptides in large yields could be produced routinely—a feat that literally made possible the peptide revolution that is exploding around us today. Merrifield's methodology won him the Nobel Prize in chemistry in 1984.

Today, Merrifield's solid-phase column is operated with computer technology and is commercially available. Any given peptide sequence can be programmed in, the peptide reproduced in a series of automatic steps during the night, and then purified out the next morning. Thanks to Dr. Merrifield, modern chemists can now spend their nights at home and sleep with their spouses.

And yet, remarkable as Dr. Merrifield's accomplishment is, our bodies are capable of making more peptides, perfectly produced in a purified state, in one night while we sleep than all the peptide chemists who have ever lived have made in all their high-tech laboratories since 1953, when synthetic peptide production began. How does the body do it? Amino acids are strung together to make peptides or proteins in minuscule factory sites called ribosomes, which are found in every cell. A double helical strand of DNA, the genetic material in the nucleus of the cell that codes for the needed peptide or protein, unwinds and makes a complementary working copy of RNA. The RNA information, which is a copy of the sequence coded by the DNA, floats to the ribosome. Every amino acid has a "triplet code" of three nucleotides that causes a given amino acid to be transferred and joined to the peptide or protein growing on the ribosome. Marshall Nirenberg of the NIH won the Nobel Prize for cracking this triplet genetic code in 1960. His work created the decoder key that has made today's mapping of the human genome possible.

THE PEPTIDE/BRAIN CONNECTION

At the time Hughes and Kosterlitz revealed their findings about the existence of the enkephalins in the brain, the field of peptide chemistry had synchronously matured to a stage where a modest number of scientists had set themselves the task of finding the peptides responsible for a wide range of bodily activities. Some were looking for the peptide factors that regulated digestion and assimilation in the gut, or the factors responsible for raising and lowering blood pressure in the body's circulatory system. Others were trying to identify the peptide components produced by the almond-sized master gland, the pituitary, found at the base of the brain below the hypothalamus. The Italian pharmacologist Vittorio Ersparmer and his fellow researchers had found and completely purified over thirty peptides in a Macbethian witches' brew extracted from the skin of frogs. All of these chemists were working heroically, if myopically, to purify and then reproduce their own chosen peptides, each of which was believed to issue from one particular source and to govern one particular activity of an organism, be it human, animal, or microbe.

The science was young enough at the time that each discovery was greeted with excitement. But none had the impact of the Hughes-Kosterlitz findings. And it wasn't simply the discovery of an endogenous substance to fit the body's opiate receptor that set the scientific world afire. No—it was the finding that this substance was a peptide, one that was not only produced in the brain but had a receptor for its functioning that was also located in the brain, that caused all the excitement. In other words, what seemed to be local pain-relief effects occurring in places throughout the body were actually mediated in the brain. This opened up the possibility that other peptides with seemingly local sources—and effects—might also be produced in the brain and/or bind with receptors in the brain. Every peptide that had been identified in the last century was now a candidate for a brain-receptor search, a task accomplished first by using the principles of the tool we had developed at Hopkins—opiate-receptor assay and the early forms of receptor visualization—and later with more sophisticated methods such as color and computerized in vitro autoradiography. Peptides could now be investigated in relationship to how they interacted within the brain to bring about many of the organism's internal processes.

It was not until much later that we realized that each and every peptide, no matter where it had first been discovered, was actually made in many parts of the organism—including, often, the brain. The pituitary peptides, it turned out, were gut peptides after all. The frog skin peptides were also hypothalamic-releasing hormones. The same peptides that bound to receptors in the kidney to change blood pressure could operate receptors in the lung and brain. Moreover, many substances not previously identified as such turned out to be peptides. Hormones, with the exception of the steroid sex hormones testosterone and estrogen, were peptides. Insulin was a peptide. Prolactin, which signals a woman's breasts to secrete milk, was a peptide. The gut cell substances that guided every step of digestion and excretion were peptides.

Although peptide structures are deceptively simple, the responses they elicit can be maddeningly complex. This complexity has led to their being classified under a wide variety of categories, including hormones, neurotransmitters, neuromodulators, growth factors, gut peptides, interleukins, cytokines, chemokines, and growth-inhibiting factors. I prefer a broad term coined originally by the late Francis Schmitt of MIT—*informational substances*—because it points to their common function, that of messenger molecules distributing information throughout the organism.

Suddenly, even before all the implications of the Hughes-Kosterlitz findings had begun to be explored, peptides were an infinitely more interesting class of chemicals than they had been before, and everyone wanted to know if their peptide was brain-involved. In my own lab at the NIH, which was where I went after leaving Sol's lab at Hopkins, I began to look for brain receptors for known peptides such as bombesin, vasoactive intestinal peptide (VIP), insulin, and a number of peptide growth factors heretofore never believed to exist in the brain. During that time, a veritable parade of these new *neuro*peptides was being reported on an almost monthly schedule. Oxytocin from the pituitary, insulin from the pancreas, angiotensin from the kidney, bombesin from frog skins, vasoactive intestinal peptide (VIP) from the gut, the impossibly named gonadotrophin-releasing hormone (I've spared you its other three names!) from the hypothalamus—all of these and more were found to be present in various locations in the brain and to have receptors in the brain.

In my lab at the NIH, we made the assumption, based on advanced methods of brain visualization we had developed, that any peptide ever found anywhere, at any time, was potentially a neuropeptide, with receptors in the brain. Adapting the new opiate-receptor technology, we went looking for peptide receptors in the brain, intending to map the location of both the actual peptide containing neurons and the location of their receptors. We were almost never disappointed. Most candidates for the search were clearly shown to both have receptors in the brain and also to be present themselves in the brain. We were even more excited—and surprised—when we found that peptides existed in all parts of the brain, not only in the hypothalamus, where endocrinologists had classically predicted they would be confined. Peptides also appeared in the cortex, the part of the brain where higher functions are controlled, and in the limbic system, or the emotional brain.

It was this mapping of the distribution of the neuropeptide receptors—showing us where they were located and where they were the densest—that proved to be the real fruit of the peptide research explosion. By understanding the distribution of these chemicals throughout the nervous system, we got the first clues that led us to theorize about peptides being the molecules of emotion. But I'm getting way ahead of myself. . . .

I need to pick up the threads of the peptide revolution story back where my own role in that revolution began, which was several years before the findings published by Hughes and Kosterlitz. Within the world of the laboratory, the receptor that I discovered in 1972 would have to wait for the discovery of its ligands—the enkephalins—to fill it. But there was nothing passive about this waiting. Indeed, shortly after my discovery, those of us in Sol's lab pitted ourselves against Kosterlitz's lab in a frantic scramble to be the first to find the ligand. They won— even though we played pretty dirty. And the story of that competition is a story typical of much of modern science.

❧

One day, after the opiate-receptor discovery, but before our first paper on the finding had been published, Sol called me into his office.

"Marshall Nirenberg wants to hear more about the opiate receptor," he announced. "Can you drive down to the NIH next week and show a few slides?"

"Marshall Nirenberg?!! The NIH?!!!" I gulped with terror.

"Don't worry. He won't bite you. Marshall's actually quite shy." Sol laughed, then stifled a yawn and began to rearrange some papers on his desk, obviously having already made his decision to send me and in a hurry to get back to the bigger fish he had to fry.

"But next week?!!" I gasped in astonishment. "I won't be ready," I stammered.

"You'll do fine," he reassured me, looking up one last time. "You need the practice. Anyway, you'll be giving a lot of talks on the opiate receptor soon, so you might as well get used to it."

4

BRAINS AND AMBITION

WINNING—it's the fuel that feeds the modern science machine. Being first with the facts is what we all strive for, but, even more, being first to publicly announce the results of the research in a top scientific journal—this is the big payoff. The history of science is loaded with tales of people who performed a "killer" experiment first, but were scooped in the eyes of their colleagues because a competitor made it into print before them.

Boldness and self-confidence are the winning scientist's defining traits. The tendency to ponder over or repeat experiments endlessly are the hallmarks of the second stringer. When a paper is rejected, as is often the case for truly creative ones, it helps when the author is willing to carry on a scathing, Solomonic, or self-righteous defense by phone and fax with journal editors—while remaining polite, of course, with just a hint of superstar arrogance.

GLORY

Less than two months after the success of the opiate-receptor assay, Sol summoned me into his office. Adele was repeating the experiment and getting good data daily, and I had accumulated a lovely little pile of graphs and tables that were carefully scrutinized for inconsistencies and thoroughly massaged for clarity. I think I expected that this process would go on forever, and I suffered quite a jolt when Sol announced that it was now time to write a paper and report our results—and that it had to be done now. Immediately.

Sol didn't know the meaning of writer's block, and had no patience with people who suffered from it. His strategy was to speedily dictate a complete first draft, no matter how crude, with the person who had done the most work present in his office at the time. The first author, usually the one who had done the work, would take over from there, editing the transcript, filling in the usual huge holes, explaining the methods, checking every number and factoid in fine detail, and finally returning the manuscript to Sol for a final read.

Now it was my turn to be first author. The first thing Sol directed me to do was boil down our data for summary in two tables, as required by *Science,* the highly prestigious journal he was targeting. Then he had me lay out my pages between us on his desk. Sol studied the data intently, all the while yanking at his hair, stroking his face, and emitting several involuntary snorts, as was his habit during heavy concentration. Suddenly, he grabbed his handheld tape recorder, leaned back in his chair, and began to dictate: "An abundance of pharmacological evidence suggests the existence of opiate receptors. . . ."

Once the rough first draft was set down, Sol was ready to teach me the further art of writing a scientific paper. Basically, he explained, the report must be simple and clearly written. Anything overly elaborate, containing too many ideas, will be rejected by the top publications, he warned. The ideal, he emphasized, was a paper of such pristine simplicity and crispness that anyone—even the dullest of technicians—could use it to repeat the work and achieve the same results.

Together, Sol and I refined months of mind-breaking labor into just fifteen elegantly written paragraphs. The title was simply "Opiate receptor: demonstration in nervous tissue," followed by our names, Pert, C. B. and Snyder, S. H. This order was a matter of scientific-paper-writing tradition, which Sol always followed: first the name of the person or persons who did the bulk of the actual work, last the name of the "senior author," who had raised the money to make the work possible, with the names of other contributors, if there were any, distributed in between.

As soon as we'd written the paper, I had so many nervous student qualms that I recruited Pedro Cuatrecasas and the department chairman, Paul Talalay, to scrutinize the manuscript before we submitted it. I was glad that they caught a number of careless errors, but it seemed that no one but me was bothered by the fact that we had failed to cite over twenty years of published data suggesting the existence of opiate receptors. How else could my research be explained? I risked making a brief but heated argument that we should at least credit Goldstein's idea in the introduction, but lost on the basis that his actual method was unrepeatable by others. Instead, we devoted a whole paragraph of the summation to making it clear that the results we had achieved with our new method bore no resemblance whatsoever to Goldstein's findings.

Now the rush was on. In early December 1972, barely six weeks after I'd completed the first successful experiment, we submitted our paper to *Science*. It was accepted immediately and scheduled for publication the first week of March 1973.

A DAY OR so before the publication date, Sol called me into his office. This time it wasn't about the data.

"Look at this. It's totally worthless and boring!" he said, angrily pushing a three-page document headlined with a Johns Hopkins logo in my direction.

I gathered that "this" was the press release prepared by the Hopkins media office to announce the opiate-receptor finding. Though it seemed fine to me, Sol clearly found it unacceptable. He abruptly turned to a dusty typewriter in the corner of his office, scrolled a piece of paper into it, and with total concentration began to pound the keys. Within a few minutes, he yanked the paper from the machine and, with great relish, handed it to me.

"Now, *this* is a press release," he announced. "Please get it down to the media office right away."

By the way he glanced at his watch, I knew I had to hustle. Clutching the press release in my hand, I tore through the hallway and flew down the stairs to the first-floor office, skipping every other step as I went.

Sol's press release was a surefire bull's-eye, and got the attention of more than a few people. A press conference was scheduled for the very next day, and I was about to experience science as a media event for the first time. That night, I tried to roll my hair in curlers and prepare for the conference, stopping every few minutes to answer the phone and talk with a reporter from UPI or the Knight-Ridder newspapers. The next day, I joined Sol and William "Biff" Bunney, a research psychiatrist, along with a handful of functionaries from government offices who were interested in showcasing the discovery as a major step toward a solution to the addictive drug problem. When we arrived at the Hopkins official press-conference room, we were greeted by dozens of reporters and photographers, their flashes lighting up the space. I remember being nervous and thinking the whole thing was a bit overblown, and was grateful when Sol and the others did most of the talking. What I didn't know then was that Dr. Bunney was about to become the first head of the National Institute of Drug Abuse; Sol was about to become one of the best-funded scientists in the world; the White House was about to be acknowledged for funding what was heralded as a giant step in the war against drugs; and I was about to become famous at age twenty-six.

The reporters grumbled a bit when it turned out that we hadn't actually found the cure for heroin addiction, which had been their expectation. But it was a good story anyway, and it instantly flew around the globe, alerting the world that a tiny bit of the human organism had been discovered, one long theorized about but never before demonstrated. A new molecular sensor too small to see, like tiny eyes or ears or taste buds, had been found in the brain. And what it sensed were the drugs of the opiate family—morphine, opium, heroin—causing the organism to be "turned on" and inducing the "high" that users of these drugs often experienced. The hope that one day this discovery might help the desperate heroin addict did not seem so far-fetched, after all.

The story was widely covered by the popular press: *Newsweek, U.S. News,* the *Washington Post,* the *New York Times*—they all picked up on the story and ran with it. When *Newsday,* my hometown Long Island newspaper, featured me in the story, I got clippings in the mail for weeks from people I hadn't seen since I was eight years old. The *Baltimore Sun* did an in-depth follow-up, complete with a large display photo of me and Sol in our lab coats. Pert and Snyder, the dynamic scientific duo, the winning team, were appearing on front pages everywhere.

It didn't take me long to get the hang of being in the media spotlight, and I must admit that I quickly came to enjoy it. But even more exciting was the chance to explain the work to our peers at the many scientific conferences that year. Sol, who generally disliked going to the more general open conferences because the specialized psychiatric and pharmacological meetings he routinely attended were more important to him, sent me out to schtumpf in his stead. Perhaps he felt uncomfortable appearing before members of the "Opiate Club," as the researchers who had been working in this field for years were called. They weren't his crowd. Sol was a newcomer to their field, and could be seen by them as swooping down to take the prize out of the hands of the "experts." So I did the traveling road show, and the more I presented, the more I felt an owner's sense of pride in this discovery.

Everyone, I found, was very excited about the news and wanted to hear more. A pivotal meeting, sponsored by the International Narcotic Research Club in Chapel Hill, North Carolina, took place a few months after the *Science* paper was published. I remember my sheer terror when I found out I'd be presenting to Avram Goldstein from Stanford, Hans Kosterlitz from the University of Aberdeen, Albert Herz of the Max Planck Institute in Munich—all solid members of the Opiate Club—along with many Europeans who had been meeting informally for years. I arrived at the podium after laboring long hours over my talk, loaded with forty or more hot-off-the-press slides. People whose papers I had read and whose work had inspired me were sitting in the audience. Now I was standing in front of them, about to reveal what they'd been searching for for years but been unable to find. My heart was pounding, my mouth dry. I fumbled with the clicker as the lights went down, hoping that when I opened my mouth, the words I'd rehearsed for hours would be there.

One of the slides I'd laboriously prepared had been taken from Goldstein's classic *Principles of Drug Action,* so I was surprised when he jumped up at the end of my talk to announce that the slide—which compared the three-dimensional chemical structure of levorphanol and dextrophan—was incorrect. Apparently, the two images I'd copied had been accidentally reversed.

"We caught the publisher's mistake," the grand old man of pharmacology intoned, "but we left it in the second edition to trip up unsuspecting young graduate students like yourself," he said with a half-smile.

A resounding hiss arose from the audience at this blatant swipe. Goldstein was obviously still smarting over the *Science* paper and was getting in his licks at the upstarts, Snyder and Pert. But I knew that I had won over the formidable crowd when, at the close of my talk, a number of people in my audience bounded up onto the stage to shake my hand and introduce themselves. It was ecstatic, this moment of instant acceptance from colleagues I admired immensely, who were so excited about the discovery that they were even willing to overlook the fact that I was a very green, very nervous, twenty-six-year-old female graduate student.

My euphoria was tempered by an excruciating migraine headache that began just as the crowd started to disperse. It was brought on, most likely, by the sudden letdown I experienced after weeks of prelecture preparation. But the mood around me was jubilant, and while I was tempted to go back to my room and lie down, I didn't want to miss the chance to be escorted around town by Hans Kosterlitz, who was calling for a celebratory meal. Hans, who was short and vigorous for his seventy years, was clearly taking me under his wing. We marched around the quaint college town, followed by a parade of a half dozen pharmaceutical company chemists, stopping in at various drinking establishments to toast the momentous occasion. I could barely keep up with him, as he easily downed Scotch after Scotch, but I did lose my headache in the exhilaration. We wound up at a steak house where we feasted on prime rib, paid for, of course, by the drug company boys, who were drooling even more over the idea of exploiting our science than over the huge slabs of beef on our plates.

It was in the inebriation of the moment that Kosterlitz confided to me that his team back in Scotland was looking for a natural substance, extracted from pig brains, that acted just like morphine when sprinkled over a certain novel smooth muscle preparation, which he, despite my excited cross-examination, refused to divulge. Leaning over close to me in order to elude the drug company boys, he whispered, "I've got a new man coming into my lab, John Hughes is his name, and he's bright, very bright! We're going to find it, we've got a way!" he boasted excitedly. In the next moment, thinking better of his indiscretion, he repeatedly swore me to secrecy.

Back in Baltimore, all promises forgotten, I told Sol about the encounter with Hans. "I think he's onto the endogenous ligand for the opiate receptor," I told him.

Several months earlier, Sol and I had abandoned a perfunctory run at finding the ligand ourselves. I had brought him some promising data that pointed to a possible ligand, but after a careful review of my data during a long meeting in his office, Sol had finally made up his mind.

"Drop it," he said. "It's too iffy, and you've already got plenty to follow up on with the opiate receptor."

But I could tell he was now more interested, and although he didn't say much, I suspected he was planning to find out more about what Kosterlitz was doing.

I continued to represent the Hopkins team at conferences around the world, and although I learned to adapt the humble demeanor at the lectern expected of someone presenting a very important finding, my budding scientist ego thrived on the strokes it received. Later, I realized that by having me appear so often and so quickly after the discovery, Sol was ensuring that we were staking an effective claim to the opiate receptor, for, as it turned out, there were other claimants. One of these was Eric Simon, a professor and researcher at NYU Medical School, who had been searching unsuccessfully for the opiate receptor for years. Recently, he'd been experimenting with a radioactive form of etorphine, a highly potent morphine analog used in tranquilizer guns to stop rampaging rhinoceri and other big game. The resulting data had shown some promise, and he planned to present it at the huge Federation of American Societies of Experimental Biology Conference in April, the month after our paper was to come out.

Missing no chances, Sol had managed to get himself a slot on the agenda that followed Simon's presentation at the meeting. Breathlessly, he showed slide after slide that I had made for him, each displaying data from the *Science* paper and from the follow-up research that Adele and I, working furiously, had produced in record time. Simon watched, taking notes, pleased that the symposium was going well.

From Simon's point of view, it was clear that he was a codiscoverer of the opiate receptor. When Sol failed to cite him as such in the early follow-up papers, he was offended and hurt. But Simon's very first paper not only cited ours, it reported the identical assay system I had painstakingly developed, using the hot etorphine instead of naloxone. I had shown him how to use the rapid filtration Triple M machine when he had visited me at the Hopkins lab.

Part of the education I received from Sol, for better or worse, was not only how to effectively beat out a competitor but also how to let the world know that you had won the race by strategic paper citation—and omission—a point of gamesmanship he knew so well. After all, we had published first, and that made all the difference.

EXPLOITATION

Once the initial hoopla died down, we entered a period of intense scientific activity. Each night, I designed experiments for Adele to set up the following morning, all attempting to answer the many questions that the discovery now allowed us to ask. Exactly where in the brain were the opiate receptors located? What part of the cell did they occupy? How simple or primitive could an organism be and still have the opiate receptor? Now I was a frequent and welcome guest in Sol's office as I regaled him with my abundant new data, and together we worked long hours to prepare papers on the results. Later, to my disgust, I learned that a few jealous postdocs were circulating the rumor that Sol and I were having an affair. This was a classic slur, one I was to hear again and again in succeeding years whenever a female colleague, particularly an attractive one, made a significant contribution and rose in status.

What was going on between Sol and me was a far cry from the illicit dalliances our colleagues were imagining. Behind the closed doors of Sol's office, I was getting trained in how to exploit a major discovery. In the scientific world, there's no time to stop and smell the roses after making a big breakthrough or developing a new technique, because someone else will come along and pick the next bouquet. The window for doing the follow-up work closes quickly, as peers join the race and gain ground rapidly. Sol and I were out in front, and we were planning on staying there for a while.

While I was working out the kinks, Sol got some of the other people in the lab to try out the new technique to see if they could use it to look for additional neurotransmitters. Anne Young (then a medical student, now the head of neurology at Massachussetts General Hospital) was the first to hit paydirt. She used the rat poison strychnine, which causes convulsive muscular contractions, as a radioactive antagonist to find the receptor for the neurotransmitter glycine, which causes muscular relaxation. Instant Eureka! Immediately, Sol switched all his postdocs over to our method and directed them to use it to scan for receptors for all the known brain chemicals. When I showed signs of getting possessive about my hard-won methodology, Sol ordered Adele to show everyone the ropes—how to make the "magic membranes," as I called them, when to mix the test tubes vigorously, how to filter—all the little tricks of the trade Adele and I had evolved to guarantee good data every day.

Sol ordered a dozen new Triple M's and thousands of dollars' worth of radioactive ligands. Like manna from heaven, successful binding assays for the various neurotransmitters began to rain down on the blessed lab in the heart of inner-city Baltimore. While it had taken me months to work up the procedure from scratch, positive results from new receptor assays seemed to fall from the sky on the first or second attempt. The norepinephrine receptor! The GABA receptor! The dopamine receptor! We found them all.

We were learning that each receptor has its own special requirements for the conditions—the soup du jour—in which it will reveal itself. One receptor might show binding when its soup was loaded with sodium, while another preferred a heavy dose of chloride. Getting it right might take the postdocs hours of fiddling, but their job was nothing like the seemingly Sisyphean chore I had done to get it straight the first time, when everyone around me believed it was impossible!

One of the first questions we tried to answer regarding the opiate receptor was why some drugs like morphine and heroin fit the receptor and caused enormous behavioral changes, while their antagonists, like naloxone, nearly identical in chemical structure, fit the receptor and resulted in no change, in effect blocking or "antagonizing" any further activity. Moreover, if an antagonist like naloxone was put in competition with morphine, it would move in and bump the morphine right off the opiate receptor, which was why it was such an effective antidote for heroin overdose. But how did this happen? A clue to this mystery came from an observation I'd made in my original opiate-receptor assay, which was that naloxone required sodium to perform its blocking action.

My first big follow-up finding came as the result of a turf battle that I fought on two fronts, playing hardball to keep the opiate receptor from slipping out of my hands. In the paper he finally published, close on the heels of ours, Eric Simon pointed out that etorphine, the big-game tranquilizer he used in his experiments, was weakened when sodium was added to its soup. The only difference between his results and ours, he reported, was that his etorphine binding was diminished by the presence of sodium, while our naloxone's was boosted. I wondered if this sodium difference could be the clue to one of the biggest mysteries in pharmacology: What makes one drug (like etorphine) an agonist, and another drug (like naloxone), which is almost identical except for tiny molecular differences, an antagonist? Why does etorphine mimic morphine in all its effects from euphoria to muscle relaxation, while naloxone blocks all the effects of these and other opiate drugs? Both agonist and antagonist were believed to bind to the same opiate receptor, but somehow their "intrinsic activity"—the effect they had on the cell—was different.

As soon as Sol spurred me on with a preprint of Simon's paper, I raced to set up an experiment to prove that sodium was the decisive factor, the one that could be used to tell the difference between an antagonist and an agonist—and not just between etorphine and naloxone, but between the agonists and antagonists in a whole cornucopia of opiate analogs we had by now accumulated. As I concocted a nifty little system for testing this "sodium shift" on all of these opiates, it was easy to stay two steps ahead of Eric Simon. But I was facing a new challenge from Gavril Pasternak, a medical student who was spending some time in

Sol's lab, and who, behind my back, had been steadily encroaching on Adele.

Sol had put Gavril on a project that involved the purification of the opiate receptor itself, a problem he was not able to crack with the pre-historic methods available in those days. Hitting a dry hole, Gavril had started to explore how some of the chemicals off the shelf affected opiate-receptor building, and so had reason to commandeer Adele when she wasn't busy with the experiments I had assigned her.

At first, I tried to ignore this infringement on what I considered my territory and concentrate on validating my method for discriminating agonists and antagonists in the test tube. My results were good, and, once again, I was making Sol jump for joy. The applications for this new testing system were enormous. It meant that a tiny quantity of any untested new chemical could be screened for its ability to be an agonist or antagonist in a day instead of the weeks or even months it had taken previously. Very quickly and very precisely, I could point to where in the spectrum between agonist and antagonist any given opiate was to be found.

The drug companies soon caught wind of what we were doing and were agog, since, at the time, they were looking for "mixed agonist-antagonist" drugs, that is, drugs that acted like agonists in one test, antagonists in another. Such drugs, they believed, would have an incredible potential as nonaddictive opiate pain-relievers. A dream come true for a drug company! I loved seeing Sol dance with pleasure as I handed him slides showing the intrinsic activity and potency of these substances, samples of which the pharmaceutical industry had given us to test with my new techniques.

But at the same time, part of me was distracted, feeling I couldn't turn my back on Gavril as he scurried around with his test tubes doing God knows what with *my* opiate receptor. I knew something was up, and it made me nervous. More and more, Adele was doing his bidding, and when I complained, Sol shrugged and offered no response. I tried to convince myself that it didn't mean anything. After all, Adele was so efficient, and the opiate-receptor assay so simple, even sharing her with Gavril couldn't possibly fill her days. But no, there was something more, and my suspicion continued to grow.

It wasn't long before I got the news. Now it was Gavril, instead of me, locked up with Sol in his office for hours, supposedly writing a paper on Gavril's findings. When they emerged, Sol asked me to give the draft a quick critical read, because they were expecting to submit it to *Science* the next day. A quick glance told me Gavril was claiming that EDTA, a component of his assay solutions, acted in the same way as sodium did, and had an equal ability to discriminate between opiate agonists and antagonists.

I took the draft of the paper home and pored over it that night, sensing that something was wrong, but unable to put my finger on it. The next morning, I was still racking my brain as I made the forty-five-minute drive to Baltimore. It hit me just as I was making my exit—EDTA had a negative charge and needed a positively charged ion to balance the crystal. The balancing ion must have been sodium! I could hardly wait to get to the lab to check the reagent bottle to confirm my suspicion. I was right—the label said *sodium* EDTA. Gavril had mistaken the EDTA as the agent of action when it was really the sodium in the solution that was responsible. He had inadvertently proved my thesis!

If I'd had less of a competitive streak, I might have kindly offered a suggestion: "Hey, you guys should double check this, I suspect that it's the sodium, not the EDTA, that is responsible for discriminating the binding action . . ." But instead, with an evil glee, I grabbed Adele and got her to do a quickie experiment, comparing the discriminating abilities of sodium chloride, sodium EDTA, and a nonsodium EDTA. The EDTA alone did nothing, while the sodium soups were the clear winners. I walked into Sol's office and cockily slapped my data down on the desk as if it were the ace of spades.

"Boy, you'd better keep closer tabs on Gavril," I announced. Sol looked up at me, clearly puzzled. "You guys almost humiliated yourselves with that paper."

After that, the opiate receptor was mine. I had won the battle, but Sol never treated me quite the same. I had gotten down in the dirt to scrap with the boys and had emerged victorious, as a new strength to be reckoned with. From that moment on, I was no longer the innocent "sweet baby girl" in my mentor's eyes.

THE UNRAVELING of the many mysteries about the opiate receptor continued to occupy our attention. Although the dominant image of the receptor was one of a lock that opened when the right key, or ligand, fit into it, I was beginning to understand that this metaphor was not an accurate one. The idea of a lock and key was much too static, not nearly dynamic enough, a description more appropriate to the older, more mechanical Newtonian paradigm than to the facts as we were seeing them. I was starting to realize that the receptor changes shape, switching back and forth between any number of predominant configurations, all the while vibrating and swaying rhythmically to some as yet unknown melodic key.

In addition to studying the action of the receptor, my other preoccupation in the lab was gathering data to show the distribution of opiate receptors in the brain. In what locations were they the thickest, where the sparsest? I was also curious about how the opiate receptor had evolved over time, so I attempted to measure them in brains of the vertebrates, starting with the gruesomely ugly hagfish, the lowest true vertebrate, and moving painstakingly up the evolutionary chain through snakes, birds, and rats, eventually reaching monkeys. They all had opiate receptors, which meant that this molecule had been conserved over time, through eons of evolution, and therefore probably had been of great importance to the organism's survival.

I knew the day would come when I'd have to go looking for the opiate receptor in the human brain, but I was completely unprepared when Sol called me into his office one spring morning in 1973 and told me to contact the Baltimore city morgue, pronto. He'd heard that a competitor was planning to publish data from a study of opiate receptors in the human brain, and our latest paper, which we had just prepared for *Nature*, contained only monkey-brain data. Sol wanted me to get some human brains, run them through my assay, and quickly assemble some data to add to the paper before it went off to the journal. I called the morgue every day at dawn for a week until, finally, I got word that three still-warm human brains were ready to be picked up.

When I arrived at the morgue, the pathology clerk sent me to a room where I saw three naked bodies lying on three separate tables, their brains not yet having been removed. One, I was told, was a man who had dropped dead playing tennis that morning, and the other two were a liquor store owner and the young man who had attempted to rob him. The ensuing crossfire had cost them both their lives, but provided me with the materials I needed to do Sol's bidding. My heart was pounding as the pathologist went to work, eventually placing a brain in each of my three ice buckets. I thanked him coolly, as though I saw brains being removed from naked dead men's bodies every day.

After the retrieval, we followed a set routine. Mike Kuhar, a former graduate student of Sol's and now assistant professor at Hopkins in neuroanatomy, dissected the brains in the cold room. I watched as he scissored out chunks from each of the major sections—frontal cortex, hypothalamus, visual cortex, cerebellum, amygdala, etc. It was then my job to weigh each chunk and place it in a test tube, adding enough liquid so that Adele could whip the mixture into a frothy milkshake on the Polytron, a fantastically expensive machine that made a deafening noise. Once liquified, the mixture was dosed with radioactive naloxone, incubated for an hour, and then put through a filtering process. The brainladen filters were then placed in the counter to determine how much radioactive naloxone actually did bind.

I remember sitting in the counting room late into the night, listening to the crunch and tinkle of the machine. When I emerged, the lab was silent. Everyone had gone home, and it was my turn to do the clean-up.

Many times in the laboratory I've felt I was moving close to the mystery, but never more powerfully than when I walked back into the cold room that night and saw the remains of those three human brains— three-pound universes when alive, in death looking like nothing so much as half-eaten turkey carcasses—waiting to be swept into the garbage. The fragility of life, the ruthlessness of science, the folly and beauty of it all moved through me, striking an emotional chord so powerful I can still feel its vibration.

I finished the clean-up, closed the lab for the night, and went home. The next morning I was at my bench, entering the numbers from the counter into my notebook, when Mike waltzed in and slapped a *Baltimore Sun* down in front of me. He pointed to the lead story, which gave the details of the previous day's liquor store robbery and described its owner, along with quotes from grieving relatives and a photograph of him in happier times. It was difficult for me to do what seemed so easy for most of my colleagues, to distance myself from the human element. I looked at the picture of this man and then at the numbers in my notebook, wondering how he would react if he knew we'd made a milkshake of his brain. Considering what he'd done to the guy who had tried to rob his store, probably not too cordially. Even so, I hoped he'd be glad to have helped in the fight against drug addiction.

As soon as we had gathered our data, I watched as Sol called Walle Nauta at MIT and read him our numbers. Nauta, the dean of American neuroanatomists, was able to analyze the data and let us know after a few minutes of studying the numbers in what part of the brain's anatomy we had the strongest signals.

"Walle says it's a limbic configuration," Sol informed me. The opiate receptors, it turned out, were showing up most densely in the limbic system, the part of the brain classically known to contain the emotional circuitry.

Looking back on that moment, I can see that this should have been the first clue in the search that eventually led me to a theory explaining the biochemistry of emotion. But at the time, I was so impressed by the ability of Walle Nauta to look at our numbers and translate them into an image of the brain, that I missed the significance of the limbic configuration altogether. To have such mastery over such complex data—I couldn't imagine anything more spectacular than that. My focus was so single-pointedly fixed on understanding the cellular and molecular level of the brain, that I failed to be interested in the bigger picture, the notion that the receptors might be part of a network designed to handle something so fundamental to the organism—emotion—that it must surely have profound implications for the functioning of that organism. Emotions, so often dismissed by scientists as intangible, if they were ever brought up at all, must matter somehow. But *how* they mattered I had not even begun to wonder.

One thing we knew with certainty, but had not yet been able to prove, was that the opiate receptor had quite a lot to do with the organism's pleasure/pain continuum, which in turn, we felt sure, was crucial to the survival of the organism. As far back as the 1950s, behavioral psychologists had diagrammed the pathways of the nerves that carried pain from the skin to the brain, where the information was processed at pain centers. They discovered that by electrically stimulating these centers in rats, behavior indicating pain would occur. They also found that other points in the brain processed pleasure, and if the rat was wired to self-stimulate, it would do so for hours until collapsing from exhaustion. Now we were asking what the role of the opiate receptor was in this continuum, and our hunch was that if we followed the receptor trail, we'd come upon a clear understanding of the network in the brain that controlled pleasure and pain.

One morning, as I was leaving for the lab, Agu called out to me: "Don't forget to check the periaqueductal gray for opiate receptors when you dissect those monkey brains today."

Agu had read in a journal that Chinese scientists had followed morphine to a site of action in the brain called the periaqueductal gray. PAG, as it's known in the jargon, is located around the aqueduct joining the third and fourth ventricle in the midbrain, and is a nodal point where many nerves converge for information processing. Although it was not classically considered part of the limbic system, it clearly had neuronal pathways that hooked it into the limbic system. Agu had been able to confirm the Chinese researchers' observations in the behavioral brain-mapping experiments he was doing in his Edgewood Arsenal lab, and we were also aware that John Liebeskind and Huda Akil of UCLA had published data hinting that certain types of electrical stimulation of the PAG might be causing the release of a morphinelike factor.

Sure enough, our lab tests confirmed that the PAG was an area where opiate receptors were highly concentrated. And Agu proved that the PAG was the area of the brain where the perception of pain is determined—or, as we would put it, the pain thresholds are set.

This last experiment caught the attention of a lot of people, among them an Englishman named John Hughes, who was laboring in the University of Aberdeen lab run by Hans Kosterlitz, my Chapel Hill host and secret (but indiscreet) seeker of the opiate receptor's endogenous ligand. Hughes, a bright young man Kosterlitz had told me, was new to Kosterlitz's lab, and had been spending his days trying to isolate a substance from pig brains that seemed to behave just like morphine when applied to certain tissues. He was beginning to wonder: Could we have found it? Was this the body's own natural morphine?

THE RACE

The frenzy that accompanied the search for the brain's own morphine was like what happens when you wave a filet mignon in front of a pack of hungry dogs. Before the competition climaxed in a major discovery, there was more adrenaline pumping in scientific circles than flows through the drivers at the Indy 500. Labs on both sides of the Atlantic Ocean raced around the clock to get to the finish line.

But the work Hughes was doing was immensely time-consuming and labor-intensive—about as far from the style Sol had taught me back at Hopkins as could be. Daily, he visited a local slaughterhouse and acquired wheelbarrows full of pig brains, which he took back to his lab. There he reduced them to proteins and salts by grinding up the foul-smelling mess with acetone to dissolve the fat, leaving it to evaporate and then redissolving the residue in various solvents, until, finally, he managed to extract a waxy, yellow material.

A Eureka moment came for Hughes when he was able to show that the extract acted like morphine in the organism and was blocked by naloxone. This he did by demonstrating that a smooth muscle called the vas deferens in mice contracted in the presence of his mystery material, creating spasms that could be reversed by naloxone. Hughes now had both a method for purifying the extract and an assay to demonstrate its activity. But until he could crack the molecular sequence and write the structure, the race was still on.

As recounted earlier, when I returned from the Opiate Club meeting in Chapel Hill during the summer of 1973, I had tipped Sol off about the goings-on in the Scottish lab, and so we decided to invite Kosterlitz and Hughes to a small neuroscience conference we were organizing for May of 1974 at an elegant mansion in Boston. The tiny but elite conference was one in a series designed to bring leading researchers together to discuss various subjects in a collegial fashion. Shortly after the conference was closed, a pamphlet summarizing the proceedings would be published in the *Neuroscience Bulletin*. Though not a regular scientific journal, the *Neuroscience Bulletin* was considered a legitimate enough forum to establish a solid claim, should Hughes reveal his work and wish to have it in print. In my correspondence with him beforehand, I assured Hughes that if he decided to tell all about the morphinelike substance at the conference, he could do it safely without fear of being scooped, because the pamphlet would establish his primacy in the field.

Hughes had good reason to be hesitant. Calling his new substance enkephalin, he had been able to identify part of the chemical structure, but hadn't gotten far enough yet to crack the whole formula. It could have been a disastrous move for him to present his most recent but incomplete findings at our meeting, considering Sol's reputation as a shark and the fact that Avram Goldstein, who'd been on the trail of the body's endogenous morphine for years, was planning to attend.

Trusting me, Hughes decided to present his findings at the Boston meeting. In his talk, he revealed that although he hadn't been able to determine the complete structure of the substance, he had done enough work to know enkephalin was definitely a very small peptide.

The minute he stepped down from the podium, an exodus began from the hall, as people scrambled for phones to call their labs and spread the news. A peptide! The revelation that enkephalin was a peptide enabled all kinds of clever end runs to be attempted. Avram Goldstein, for one, was eager to find his own source of the mysterious ligand,

and, knowing that the pituitary gland was a rich source of many peptides, began to stock up on pituitary extract, which he acquired from commercial meat-packing companies.

I felt terrible as I watched all this, thinking I'd led Hughes like a lamb to the slaughter. But at the same time, I understood that this was the way the game was played. And why not? Why should Hughes be permitted to take his time in making a major discovery that could potentially benefit millions? In fact, Goldstein's approach ultimately led to the discovery of several completely new and important forms of natural opiate peptides.

The day after we got back to Hopkins, Sol lined us all up in battle formation for a run at the still-to-be-cracked molecular structure of enkephalin. As I sat in a meeting and listened to Sol map out his strategy, my inner conflict grew. I couldn't ignore a gut feeling that something was very wrong. Certainly I could sympathize with Sol's desire to win this race, but what seemed to me a clear lack of respect for the integrity of Hughes's work left me sick to my stomach. I was too upset to say anything, and kept my feelings to myself.

Some of my revulsion was the result of being pregnant with my second child, Vanessa. It wasn't morning sickness that was the problem, however, but the fact that the hormones of pregnancy seemed to have caused me to lose my macho-competitive tendency. What I really wanted to do was take a break from the fast lane and do something else until the baby came. The day after our meeting, I told Sol I had decided to pass on this one, even though I knew that would mean giving up Adele. My Ph.D. had been granted on the basis of the opiate-receptor work, and the next thing for me to do was begin my postdoctoral training. One project that interested me was working with Michael Kuhar to develop a method that would allow us to see the actual distribution of opiate receptors in the brain. The possibility of getting a visual image for what Walle Nauta had seen in the numbers—the exact locations of the opiate receptors in the brain—fascinated me, and Sol agreed to let me pursue it.

But the days of basking in the warmth of Sol's attention were at an end. Now the hottest project in the lab was the race to be first to find the chemical structure of the brain's natural opiates, and I had dropped out. Within weeks of Hughes's announcement at the Boston meeting, Sol and his student Gavril used peptide procedures to extract a brain substance they called MLF, for morphinelike factor, although they couldn't yet write the formula, either. This work, done in Sol's lab weeks after the Boston meeting, now appeared in the same *Neuroscience Bulletin* where Hughes reported his findings, making it appear that the two labs had done their work simultaneously. Again, I was embarrassed for having urged Hughes to openly announce his findings, and shocked that Sol would go to such lengths to make it look like he was neck and neck with Hughes. But that's as far as Sol and his students got. All that year and into the next, they hit MLF hard, but were unable to crack its molecular structure.

THE HARE TURNS TORTOISE AND DROPS OUT OF THE RACE

My work seemed much less exciting, having none of the high drama that accompanied the search for the endogenous ligand, but I was content, in my pregnant bliss, to plod along at my task. Straightening out the many technical glitches involved in getting a clear radiological picture of opiate receptors in the brain was by far the most tedious and exacting job I've ever done. It was a project perfectly suited to a very pregnant woman who had all the time in the world, and plenty of patience to spare.

Autoradiography, as the technique was called, had been around since the 1950s, and essentially involved injecting an animal with a radioactively labeled substance, such as naloxone, killing the animal, and then removing a sample of the desired tissue for study. On exposure to film, the radioactive substance would appear as a bright spot of light in the tissue. The challenge was to figure out how to get enough of the radioactive naloxone to stick to the receptors, and thus give a precise image. I worked meticulously at ironing out all of the technical niceties for five months alongside Mike Kuhar, whose knowledge of neuro-anatomy contributed immensely.

As soon as we had the method perfected, we began constructing a map of the opiate receptors throughout the brain. Using chunks of brain from rats that were nine, fifteen, and twenty days old, I worked with Joe Coyle, another assistant professor in the department (now the chairman of the Harvard psychiatry department), to slowly build up a picture of how opiate receptors developed in the brain. We saw how they were concentrated in areas that were classically associated with emotion, pleasure and pain, and sense perception. This confirmed the work I'd done earlier, locating the receptors in the vertebrates that ranged evolutionally from hagfish to monkeys, once again demonstrating to me that the system we were looking at was one that had been conserved for eons of evolutionary time, thus one that had to be very basic and fundamental to the survival of the species.

In a classic example of acquiring a scientific rhythm, we had developed a technique and were exploiting it fully, building up a database without paying too much attention to what it all might mean ultimately. I was learning that life in a lab can be a very left-brain exercise, much of the time spent doing endless variations to try to make an experiment work, hoping to extract a signal from an ocean of noise. And then once a signal is heard, the rest of the time is devoted to asking all the questions the new finding allows you to ask. This can go on for years, and often does, keeping us scientists occupied and busy in our laboratory worlds.

VANESSA WAS born in the spring of 1975 and, after spending a brief time at home with her to bond and establish a breast-feeding, breast-pumping routine, I returned to Sol's lab to finish my postdoctoral work over the summer. That June, I accompanied Sol and Gavril to the annual Opiate Club gathering, now being officially called the International Narcotics Research Conference (instead of Club), which was held at Airlie House in a suburb of Washington. I arrived for the two-day affair with the slides I'd prepared to present opiate-receptor autoradiography for the first time, and with a batch of empty Playtex baggies that I planned to fill with breast milk for the infant I had left at home.

The decision to leave Vanessa with a nanny had been a difficult one. I knew the meeting would be tense, with Hughes's report of the molecular structure for enkephalin imminent, and a throng of competitors poised to spring and claim primacy. I envisioned the testosterone frenzy that would surely be the mark of this meeting, and I didn't want to expose my new baby to the harshness of such an environment. Alternately, and somewhat paradoxically, I wanted to be ready to leap into the fray myself and do battle with the boys. But if I wanted to present a strong front, having a brand-new baby suckling at my breast wasn't the way to do it. That was a scenario I couldn't quite imagine.

At the meeting, I watched from the sidelines as the clash of the titans got under way. Each of the competing researchers presented his version of the endogenous opiate ligand, all of them vying furiously for first place in the race. To say the boys were getting edgy as they bore down on the finish line would be a vast understatement.

Gavril gave his talk first, revealing his and Sol's findings without giving credit to the pioneering efforts of Hughes and Kosterlitz. He was quickly upbraided by an irate Kosterlitz, who leapt to his feet, demanding an apology and a corrected statement of the facts. Gavril, his face turning bright red, responded that there had been no time in his talk to mention the Scottish team, but that he had cited them in the actual paper he had submitted for the proceedings of the meeting. I took pleasure in seeing my old rival humiliated, so caught up was I in the drama of the competition, the lust to win. Yet I also knew my own role in trying to steal Kosterlitz's thunder.

The climax of the meeting came during John Hughes's talk, when he dramatically ripped open the envelope of a freshly arrived telegram (these were prefax times) to triumphantly announce the latest analysis of the amino acid content of his enkephalin extract. But he still didn't have the much-coveted sequence needed to write the formula. For that we would have to wait six more months.

NEW HORIZONS

The day arrived when, having completed my postdoctoral work, I was expected to leave Hopkins and find myself a real job. Sol, who, unlike many lab chiefs, prided himself on finding the best possible positions for his students, was tremendously supportive. He used his influence to get me an offer from the National Institutes of Health, for which I was very grateful. But because of the opiate work, I soon found out, I was a bright star in demand. A dozen universities offered me faculty positions. In the end, I decided to accept the NIH offer, partly because Agu was offered a position there as well, but more because my burning desire was to do pure research, and at NIH I wouldn't be required to teach classes or write grants or advise students.

It was time to leave the nest, never an easy transition for child or parent. I felt like a bright but awkward adolescent pushed a bit prematurely from my scientific home, but eager to embark on the adventures that lay ahead. It was a classic mentor-disciple, father-child transition that was taking place. As the time for my departure drew closer, there was a palpable tension between Sol and me, one that seemed to be exacerbated by the fact that what he and I had in common was not a genetic but a scientific bond. But there was more to the tension than that.

When I went to say good-bye to Sol in his office on the last day, I remember how awkward we both were, exchanging platitudes, neither of us saying what we really felt. But suddenly he said to me, in a tone of real finality, "Candace, I want you to promise you won't work on the opiate receptor in your new job."

My heart sank. I was stunned. Even though I hadn't really thought about what I'd be working on, this request seemed unfair, even cruel. I grunted and mumbled incoherently, which seemed to satisfy Sol, and made a hasty exit before he could ask me to sign on the dotted line.

Later I wondered, Why didn't Sol want me to work on the opiate receptor? Had I upstaged him by my many conference appearances, making myself a star while he stayed behind in the shadows? Suddenly, I remembered an incident that had made no sense when it had happened a few months before. I was sitting on the centrifuge in the counting room and talking to Sol, who had a habit of contorting his body and resting his elbow on his knee, chin in hand, presumably so he could think better, when there was a pause and he stared at me intently for a full minute. As if he had just seen me in a new and puzzling light, he said, "Have you ever heard of *The Prince* by Machiavelli?"

Now, political science had never been my forte, but I vaguely knew that *The Prince* was a classic work that had been written in the sixteenth century to instruct princes of the time in how to use any means at their disposal to gain power and manipulate the masses. Why Sol would be mentioning it to me then I had no idea.

"You really should read the chapter about killing the king," he said dryly, straightening up. Then, looking me straight in the eye: "If one is going to kill the king, then one should never wound him, but finish the job and be done with it."

I stared at him blankly, having no idea what he was talking about. Later, when I pondered the incident, I wondered if I'd been too aggressive when I asked to be listed as coauthor on an article summarizing our work together that was about to appear in an upcoming issue of *Scientific American*. I was left with the disturbing thought, Was my growing ambition becoming a threat to Sol?

5

LIFE AT THE PALACE

THE PALACE

I arrived at Bethesda in September of 1975 to begin work as a staff fellow in the biochemistry and pharmacology section of the biological psychiatry branch of the National Institute of Mental Health (NIMH). This was a subsidiary of the larger National Institutes of Health (NIH). During my stay there, from 1975 to 1987, I published over two hundred scientific papers, and, for a while, was the most cited scientist at the NIMH. Although much of my success can be traced to the methods and techniques I invented, some credit must be given to my having been in the right place at the right time. The field of receptor science was exploding in the late seventies, a time when new neurochemicals, most of them peptides, and their receptors were being identified practically every month.

Spread out across hundreds of lawn-covered acres in a suburb of Washington, D.C., the National Institutes of Health is the home base of the United States government's premiere biomedical research establishment, which puts our tax dollars to work to support research in pursuit of answers to all the big questions of health and disease. Though most of the NIH's budget is deployed through universities and research institutes scattered throughout the country, the headquarters are here, where the laboratories and offices are housed in sixty-five all-brick buildings, and roughly 13,000 employees are divided among eighteen subsidiary divisions, one of which is the National Institute of Mental Health.

During the thirteen years I spent at the NIMH as a research scientist, I referred to the entire conglomerate as the "Palace"—partly out of affection, because it truly was a dream kingdom come true, a veritable Versailles, opulently funded with seemingly endless freedom to do research, partly with irony, because of its rigid political hierarchy and the sharply cut boundaries that divided the disciplines, carving up the turf. It seemed as if the old paradigm's insistence on the separation of one biological system from another, as well as mind from body, was as solidly entrenched as the old brick buildings in which the research on those systems took place. Science as an interdisciplinary, interdepartmental phenomenon was an idea whose time had not yet come to the NIH.

Even today, a casual visitor would be instantly aware of the hierarchical compartmentalization of the Palace, a quality made visible by the subtle but pervasive dress code. At the bottom of the hierarchy are the muscles, who travel in packs, dressed in bright blue or orange jumpsuits, passing through the hallways at all times of the night and day, repairing the many vital systems of the infrastructure. Above them are the lab coats, the technical assistants, like Adele, who serve the mostly male postdoctoral students, who form the next category up the ladder and are invariably marked by their jeans-sneakers-T-shirt uniforms. These last two categories are the worker bees, dominating the general populace and forming a huge pool of willing arms and bright minds. Above them in the hierarchy are the alphas, the permanent senior scientists and all those who are poised to move into tenured positions. These folks all dress distinctively, expressing their unique and privileged individualism. And, at the very top, are the brass, the reigning princes (no princesses among them!), and the lesser barons, who administer the complex beehive of laboratories, offices, and institutes and control the resources. All the top brass are medical doctors, and they all wear suits and ties.

And then there are the gypsies, whose place on the ladder is rather vague, somewhere between worker bees and alphas. This is a group of older researchers who never made it to alpha status but are addicted to the Palace, research junkies who roam freely from laboratory to institute, valued for their knowledge of the system, which, over the years, they have learned to manipulate. Some of them may have left for a while but felt compelled to return, so attractive is the atmosphere of the Palace, offering them a kind of energy and excitement to be found nowhere else in science.

At the NIMH, where I worked, the brass were all psychiatrists, medical doctors whose territory stops at the neck. The alpha scientists, who had Ph.D. degrees like myself, work for the medical doctors, feeding them data to present at the many conferences they attend around the globe. A wise scientist will seek out a niche protected by a powerful M.D. and be content to stay there. No matter how smart or productive he or she may be, the scientist with a Ph.D. has absolutely no chance of ever rising to a position of controlling resources. M.D.'s only need apply.

This intellectual imbalance creates a certain amount of friction between the two categories, scientist and medical doctor. Success often depends on a certain amount of sucking-up to your superiors, something that doesn't come easy for a lot of brilliant scientists. During my years there, I saw more than a few who weren't willing to play the game, usually because they considered their boss an imbecile who wouldn't know an experimental breakthrough if it strolled into his office and burst into song. While most of the doctors high up in the Palace had a passing familiarity with experimental science, few were experimenters, and they often had a difficult time evaluating data, particularly when two experiments conflicted. But they were in charge, the princes, the power-boy doctors, and I saw more than one frustrated scientist return to his basement lair to bang his head against the walls over a discrepancy between what his boss had learned in medical school and what he himself had just seen under his microscope.

During my time at the Palace, initially as a staff scientist and later as a lab chief in the brain biochemistry section, the psychiatrists who were my bosses were men I would describe as medical doctors with people skills. By temperament, they exuded a kind of charm that, coupled with their keen insight into the human dynamic, allowed them, if they so wished, to zoom up the professional ladder. In jest, I used to refer to them collectively as the Slick Boys, because of their expensive suits, smooth manners, and elegant office suites.

When I first arrived at the Palace, I was glad to see that there were as many women as men passing through the halls, but I soon realized that most of them clearly belonged to the technician class. Even today, few females ascend to alpha status, where tenure is the prize, instead remaining down in the trenches, lowly serfs toiling away at the donkey-work. The unspoken belief that women lacked the right kind of mentality to do science because they were too emotional was a bias solidly in place at the Palace.

But in spite of the politically charged and socially stratified atmosphere, there was an undeniable energy and excitement at the Palace, unique in all of science. This is bound to happen when the sharpest minds are brought together with nearly unlimited resources, and the creative sparks are allowed to fly. In those days, something in the Palace air called forth the very best from a person.

LANDING

My first year whizzed by, a blur of scrambling to get a lab set up, recruit a staff of technicians and postdocs, perform experiments, and publish papers. I quickly found out that my greatest challenges were less about science than about learning to deal with the gigantic, overreaching bureaucracy and finding a place in the complex sociopolitical scene of Palace life. By the time I reached my first anniversary, the overall feeling I had was one of pure, abject gratitude for having survived.

I was very nervous that first year, unsure that I would be able to continue to be the "star" I had become under the tutelage of my mentor, Sol. How much of my success had been due to his support, and how much had I genuinely achieved on my own? While my new colleagues received me positively, treating me somewhat like a prodigy from whom much greatness was expected, still I was uncomfortable. One of the seemingly trivial concerns that stemmed from this insecurity was about what clothes to wear to work. None of the standard uniforms seemed quite right for me. The few women who were at my level in the hierarchy were older and belonged to the science-nun generation—their uniform was not mine. The younger women were mostly technicians or postdocs, and I knew I had to distinguish myself from them if I was going to take any kind of leadership role befitting my status. I conferred with another newly hired female colleague: How could we dress comfortably, retain a modicum of femininity, and still be taken seriously by our fellow scientists? Together, we came up with a totally new style consisting of designer jeans topped off by a fashionable, obviously pricey blouse.

But dress code quickly became a subsidiary issue, replaced by what became my first Herculean labor—to secure a lab in which I could begin to act like a scientist. I'd assumed when I'd accepted the job at the Palace that a lab had already been designated for me. It was a shock to find out I was expected to literally create one from nothing, as territory was precious at the Palace, and, at any moment, there were scientists stacked up like airplanes over JFK, waiting to land in a working lab. I was told that work orders for my space had been written some time ago, but that the renovations had not yet begun.

In time, I came to realize that my situation was not unusual, and that most new scientists were expected to set up their operations without a great deal of guidance. The implication was that now you were on your own, and the only cost of your freedom was the initiative required to get your own projects rolling. It was a kind of Club Med for scientists, providing an atmosphere totally insulated from concerns about money, with a rare and incredible freedom to do what we came here to do: pure, unadulterated research!

While I waited for my lab, they parked me in an empty library room. I was able to push some tables together to create a makeshift bench, but was stymied by the challenge of how to carry out my experiments without access to any running water. My solution was to lug my extremely heavy filter machine down the hall to the ladies' room to empty it whenever I had to do any filtering. Under these conditions, most of my early experiments crashed, and in those first months, I had some real moments of frustration and despair.

I found myself wandering up and down the halls frequently during this period, checking out the labs of the Palace bigwigs, the senior scientists who were internationally recognized leaders in their fields. I was thrilled to be invited in occasionally for a friendly chat, and to be treated by these top scientists as a colleague and peer. I tried not to be too bothered by the fact that so few of them were women, reassuring myself with naive idealism that science was truly a meritocracy, and that if I produced good work and did great science, someday I'd rise to the top, too.

IN DECEMBER, my laboratory was ready, though it was a far cry from the fancy labs I'd been visiting. The office was so narrow that if I closed the door, I could barely fit a desk and chairs for two people in it. Agu, who had also been in lab limbo for the past four months, occupied an equally tiny office next to mine. Even though we were hidden away at the far end of Hallway 2 North, Building 10, we liked to think that together we generated a synergistic energy that attracted the younger and more open-minded researchers in our direction.

It was a potentially fertile spot, our little corner of the Palace, where the boundaries of two separate disciplines, Agu's psychology and my neuropharmacology, touched. Here there was a promise of the kind of interdisciplinary research that was relatively rare in the compartmentalized environs of the rest of the Palace. I could walk next door and tap into a worldview that focused on behavior and even "mood," a term used in experimental psychology that hinted at "emotion" or "consciousness," domains my own field was in no way comfortable with. It was an arrangement that encouraged my bent toward collaboration and boundary-crossing, as well as my interest in the psyche, the mind.

INTRIGUE

I had been in my new lab for only a few weeks when Les Iverson, a visiting English researcher and one of Julie's boys, dropped by for a visit. He brought with him the groundbreaking paper that Hughes and Kosterlitz were about to publish—the one described earlier in my lecture—where they revealed the chemical structure of the mysterious peptide they had christened enkephalin. This was the same substance that Sol had isolated and was calling MLF, but for which he was unable to write the chemical formula, even though he and his lab were laboring frantically to crack the code.

Les had been given a preview copy of the paper because he was an advance reviewer, but of course he wouldn't let me see the contents, showing me only the tantalizing title: "Isolation and chemical characterization of enkephalin—the brain's own morphine, a pair of pentapeptides." I would have to wait, along with everyone else, until the paper came out in the year-end issue of the prestigious British journal *Nature* in December 1975. Hughes, who had gotten a lot cagier since the Boston meeting, was following a common strategy, which was to publish at the very end of the year, thereby preventing any particularly facile competitor from seeming to scoop the discovery by publishing his own version in a different journal during the same year. Any subsequent publications on the subject would bear a post-1975 date.

Iverson was staying up the road a bit, a houseguest of Sol's, and as soon as he left to return to Baltimore, I trotted over to Agu's office. I knew that once the structure of the substance was revealed, there would be a rush to do the confirmation experiments, and my lab, together with Agu's, was perfectly set up to get a jump on this important phase of the discovery. What I needed was to somehow get ahold of the chemical formula for the structure as soon as possible, and then have a chemist manufacture the substance for me so I could use it in an experiment that proved, yes, indeed, that what Hughes had found was the real McCoy.

But we had to move fast. If everything went right, we'd submit our paper to the highly visible but politically treacherous journal *Science*, a risky choice because the old boys had a firm grip on the review process, and, for political purposes, might choose to ignore our submission.

I placed a call to my friend Dr. Jaw-Kang Chang, who, with his wife, Eng Tau, had just opened a peptide "boutique" in the garage of his home near San Francisco. At that time, they were the first of what would be many small commercial labs springing up all around the country equipped with the machinery and manpower to manufacture just about any kind of peptide in the growing field. Chang thought he could find a leak from a secret source, and as soon as he had it, he'd make the peptides and send them to me overnight.

It was a great plan. Chang would make the peptides, and Agu and I would do the experiments. Agu would prove that the peptide caused pain relief when dropped into the brain at the analgesic site, the PAG, then I'd demonstrate how it blocked binding to the opiate receptor in a test tube, just as we'd seen done with morphine and other synthetic opiates tested in a similar fashion. I hoped to bask in the reflected glory of the enkephalin discovery, reminding everyone of my earlier contributions, and Agu would have a chance to show off his finely honed brain-injection techniques, reminding everyone of his work showing how morphine acted in the PAG to relieve pain. Chang and his partners would have a chance to put their start-up company on the map and receive multitudes of orders for the synthetic version of this hot new substance enkephalin, which was sure to be in global demand very soon.

But it didn't work. At least not the way we planned.

The very next day Chang got the structure via a transatlantic phone call from a Chinese chemist in England who teched in a commercial lab that Hughes had consulted. Working night and day with all hands and machines on deck, Chang's group produced two test tubes of the enkephalins in less time than it took to mail the stuff from California to my lab in Maryland. All this happened within forty-eight hours of our first tip-off that the paper revealing the structure was about to be published.

We ran the tests and Eureka! It was the real thing, a substance identical to that made by the brain, which acted just like the opiate drugs, blocking pain and binding to the opiate receptor. Here was what everyone had been looking for—something that chemists could manufacture for use as a harmless, nonaddicting painkiller. I quickly wrote up our results and mailed the manuscript to *Science*. We were disappointed but not surprised when they rejected it. In order to get it printed as close on the heels of Hughes's paper as possible, we repackaged it and sent it out to the low-rent, quick-turnaround journal *Life Sciences*, where it was published in the late-January 1976 issue, one month after Hughes's paper appeared in *Nature*. Our timing had been good, but a competitor managed to get his paper into the more highly visible *Nature*, and, even though it appeared a few months later than ours, it is his work, not ours, that today is cited for the confirmation experiment that proved enkephalin was the body's natural opiate drug.

There was, however, some poetic justice in the end. While not advancing our reputations as planned, our research had led us to unearth a rather fishy discrepancy. We had noticed that the enkephalin was as potent as morphine in the cold test tube experiment, but surprisingly weak when dropped directly into the brains of Agu's rats. We surmised that it must be getting chewed up and rendered inactive in the warm brain by enzymes that normally go to work at body temperature. Once again working with Chang, we designed and constructed an analog of enkephalin, replacing one amino acid, L-alanine, with its mirror-image amino acid, D-alanine, thereby making a longer-lasting form of enkephalin, which was shown to be more resistant to the enzyme's action. This time our paper was accepted by *Science*, where it appeared only a few months later, in September 1976.

We got plenty of credit for that finding, more, in fact, than anyone bargained for. I presented my "super enkephalin" at a meeting in Scotland that summer, which sent the pharmaceutical company scientists in my audience rushing to the phones to call their labs and mobilize their companies' patent lawyers. These giant biotech department stores had been very busy manufacturing and testing all kinds of peptides for commercial potential. Now they thought they had the Holy Grail, the natural-style morphine, the nonaddictive painkiller (and possible antiaddiction drug), for which the industry had been doling out dollars for years in order to defray the high costs of the fabulous Opiate Club meetings around the world.

Shortly after I returned to the States, I was visited by a Justice Department lawyer who showed me how to write a patent before "disclosing your invention," as she diplomatically put it, implying that I'd best not go shooting off my mouth to a bunch of industry guys about an invention the NIH funded and owned—a little detail I had never

thought about. A royal battle ensued among ten pharmaceutical companies, each one claiming they had the D-alanine enkephalin first. The
resulting federal lawsuit, *U.S. for Pert* v. *Burroughs Wellcome, etc.*, took
several years of Justice Department involvement, and in the end we
won, but it was a Pyrrhic victory. The new enkephalin turned out to be
just as addictive as the original one, and even more expensive to make at
the time, so it was of no use to the industry.

But back to the early days of December 1975. After Les's visit, but
before Hughes's paper hit the stands, word drifted down from Baltimore
that Sol and his lab had been able to crack enkephalin's structure. I knew
that Sol had been working relentlessly with his most aggressive postdoc,
a hard-working Israeli named Rabi, but that they hadn't been able to get
enough of the purified peptide out of the pig brains to run a reliable test.
But now, apparently, they'd done it. The very day after Les Iverson flew
back to England, and around the same time John Hughes sent Sol his
Nature preprint, Sol was reported to be racing up and down the hallways
of Hopkins, waving a long piece of data chartpaper, proudly exclaiming
to all that he and Rabi had finally cracked the enkephalin formula. And,
sure enough, there was the data, with all the right amino-acid peaks. It
was the same data they had produced weeks ago, but at the time they'd
been unable to understand it. Now they were able to interpret it and call
it the real McCoy.

Too late to scoop Hughes, but eager to try to tie him, Sol and Rabi
published a decent structural proof in the early January issue of *Life Sciences*, right before the issue in which we had published our glorymon-
gering *Science* reject, confirming the analgesic properties of enkephalin.
But it was too little, too late. Sol had run it up the flagpole, and no one had
saluted. Ever the politically savvy wheeler-dealer, Sol quickly distanced
himself from any appearance of trying to share credit with the Scottish
research team for revealing the chemical structure of enkephalin. In the
end, he even sent Kosterlitz, the senior man, a congratulatory case of
Cognac. It was this gesture, I was later to speculate, that laid the ground
for the magic three to move forward in unison to claim their future prize.

ENDORPHIN HIGH

My immediate superior in the early days at the Palace was Dr. William
Bunney, head of the adult psychiatry branch and former director of the
National Institute of Drug Abuse, whom I had first met two years earlier

at the Hopkins press conference announcing the discovery of the opiate receptor. Dr. Bunney occupied an opulent suite, furnished with furniture and fine art he himself had purchased, a few floors above my lab. In a weekly ritual, I would ascend to his spacious office and deliver the update on my research projects. Dr. Bunney had a calm and clipped style that perfectly fit the Hollywood image of the classic psychiatrist, and he listened intently as I reeled off the week's findings. He always wore a dark, pin-striped suit, which, I imagined, was one of dozens of identical ones in his closet at home.

Biff, a nickname I eventually felt comfortable using, had risen to his position within the Palace hierarchy by showing that lithium was an effective drug for manic-depressive psychosis. When I first arrived, he controlled a large corner of the funding empire through the NIH's National Institute of Drug Abuse, an organization that had come into existence simultaneously with the opiate-receptor discovery. The purpose of NIDA was to support research that it was hoped would one day lead to the development of drugs to cure addiction.

Biff's very first question took me by surprise. I was in his office for my weekly report when he leaned forward, looked me straight in the eye, and said in a flat, dry voice, "Do you realize, Candace, that for a heroin addict the first intravenous injection hits the brain like a sexual orgasm?"

"Gosh, no, Dr. Bunney, I didn't," I responded uncomfortably.

Biff explained that he believed the pleasure experienced during orgasm was accompanied by a surge of endorphins—the term that was being used to refer to Hughes's enkephalin—into the bloodstream. Now, that caught my attention, as did any idea that might explain the influence of the opiates and how they worked to produce pleasure and relieve pain. Soon I was designing a test that could measure levels of endorphins in the blood and running a series of experiments to determine which kinds of behavior made these levels go up, and which made them go down.

My investigation of this question spanned a period of almost two years. We used hamsters for one study, the classic lab animals for studying sexual behavior because of their predictable cycle of sexual behavior—two minutes of licking this or that, three minutes of humping, etc., and the act was complete. The males are extremely prolific, ejaculating about twenty-three times per cycle. Later, we were joined by Nancy Ostrowski, an accomplished scientist who had left behind her desire to become a nun and gone on instead to become an expert on the brain mechanisms of animal sex. Nancy would inject the animals with a radioactive opiate before copulation, and then, at various points in the cycle, decapitate them and remove the brains. Using autoradiographic visualization of the animals' brains, the two of us were able to see where endorphins were released during orgasm, and in what quantity. We found that blood endorphin levels increased by about 200 percent from the beginning to the end of the sex act.

With our newfound method of measuring endorphins in the blood, all kinds of other projects were now possible. We explored the question of how exercise affected the release of endorphins when I recruited twelve young Palace psychiatrists who were serious runners to let us take blood samples before and after their daily runs. The results showed a definite increase in endorphin levels, but the assay crashed at a few key moments, causing us to lose the precious samples that my subjects had literally sweated to produce. Nothing much came of these studies until Peter Farell, an exercise physiologist from outside the Palace, pulled a paper together that made use of my expertise, but was based mostly on his own efforts. He generously made me a coauthor of the paper, which was the first published study to provide the physiological validation of the phenomenon we now know as "runners' high."

My next project was the obvious biggie, the human-orgasm experiment, which presented a challenge in both recruitment and design. Since it wasn't possible to have a technician present to draw blood at the ultimate moment, we had to settle for measuring the endorphins in our subjects' saliva. Our subjects consisted of friends, as well as Agu and me, all of us agreeing to chew parafilm (which generates saliva) at various moments during sexual intercourse and then to spit into a test tube.

While enjoyable to do, these experiments were ultimately considered failures from a Palace point of view, because the results, suggestive as they were, lacked sufficient clarity to be written up and accepted by a medical journal. The work did produce a number of very interesting abstracts, which were presented at some early neuroscience meetings and were, understandably, very popular. But the idea that human orgasm is accompanied by the release of the organism's own pleasure chemicals has never quite seen the light of day in a prestigious journal.

SUCCESS

And so it goes, month after month, planning experiments, brainstorming with your postdocs, and gathering data with an eye to publication. I was usually a pushover when it came to believing that data. If it looked crisp, and if, after massaging it this way and that, my gut reaction was still affirmative, I'd give the green light for a paper to be prepared. Most lab chiefs were just the opposite, making their postdocs repeat experiments ad nauseam, terrified of attributing truth to something that turns out to be a trick of the numbers, or an artifact—an error created by the method, which leads to a false conclusion.

But when the data held true, and it looked like we'd uncovered a legitimate piece of the picture, then we'd move ahead to getting it in print. As I said earlier, we scientists measure our success in terms of papers—how many we've published and where they've appeared, in journals considered top of the line, middle of the list, or bottom of the barrel. That, in a nutshell, is what scientific life is all about. The pay is comfortable but not spectacular, and the only real glory comes from seeing your name in print under the title of a paper. Even more thrilling, at times, is seeing your work cited in another scientist's paper, which is significant because it affects your status in the professional hierarchy. Your position is determined by a huge database called the "citation index," a listing of every paper that has ever been referred to in another paper, ranking each paper according to the number of times it has been cited. For many years Sol has been the very top-cited scientist in biomedicine! For one ten-year period, my ranking was 130 among the most highly cited scientists in the world.

The percentage of papers that get cited more than a few times is very small, and for this reason everyone always refers to their own previous papers as much as is practically possible. Because appearing in print is so important, and because modern scientific projects can involve several collaborators, more bitter and intense arguments arise over the order of authorship on a paper than over the thorniest of theoretical issues. As was the case with the opiate-receptor paper that Sol and I published, the first author cited is generally the person who took the lead in designing and implementing the experiment. Then come the names, in descending order of importance, of all those who participated, either by advising or assisting, sometimes as many as ten or fifteen names. The last name to be mentioned is the person who either runs the lab or has raised the money to make it all happen. Traditionally, the names of the technicians who do all the actual lifting and hauling are left off of papers, but I always thought it was the decent thing to do to include them on my papers. I was also glad to let my postdocs take the first-place position, especially if the paper was an important one. I had learned from my work with the opiate receptor that first authorship on a key paper could go a long way toward assuring a person's career.

It was this name game that was partially responsible for my own swift rise to the top, and would soon be a pivotal element in an unfolding drama that would completely change the course of my career.

In a prophetic glimpse of things to come, John Hughes had dropped by my house for a casual visit in the spring of '78. As we sat out on the back deck sipping cool drinks, he turned toward me and asked rather abruptly, "Candace, have you ever heard about an award called the Lasker?"

"No," I responded. "What's it for?"

"Well, it's kind of an American Nobel Prize, given each year to scientists who have done outstanding medical research," Hughes explained. "In fact, the scientists who receive it usually go on to win the Nobel. It's sort of a stepping stone."

Now he had my attention. I knew the Nobel was the biggest prize in science, but I hadn't a clue as to how the scientists who won it were selected.

"What if I told you that Hans, Sol, and I were about to receive this year's Lasker for the opiate work?" John asked.

It took a moment for his words to register, but when they did, I blurted, "You've got to be kidding, John! And leave me out? Why, I'd be furious, of course!"

6

BREAKING THE RULES

INVITATION

Science in the big leagues is a lot like what goes on up at the basket at the NBA play-offs: very competitive, with sharp knees and elbows flying hard and fast. As individuals vie fiercely for credit, everyone knows you have to take care of yourself, because no one else will. The exception, of course, is your scientific family, your collaborators, whose job it is to watch out for you, scratch your back, and see that you get a chance at the basket when it's your turn.

While I found the game thrilling to play, I had not been sufficiently conditioned to accept the code of loyalty that it demanded. In a series of events that caused me much heartbreak and earned me much notoriety, I broke the rules and was dealt the cruelest of punishments, alienation from my scientific family. Later, in a popular book entitled *Apprentice to Genius*, Robert Kanigel would make a dramatic case for how I'd embarrassed a most royal medical dynasty, although this had never been my intention. In retrospect, I can see how my actions were part of a greater force at work to bring about a major shift from old-boy rule to a more egalitarian system.

It all began in the fall of 1978, when Sol Snyder, John Hughes, and Hans Kosterlitz were recognized for their opiate receptor/endorphin research and received a prize nearly as prestigious as the Nobel, the Lasker Award. My name wasn't mentioned.

John Hughes had tried to tip me off when he paid a friendly visit to my home the summer before the awards ceremony in October. But I was so stunned when he mentioned that the Lasker might be given to him and the other researchers, that I blocked all further conversation on the topic. Later, I realized that John had tried to alert me to what was already in the works, so that if I had wanted to do something about it, I still had time. But I was so naive politically that I chose to stick my head in the sand and reject the possibility that such a scenario would ever take place.

I didn't give it another thought until some months later, when I got a phone call from Sol, inviting me to a luncheon in New York City.

"Hello, my little baby girl," he cooed affectionately, with the usual politically incorrect term of endearment I had tolerated with mixed pleasure and horror for nearly seven years. We chatted briefly, and then he announced, an audible tension coming into his voice, "Candace, I'm receiving an award in New York City next month, and I want you to be one of the five invited guests I'm allowed to bring."

Although I'd been away from Hopkins for over a year, I was still flush with gratitude for Sol's help in procuring for me a lab and a position as staff scientist at NIH, and I was thrilled that he would think to invite me to this obviously important event. He then mentioned that two other scientists were receiving the award with him, but failed to mention what it was for, or who the other two were.

I hung up feeling pleased, but in the back of my mind an uneasy question surfaced. Sol had been unmistakably uneasy in extending this invitation, and I wondered why the Golden Boy, the wunderkind of neuroscience, would be afraid of asking his former graduate student to a luncheon where he'd be receiving an award? It took only a few seconds for the answer to hit me square between the eyes. The Lasker Award! The one John Hughes had talked about! This is what he'd meant, that he and Kosterlitz were to join Sol in receiving the Lasker Award for the opiate receptor and endorphin findings, the very work I had played such a pivotal role in. And now Sol was inviting me to the awards luncheon as his guest!

My heart had already begun to pump furiously, even before my brain got the full news. I picked up the phone and dialed Sol back.

"Sol," I said, barely concealing my anger, "do I understand that you, Hughes, and Kosterlitz are accepting an award for our work on the opiate receptor, and I'm not included?"

Taken off guard by my rawness, he admitted, half-apologetically, that, yes, it did seem strange, but that's how these awards things went, they were unpredictable, not always going to the person you'd expect. And anyway, he assured me, it was too late to do anything about it now. To make it up to me, he would see that I stood up for a bow at the awards luncheon, which he thought I would enjoy, he told me, since Ted Kennedy would be presiding over the ceremonies at the famed Rainbow Room, and I'd have a chance to meet him.

I hung up again, and tried to see things from Sol's point of view, but I could not put my rage to rest. The idea that I would be sitting in the audience while he and the others were honored for the work that I had played such an important part in seemed blatantly unfair. Was I really expected to stand down for my part in what had turned out to be a tremendous discovery, one that in the few short years since it occurred had been reshaping the entire field of the neurosciences? No, I resolved, this was not something I could watch happen. But what could I do?

DEEPER ISSUE

At the time all this took place, I had been reading several books that greatly influenced the development of my thinking and feeling. They were biographies of Rosalind Franklin, the brilliant scientist who had provided the critical link in the chain of reasoning that allowed Francis Crick and John Watson to show that the DNA structure was a double helix. As a result, the two men beat Linus Pauling to the punch and bagged themselves a Nobel in 1962. Franklin was a classic science nun whose life was totally given over to her work, and in Watson's book, *The Double Helix*, we get to see how these women were viewed by their male colleagues. The contempt all but bleeds through the pages, as Watson justifies his and Crick's actions in a graphic example of unabashed sexism in science.

But the truth of the matter is, the two men visited Franklin's lab when she was out of town and persuaded her boss to let them take a peek at her data. In what must have been a moment of incredible rationalizing, they stole Franklin's findings and got away clean, tossing her a bone of acknowledgment in their seminal paper, which won them the biggest award in science. In his best-selling book, Watson actually boasts about the theft, deriding their colleague for withholding her findings for publication in her own paper, which came out in the same journal—*Nature*—where theirs had appeared just a few months later. At the time, no reviewer, to my knowledge, cried foul, although in later years some heroic attempts at correcting the record were made.

To say I was livid about Rosalind Franklin's plight is an understatement. Even more, the tale of this outrageous deception deepened my appreciation for all the women who had been my teachers. I no longer saw them as second-class scientists who had failed to achieve status as lab chiefs and had settled instead for academia. I now realized that they were the pioneers who, if I were ever to run a big lab, I could thank for trailblazing the way, enduring the rampant sexism of their male colleagues in order to level the playing field for future generations of women in science.

But even while I acknowledged the debt I owed to these women, I was appalled at how little had changed since their time. The deeply ingrained bias against women surfaced often at meetings, especially at the quarterly study section, where we would review grant applications. Whenever my male colleagues came across a grant whose principal investigator (the "PI," as we call the controller of the government green) was a woman, they would unfailingly amuse themselves by subtly conjuring up an image of the eccentric, asexual lady scientist. This was followed by a silent consensus concerning her scientific untrustworthiness, an assumption that reverberated around the conference table until, finally, the application was given an inappropriately low rating. Most interesting was the discussion about the budget section of these female-initiated grants. While grants with a male PI requesting twelve postdocs never raised an eyebrow, the grants with a woman PI asking for a secretary and extra technicians would be chewed over ad nauseam. Like the attack cry of a flock of ravenous birds, the word *she* was repeated incessantly during the deliberations—Why couldn't *she* manage with less? I once amused myself by scientifically documenting this male pack behavior, carefully marking the number of times the pronoun was mentioned during one long afternoon session. My record showed that "she" was used nine times more often than "he," in spite of the fact that grants involving female PI's were rarer by far than their male-headlined counterparts. In as humorous a way as possible, I would try to point out the subconscious prejudice, but my words fell on deaf ears.

These, then, were the thoughts and feelings that were occupying my attention when the challenge of the Lasker landed in my lap. I woke up one morning and looked in the mirror—only to find Rosalind Franklin looking back at me.

Whereupon I took to the phone and solicited all my friends for advice. Almost to a person, each advised me to shut up and stay put. "That's just the way the game is played," I heard over and over again. And certainly that's the way Rosalind Franklin had handled it. She let Watson and Crick swipe her piece of the biggest scientific discovery of the century without so much as a peep. I was sure her friends told her the same things mine were telling me: Be careful—if you upset the boys, they might not let you play with them anymore.

The more I thought about it, the more angry I became. Was I expected to shove my feelings deep down inside me, where they might fester for years, eating at me and eroding my self-respect, my sense of pride and accomplishment, my self-esteem? I knew that I had to take my chances and blow the whistle or spend the rest of my life cringing with resentment and disappointment whenever the subject of the opiate receptor came up, every time Sol's name was mentioned in connection with the discovery and mine was not.

In my reading, I had learned that a few years after the theft of her data, Rosalind Franklin died of cancer. Even then, when almost no one had done any serious work on the possible effects of the emotions on health, I believed that her disease had been exacerbated by the humiliation she suffered at the hands of these, and probably many other, old boys, and that her failure to express her anger contributed to and possibly caused her death. It was partly intuition and partly common sense, but I felt that by not speaking up, I would be sacrificing my self-esteem and self-respect, not to mention possibly setting myself up for a nice case of depression and maybe a cancer or two down the line.

I certainly had no intention of allowing this to happen, and neither could I see myself going to the awards luncheon in order to "take a bow," especially when I learned that Avram Goldstein and Eric Simon had also been invited and would be acknowledged equally for their contributions. This was too bitter a pill to swallow, to have to stand alongside those who had failed to go the full nine yards, when I was the one who had done the brain-breaking work to put the cap on the opiate-receptor search, and had done it despite the abandonment of the research by a man who was now accepting the award for it. No, I told myself, I couldn't let this happen, to be forgotten and ignored by history, while the boys waltzed away . with the prize.

DEFENSE

When the official invitation to the Lasker luncheon arrived, I knew I could never accept. I decided to tell the truth.

"Dear Mrs. Lasker," my letter began. "I make it a point never to tell white lies concerning social occasions. Therefore, I want you to know that the reason I will not be attending your luncheon is that I'm upset and angered at being excluded from this year's Lasker Award for research I conducted in collaboration with Dr. Sol Snyder. So often women and other people who are low on the scientific hierarchy are excluded from receiving recognition for work which in truth they did, while the senior male scientist on the project is rewarded in their place."

That was the first statement about the incident. The second and much more public one was the result of a copy of that letter my husband Agu mailed to Jean Marx, an editor at *Science* magazine. Agu was even angrier than I was. Being male, he knew that the boys played a nasty game, and that if I let myself be shuffled out of the spotlight, no one was going to come to my rescue. While I was practically debilitated by my anger and confusion, Agu didn't think twice about shoving a stick in the spokes of the scientific wheels as they spun on. He dropped the letter in the mailbox, and life went on.

I never spoke personally with Jean Marx, but the letter caught her attention, especially the part about women being routinely excluded from scientific prizes. Her journalist's eye had picked up on the larger issue of scientific credit, and certainly the Pert-Snyder Lasker flap was a guaranteed hook for her readership. Her lead editorial in the very visible *Science* magazine, "Lasker Award Stirs Controversy," appeared in January 1979, complete with a large picture of me that the NIH had conveniently provided while I was at a conference out of town. My fear of being forgotten was instantly replaced by a feeling of pure fear.

The history of science is full of tales about feuds over who deserved credit for what. One classic brouhaha had erupted right in my own scientific family, when Julie Axelrod broke with his mentor, Steve Brodie, over the microsomal enzyme discovery. Brodie, the senior man at the time, tried to monopolize the research, and Julie, the underdog, had refused to lie down and roll over. But more typically, the scenario is that the junior man puts up with the injustices in the hopes that someday he'll be on the top.

When reporters from various scientific publications questioned Sol in the days following the *Science* editorial, he claimed that he had contacted members of the Awards Committee and asked them to include me in the award. He readily confirmed my written statement to Mary Lasker that I had indeed played a key role in both initiating and following up the research. But that was as far as he would go. In the month-long interval between Sol's telling me about the award and the actual ceremony, I called him repeatedly to ask him to make a statement on my behalf—either to the Lasker committee itself, requesting that they include me in the award, or at the ceremony when he went up to accept the award. Sol refused. In a last-ditch attempt to salvage something from the situation, I asked if he would at least agree to donate half the prize money (which I've since heard was quite modest, perhaps as little as $15,000) to a Bryn Mawr scholarship in my name. To this, too, the answer was no.

Meanwhile, word of my plight was picked up by Joan Arehart-Treichel, a biomedicine editor at *Science News* who had followed the field of peptides and receptors closely. In her February 1979 article about the Lasker flap, she stated her initial view that I had been excluded from the award because only three scientists could be nominated for the Nobel. She wrote: "I found myself outraged when America's most prestigious medical award, the Lasker, was handed out in the brain peptide–opiate field to only three of the four scientists whom I had expected to receive it. The three who were acclaimed were men; the one who was excluded was a woman."

Assuming it was the numbers that were the cause of my exclusion, Arehart-Treichel went on to say that her investigation had revealed a deeper, perhaps nastier truth. The reason I had been passed over was because no one had considered putting my name up for nomination. Neither Sol nor any of his buddies had ever mentioned my name as someone involved in the discovery. My displeasure at not being included, it seemed, had come as a total shock to the Awards Committee. "Pert? Who's she?" they responded when the editor called each of the committee members individually. When she pointed out that Pert was the name of the first author on the opiate-receptor paper, they squirmed uncomfortably, and they were further chagrined when she mentioned that Hughes, in a directly analogous situation, had been the first author on the endorphin paper. The article ended on a speculation about whether it was still possible for me to receive a Nobel for my opiate work.

"I'm not very optimistic," she wrote, "since the same informal roster of predominantly male scientists, sometimes referred to as the 'old boys club,' seems to be responsible for nominating scientists for both the Lasker Award and the Nobel Prize in Medicine. . . ."

Also appearing in print at the time were the comments of Eugene Garfield of the Institute of Scientific Information, an organization that ranks scientists according to how often peers have cited their papers. Garfield had developed ways to interpret this citation data to give a sense of a particular research dynamic.

"Since the Lasker committee deliberations are confidential," he wrote in his article, "we do not know if the members used citation data." He then went on to identify seven other scientists whose early work on the opiate receptor was well documented in the literature. His recounting of my work served as a summation of my entire scientific career up to that time:

> From 1973 to 1976, Pert and Snyder co-authored 17 journal articles on opiate receptors. These papers have received to date an average of 87 citations per article. During the same period, Snyder and other collaborators published 23 papers in the opiate receptor field. These papers have received an average of 37.5 citations per article. Of Snyder's papers on opiate receptors, Pert co-authored five of the six which received over 100 citations. She co-authored 10 of his 20 most-cited opiate receptor papers. Since leaving Snyder's lab, Pert has published 18 articles. Seven of them appeared in 1978 and have had relatively little time to receive citations. Yet these 18 papers have received over 300 citations, or an average of about 16 citations per paper. And one of her 1976 papers (the discovery of the long-lasting enkephalin analog with D-alanine that started a pharmaceutical stampede) proved to be among the one hundred papers most cited in 1976–1977. Thus, Pert's work at NIMH continues to be significant to her colleagues. . . .

Garfield's meticulous analysis settled the most persistent argument used against me, that as a graduate student I'd only carried out the orders of a senior scientist, and had done nothing on my own worthy of a mention since leaving my mentor. This attempt at exoneration made me feel better, but in the bigger picture, I soon realized, the major damage couldn't be undone. By involving journalists in what should have been a private spat within my own scientific family, I'd stepped too far over the line.

BETRAYAL

A few weeks after the first wave of publicity had passed, Julie Axelrod called me into his office. Still not realizing the impact of what I had done, I half-expected to be given a special project or other favor—after all, he'd been so positive and enthusiastic in his recruitment of me to my NIH post. Instead, Julie asked me to help him fill out the form he was submitting to nominate Snyder, Hughes, and Kosterlitz for the Nobel Prize. He hinted that my cooperation would go a long way toward soothing the bad feelings that still lingered as a result of the Lasker publicity. And, of course, he emphasized, I must know that the committee that made the final selections for the Nobel Prize hated any kind of scandal.

Without any hesitation, I shook my head and flatly refused. Julie's mouth dropped open.

"You've got to do this. You're the only one who knows everything that happened," he pleaded.

"Exactly," I responded, "and it's precisely because I know everything that I can't do it."

"Don't you love Sol?" he insisted, his voice rising. "Why are you doing this? You help him now and he'll help you later. This is the way it works. How old are you anyway? C'mon, you're a nice girl, Candace . . ."

He was shouting at me when I stood up to leave.

Unquestionably, Julie was correct about one thing: That's the way it works, the game of scientific recognition. If I helped Sol now, he'd help me later. And if I didn't, well, I would soon learn what would happen then.

Partly it was my still-raw anger that kept me from having any broader perspective, and partly it was a new heat arising in me from a recent revelation. During my own private probing of the Lasker nomination process, I'd learned a shocking piece of information from two highly reliable sources inside Sol's lab. Evidently, Sol had been officially nominated by the chairman of the pharmacology department at Hopkins along with Hughes and Kosterlitz, but the forms had been seen on the desk of Sol's secretary, being typed according to his instructions. I now knew that Sol had prepared the nomination documents himself, intentionally leaving my name off them. Perhaps there was nothing more nefarious going on than that the chairman, a new appointee at Hopkins, had told Sol to fill in the specifics, trusting he would know the names of those involved and the details of the actual work. But knowing Sol's love of applying for grants and awards that promised a fabulous prize, I chose to believe that he had initiated the nomination process and then sent the forms along to the chairman for a signature, so that the chairman would appear to be the source of the nomination.

This, then, if true, was the final twist of the knife. Was it possible that Sol had been blaming the Awards Committee for my exclusion when he himself had cut me out? Had he counted on my devotion, on the eager-to-please, self-sacrificing temperament I had displayed as his apprentice, to get me to go along with his plan? I was convinced this was the case. No male, regardless of status, would ever have accepted the situation, but because I was a woman, Sol had taken a chance. In my bitterness, I vowed to put behind me all traces of the naive female I had once been, Sol's "baby girl."

I exited Julie's office in tears, his angry, demanding voice reverberating in my ears as I slinked miserably along the Palace corridors to my office. Once there, I closed the door, slumped to my desk, and released my feelings in several deep sobs. By now I felt my defeat was complete. Pulling myself together, I called Dr. Bunney, my immediate boss, who descended from his office to meet me in my tiny cubicle—the first, last, and only time he ever made that particular journey. Sitting across from me, so close our knees were almost touching, he listened in silence to my mixture of grief and rage. "The truth is, I did ninety-nine percent of the actual scientific work on the opiate receptor, both inspiration and perspiration, even though Sol got the grant, slipped me Goldstein's original, if wrong, paper, and counseled me when the going got tough," I complained tearfully. "But when it came down to it, he pulled the plug and ordered the entire project closed down, kaput! I took a chance and continued, behind his back, because I believed it could be done. Oh, I really do love Sol!" I wailed on, while Biff practiced his best listening technique, and handed me a Kleenex.

Biff reminded me that no lab chief at the Palace would ever let me become so closely associated with such an important discovery, as Sol had done. A Palace Slick Boy would have seen the discovery for what it was and cut me out mercilessly, regardless of my contribution, as soon as it surfaced for publication. But the bitter irony was that by letting me be first author on the paper reporting our findings and then sending me on the road to present them for all to see, Sol had assured my career when I was only a graduate student—and assured himself a huge heartache by losing maybe his only shot at the Big One, the Nobel.

When I calmed down, Biff looked directly at me and said, "Do you realize now how important you are? If you don't sign off on this nomination for the Nobel and smooth over the Lasker controversy, it'll cost them all the prize."

I was beginning to understand what was going on. Biff later showed me a copy of Nobel's will, which stipulated that only three living scientists could win the prize at any one time. Someone had to be axed, and Sol had expected, again, that I would take the fall gracefully. I realized that my support, in spite of my earlier attempt to blow their cover, was their last hope for the Nobel, making me a potential and very important conspirator in their plan. But, even then, I wouldn't budge.

Moving quickly, and unbeknownst to me, Biff drafted a statement nominating me for the Nobel, along with Sol and Lars Terenius, whose brief paper in an obscure Scandinavian journal had reported findings parallel to our own, but in so densely technical a fashion, and with so little fanfare, that it had gone almost unnoticed. I can only surmise the effect this had on the nominating committee, no doubt confusing them to the breaking point, since Sol's name also appeared as part of a different trio in a separate nomination.

While a large percentage of Lasker recipients do go on to win the Nobel Prize, Snyder, Kosterlitz, and Hughes did not get that year's prize. After a long and reportedly heated debate, the 1979 Nobel was handed out to a completely different trio of male scientists for another discovery altogether. Within the cozy world of the Palace, word spread rapidly that my uprising had cost them the Nobel. And perhaps it had.

SURVIVAL

As a result of the Lasker flap, I developed a somewhat notorious reputation. Some of my closer friends would jokingly introduce me at meetings as "the scarlet woman of neuroscience," in an attempt to make light of what they saw as an overreaction by others. But it was not amusing when people I barely knew, senior male scientists, would cautiously move to one side of the corridor when they saw me coming. The rumors, as is typical in these kinds of affairs, were far worse than the actual incidents,

and many people had the illusion that I had called a major press conference to publicly deny Sol's role in the opiate-receptor discovery. I hadn't,
of course, but people chose to believe in such a dramatic and fabricated
scenario for their own reasons.

Even more upsetting than the gossip and the shunning was the
blackballing that went on in regard to important meetings and forums.
My invitations plummeted. I was no longer asked to address my colleagues at the top-of-the-line symposia, and was now lucky to receive
even the most peripheral spot at the low-rent affairs. My response was to
accept whatever came my way, hoping that by doing so I would be able
to hold on to some of my previous visibility. I was determined not to disappear.

The reality of my new in-house status was painfully driven home
when the Palace hosted a widely publicized opiate meeting later in 1979.
Hughes and Kosterlitz were flown in from Scotland to headline the
event, and Sol and other Opiate Club members were invited to present.
For some unspoken reason, the organizers were unable to find a spot on
the roster for me to report on my new work at the NIH, and to add insult
to injury, more than a few subtle suggestions were made that I not
attend. But I went anyway, my heart in my mouth, and endured the
chilly reception that greeted me.

This was a period of time when whatever I did seemed to send sparks
flying and start fires. One day, I decided to liven up my drab, institutional
workspace by painting a brightly colored rainbow strip along the walls
and out into the hallway. I'd had a fondness for rainbows ever since my
days at Hopkins when I had surprised my labmates by painting my finger-
nails with the tiny, multihued crescents. To me, the rainbow was a profoundly hopeful symbol, separating the white light of appearances into its
multiple spectrum and revealing a hidden dimension. It reminded me of
my belief that it was the mission of science to pierce through the layers of
everyday reality and penetrate to the truth. But to my colleagues, this
innocent exercise of right-brain expression made it seem as if I were trying to drive a stake through the very heart of modern science. Many of
them were simply embarrassed, and in spite of the continued productivity of my lab, my reputation as an eccentric grew.

As a survival strategy, I garnered the support of two important
allies—other women scientists, who were a growing force in research
circles, and the media, as represented by a never-ending supply of journalists who knew a good story when they saw one.

After decades of sexism that had never been honestly confronted until recently, my female colleagues hailed me as a heroine for standing up to the alpha male scientists. Even though I'd lost, they recognized my courage in making an attempt to see that justice was done. In the early eighties, I joined with a number of other female neuroscientists to start an organization called WIN, Women in Neuroscience. We selected as our motto "WIN with Brains" and posted an announcement in the Palace's women's rooms stating our intention that at the next Society of Neuroscience meeting, WIN would convene its own symposium. Much to our amazement, about three hundred women showed up for an event that turned out to be part group therapy session, part serious scientific meeting. It began in a spirit of lighthearted camaraderie, everyone happy to be in the company of women and for once not made to feel like outsiders at the boys' clubhouse. I gave my lecture barefoot, and, since I was pregnant at the time (with my third child, Brandon), wore a wildly colored hippie gown. But underneath the laughs and good time, there was an undercurrent of anger so strong that we could do nothing but ride it for many hours. So we steamed and vented, raged and wept, sharing horror story after horror story. And even though I knew firsthand how tough it was to be a woman in science, I was completely overwhelmed by what I heard.

The purpose of WIN, as it evolved, became increasingly political. By organizing ourselves in this way, we women were attempting to change our status from that of an oppressed minority to that of a modest interest group. We intended to lobby for more women to serve as chairs of our professional meetings, as well as set up a system of mentorship so that the more successful could assist the less successful to learn the ropes of grant-writing and political maneuvering. My involvement with WIN was very healing, even energizing for me at this time. I enjoyed being in a position of leadership, and I liked making waves that rocked the boats of the power boys and their established structures.

While this was going on, the media continued to sniff around, hoping to more fully exploit the Lasker incident and catch up on my latest exploits. But I responded by throwing them a curve, saying the Lasker was "old business," and offering them instead my views on the sorry plight of women in science, as well as the exciting discoveries that were happening almost daily in neuroscience laboratories around the world. Over time, I grew to like the press and came to consider journalists my allies in what was turning out to be a personal battle to rescue my former reputation as a promising young scientist—a Golden Girl—who had done some great science, and was yet to do more.

In retrospect, I'm amazed at how brazenly I dealt with the press, often throwing caution to the winds and saying whatever was on my mind at the time. I hadn't yet learned that I could demand that my quotes be read back to me, and so made some real blunders I later regretted, especially when I was giving a lecture and journalists reported my words as part of the public record. In a most egregious example, I was once quoted as saying: "Don't get me wrong, I like men—but in their place, which is the bedroom. Let them out, and they start wars." In context—though I have long since forgotten what the context was—the words seemed appropriate, but out of context and in print, where I never expected to see them, they looked outrageous. Unfortunately, it's a quote that has followed me around like a faithful puppy dog ever since.

It took a while to develop an effective media filter, but eventually I grew savvy enough to be able to handle myself in a way that didn't cause a sensation. And once I got over the need to use the media to give voice to my plight, I began to see that they could also help me communicate with the public about the science I was doing—which would ultimately be much more helpful to both my reputation and my work.

Still, in spite of the outlets that I cultivated with women and the media, the rejection I experienced from my colleagues hurt tremendously. But like all assaults on the organism that don't destroy it, the injury turned out to be beneficial in the long run. My faith in science as a search for truth was only strengthened as I became more determined than ever to work hard and do great science. The single-mindedness of my focus on work was reinforced, ironically enough, by a near-screeching halt in my advance up the Palace hierarchy after the Lasker incident, a trend that kept me in the lab and close to the bench. Usually, as scientists rise to power, their political skills become more finely honed than their scientific ones, causing them to lose the intuitive feel that good experimentation requires. I was spared this trade-off at a time when the bench was the very best place for me to be.

By 1982 the storm had passed, and I had emerged once again as a scientist whose reputation was based more on productivity and significant advances than on rumors, prejudice, and ignorance. I had learned to put my bitterness behind me. But never again would I be the starry-eyed Bryn Mawr girl seeing the world through rose-colored glasses fixed firmly on her nose. Most of my illusions about the romance of science had been stripped away, and I was left with some hard personal choices. Would I continue to try and beat the boys at their game, becoming more aggressive, more competitive, and more ruthless as I ascended the ladder to success? Was I going to let myself be motivated to do science for the fame and status, maybe even the money? In my heart of hearts, I knew that the only payment the fast track would bring me was a migraine at the end of the day, or perhaps a coronary bypass before the age of fifty.

DOING THE WORK

During the early days of the neuropeptide explosion, when new brain chemicals were being discovered every month, the field was so hot, so terribly fashionable, and so productive that even after my disgrace I did not have to shape my lab's work along pragmatic lines dictated by economics or changing science trends. Instead, during the late 1970s we were able to follow our own inspirations, striking out in directions where the work was assuredly doable, yet without having any real global perspective to guide our work. Only later would the bigger picture emerge, and our efforts be seen as contributing to advances that were not even thought of as yet.

My lab, like most in the Palace, hummed with activity night and day. By 1978 I was co-supervising, along with John Tallman, a staff of up to ten and getting a taste of what it meant to be in charge of the hiring and firing, the managing of people as well as resources and projects. In the beginning, I had a personal preference for hiring Italian graduate students and postdocs. I felt that their warmth, spontaneity, and lust for life infused an element of fun as well as creativity into the lab atmosphere. They liked to come into the lab around two in the afternoon and work all night, often peaking in rhythm around the time the evening news came on.

But I never developed any of the rigid rules about hiring staff that most of my colleagues did. Most of the time, I relied on my gut feelings about people, sensing an intuitive connection before I made my decision to hire, and so creating quite a bit of diversity in my lab.

As might be expected, my lab had a much higher female quotient than other labs in the Palace. Quite often, I hired brilliant women, some at the level of senior scientist, who were unable to find positions that matched their experience. I remember one refugee from a lab on Long Island renowned for its sexist policies who had invented a technique to visualize a single neuron, making it light up with monoclonal antibodies. It was a revolutionary technique, one that should have skyrocketed this woman's career, but, instead, I found her knocking on my door like a beggar, asking to work in my lab on a temporary basis until she could find other work.

Sometimes, a postdoc would arrive with his or her own salary line from a grant, as was the case with my very first postdoc, Terry Moody, a long-haired, California tennis jock. When Terry arrived in my lab in 1977, I had a project ready and waiting for him. Just recently I had acquired a sample of the bombesin peptide from Marvin Brown, a California neuroscientist with whom I'd waited out a plane delay one stormy day in Atlanta. Marvin told me he'd injected bombesin, which had been extracted from the skin of a frog known as *bombix bombina,* into rats, and it caused them to scratch furiously while losing ten degrees in body temperature. As it later turned out, itching and lower body temperature were found to be among the effects in humans as well. Of course, it had to have receptors in the brain, which Terry and I set out to find. Using the same approach I had devised to find the opiate receptor, Terry was able to locate bombesin receptors in the limbic areas of the rat brain. He and a number of other postdocs were responsible for a string of such discoveries, which kept us all in a perpetual state of excitement.

Typically, a Palace lab chief might supervise anywhere from ten to thirty postdocs, each one of them assigned to a particular line of investigation. The predominant management style at the Palace is best described as a wheel, with the lab chief at the hub and the staff all extending outward as individual spokes, each totally ignorant of what the others are up to. This ensures an atmosphere that is very secretive, which lab chiefs can then exploit in their efforts to motivate the postdocs. Two or three people may be assigned to the same experiment, sometimes even doing the exact same work. The chief can then use them against one another to ensure accuracy. "Paul did such and such," he'll say to Peter. "But I don't really believe it. You think Paul's off base, don't you? Why don't you repeat the experiment and find out what's really going on." Managed competition we call it—a technique that Sol used in his lab at Hopkins, and one that a large number of other successful scientists of his generation have used. As a management style, it can be extremely productive, but not a lot of fun, especially if you're one of the spokes in the wheel.

In my own lab, I consciously tried to develop a nurturing, even maternal mode of management, motivating people by praise rather than criticism, by team spirit rather than competition with each other. The rejection I had experienced after the Lasker episode sensitized me to the pitfalls of the gamesmanship in lab politics, making me want to have as little to do with it as possible. I strived to create an atmosphere in my lab that would allow for more collaboration, opening up doors for different labs to work together on joint projects instead of viciously competing with one another to win funding and glory.

Much of the work we did in my NIMH lab, both before and after the Lasker incident, was a continuation of the brain-receptor mapping I'd been doing since my days in Sol's lab. We went on the assumption that any peptide that could be found anywhere in the body would have a specific receptor for a perfect fit somewhere in the brain—hence the term *neuropeptide*. We proved this by putting one peptide after another through assays constructed along the lines of the original opiate-receptor experiment and detecting if there was binding. Someone dubbed this process "grind and bind," *grind* for the process of pulverizing the tissue into milkshake consistency, and *bind* for the action of the peptide as it docked onto its receptor.

Our grind-and-bind assay method was soon replaced by a new autoradiographic technique, which further confirmed the distribution of receptors for all the known peptides at different locations in the brain. In developing this, I worked closely with Miles Herkenham and Remi Quirion to take the art of receptor autoradiography to the next level of precision, creating a quick and easy way to visualize the receptors in the brain, and even see their levels of density.

My first contact with Miles was on the phone, when, shortly after arriving at the Palace in 1977, he called to invite me to a talk he was giving. Miles was a neuroanatomist who had studied with the renowned Walle Nauta at MIT, the scientist who had earned my admiration for interpreting our opiate-receptor assay data and locating the densest concentrations in the limbic, or emotional, brain. Miles, like Nauta, was involved in constructing a circuit diagram of the neuronal cells and their pathways, an ongoing project scientists had been working at for almost a century in the hope that it would reveal the full extent of the electrical brain. However, their methods had not yet enabled them to determine which neurotransmitters were secreted by a particular neuron. Autoradiography had allowed Miles to map the pathways of neurons and axons, which showed up as round holes on his film, making for a kind of

Swiss-cheese effect, as if myriads of tiny islands were afloat in a sea of microscopically viewed tissue.

The work Mike Kuhar and I had done when I was a graduate student at Hopkins had yielded a map of the chemical brain on which the receptors, when visualized, looked like tiny dark patches against a lighter tissue background. Having read about that project, Miles had been wanting to ask me for some time: Could his holes match my patches? Would his electrical brain mapping correspond to my chemical brain mapping?

When I attended his talk, I was stunned by the slides Miles showed, so beautiful, the tiny neurons showing up like the Milky Way projected against a vast black universe. Rarely did I encounter a scientist who had an aesthetic sense, most of them preferring a drier, numbers-only approach. But Miles had an unusual respect for the natural beauty of what he was seeing, and I resolved then and there to work with him. And he himself was as gorgeous as his slides—a real hunk!

Our collaboration resulted in a great improvement on the tedious visualization methods I'd used back at Hopkins, and greatly advanced the art of receptor autoradiography, moving it from binding in animals (in vivo) to binding on slides of presliced brain (in vitro). "Slip and dip," as we called the new methodology, involved fixing radioactively labeled ligands to receptors and then dipping them in a radiosensitive liquid emulsion. While it was a difficult and unforgiving method, partly because much of it had to be done in the dark, it produced excellent results, allowing us to see receptors as tiny sparkling grains in a sea of colorfully stained brain tissue.

A much simpler and quicker method that we developed, which was eventually adopted as the state-of-the-art technique, involved taking the brain slice after it had been fixed on the slide, after the receptor is bound in vitro, and putting it up against radiosensitive film, which is housed in a cassette. In theory, it was a great idea, but our first attempt was a total bust, yielding black on black.

We were about to abandon the pursuit when chance brought me to a conference where I met up with my old friend from graduate school days, Anne Young. Now the head of neurology at Massachussetts General Hospital, Anne is renowned for her unpretentious, zesty humor. She graduated first in her class at Vassar, wearing a molecular model "hat" on her head, which she gaily tipped to the president on receiving her diploma. One night during the conference, we went to Anne's room and talked well into the night, catching up and enjoying the bottle of Scotch I'd brought with me, and Anne happened to mention that her lab had been working with the slip-and-dip method Miles and I had developed, and had refined it. I was all ears.

"We put the slide up against the film, in a cassette," she volunteered.

"Really," I said. "Miles and I had the same idea, but we couldn't get it to work."

"But it's so easy. You've already done the hard part, labeling the receptor in the tissue and then getting it to stick on the slide," she said matter-of-factly. "I bet you put the film in backwards—an easy mistake to make in the pitch dark."

In a stunning example of how scientific schmoozing can create a breakthrough, leading to the correction of the one "tiny" mistake in a long chain of well-executed experimental steps, Anne had given us the key. Back in Bethesda the next day, I went to work with pharmaceuticals expert Sandy Moon, an African-American woman who was one of the elegant, brilliant scientists I'd been able to attract to my lab. She put the same slides up against the correct side of the film for me, and instant Eureka! A more valuable bottle of Scotch I don't think I've ever shared with a friend.

This new technique allowed us to map our receptors in a week, rather than spending up to a year on each receptor. Now we could find receptors easily and quickly for ourselves, as well as for the many labs engaged in this kind of research and needing to identify a particular peptide. My postdoc Remi Quirion was immediately put on a project to find the PCP ("angel dust") receptor, and he found it right away. A Canadian whose parents ran a diner, Remi was a brilliant short-order cook in the lab who turned autoradiography into a process that resembled a fast-food operation. By the time he was ready to leave my lab, Remi was using his autoradiography technique to collaborate with several other labs in the Palace.

While the new method was incredibly efficient at locating the receptors, it also allowed us to expand aesthetically and bring color to autoradiography. Up until then, the different densities had appeared in black and white only, making it difficult to discern subtle fluctuations. Now, because the image was on film, we could employ the computer to use color coding, giving us quantitative receptor autoradiography. This made the density of the receptors in the brain visible by giving us a picture that resembled a modern-day weather map showing temperature gradients for different geographical areas. Areas that showed up as yellow might indicate a certain number of receptors, while areas that were orange or purple might indicate greater or lesser numbers. We came to call our cassette and computer combinations "the rainbow machine," evoking for me my favorite symbol of the promise of science. When I looked at the images, I thought they resembled colorful butterflies, and I couldn't resist making some of them into posters to grace the otherwise sterile hallway walls. I fantasized that we'd invented a new art form called "photoneurorealism" that would be shown one day in a New York gallery—exquisitely beautiful and full of scientific data.

Now we went to work on our goal of connecting the neurochemical maps of the peptides and their receptors with the circuit diagrams that the anatomists had been working on for years. Miles's circuit diagrams showed the actual wiring of the brain, its electrical reality, and marked out paths of communication between the nerves, axons, and dendrites. By overlapping our autoradiographic maps with his diagrams, we could see which nerve pathways had receptors for endorphins and which could receive messages carried by other peptides. At the time it was my dream that we'd soon have a large color-coded map of the brain showing the chemical system as it interacted with the electrical, a veritable rainbow of color and information.

Another finding that emerged from our color autoradiography was just how rich the emotional, or limbic, areas of the brain were in the receptors for peptides that regulated the pleasure/pain continuum. Miles had already mapped some of these limbic connections anatomically, in particular the thalamus-to-cortex pathways of neurons in the brain. Now we could see how the opiate receptors were following the exact same pathways as the subgroup of thalamic neurons connected to the limbic system. We jokingly called our newly revealed spots "love patches," because they fit precisely over Miles's floating island holes. At last we had confirmed that the holes on his film, which were formed by neurons projecting to the limbic system, were a match for my opiate-receptor patches, and in August 1983 we had a gorgeous, full-color cover of *Nature* magazine to illustrate our findings.

In my more leisurely moments, when I could take time away from the minutiae, the day-to-day routine, I dreamed of a grand mosaic design, one made up of all the neuropeptides and other messenger molecules being discovered by my lab and others. Surely what we were seeing was the basis for some kind of complex communication within the brain. And since the brain peptides and their receptors were showing up throughout the far-flung systems of the body, perhaps this was an indication that communication was taking place not just *within* the brain, but *between* the brain and the rest of the body. I was beginning to wonder: Did all these systems hook up somehow? And if so, why?

PROCESS

By the early eighties, although the Palace air was still charged with competition, our lab had become a hotbed of collaboration and exchange. In 1982 I was promoted to section chief in brain biochemistry, and with a staff sometimes numbering as many as fifteen, I was in heaven. Scientists from within and without the Palace walls were calling my lab every day for appointments to have their peptide and receptor findings autoradiographed for possible measurement and location in the brain. A researcher might drop by to show me the data for an experiment using naloxone to control the eating behavior of his overweight rats. He had found that when the opiate receptors were blocked, the rats ate less, suggesting that endorphins might be somehow involved in obesity. Could I measure the output of endorphins in the rats' pituitary glands? Another researcher might call to tell me of her study that correlated mood swings to the menstrual cycle. Could the emotional upswing occurring between days five and seven be explained by endorphin release? A psychiatrist might come by to discuss the work he was doing with patients extremely sensitive to pain and unable to get relief from even the strongest painkilling drugs. Could I measure the endorphin levels in these patients' blood? The stream was endless, and I loved every moment of it.

Things were moving so fast during that time period that I often felt like a juggler adeptly spinning plates on the top of two long poles held in each hand. Getting the plates to spin was only half the job. Keeping them in motion—that's the art. If one plate falls to the ground, if an experiment crashes and must be shut down, even for a few weeks, the task of getting it up and spinning again might be impossible. The technology seems to have a life of its own—once it ceases to work, it's dead, and it can take months, sometimes years, to bring it back to life again.

The assay to measure endorphins in the blood, the assay to locate receptors in the brain, and autoradiography—these were the plates I attempted to keep spinning. Every day in my lab my postdocs would use these techniques, training their successors to do the same when they moved on. We might ask different questions, but the technology remained the same. For example, we wanted to understand how the opiate receptor and endorphin system had evolved over time, and to map this progress using autoradiography. To do this, we would examine the fetal rat brain at three days in utero, then seven days, then fifteen days, patiently recording the changes in cells and structures. Then we used the same technique to see how these systems had developed in the brains of monkeys. We'd ask: Were the receptors densest in those locations where sensory input was received from the ears, eyes, nose, mouth, and skin? Were they thickest in the cerebellum or in the neocortex?

My method was to develop a technique and then ask all the questions to which the technique could supply an answer. This process can go on for years as you keep gathering pieces of a puzzle that gradually add up to a big picture of which you catch only fleeting glimpses along the way. And then one day, when the light flashes on in your brain, you see the big picture, the grand scheme, and it all comes together; all the data you've been accumulating for years starts to make sense. Or maybe you never get to that point and, instead, continue to create new techniques, brainstorm more questions, create more data, massaging it all into shape for publication. Then maybe someday someone else will see it, and it will be the missing piece of a puzzle he or she has been working on for years.

Most people think of science as a series of dramatic results, breakthroughs, advances, but science is really first and foremost a process. You start out on one path and then take a sudden turn to find yourself going down a totally different road. Sometimes the steps are small and the progress incremental. In my lab, when nothing was working, my postdocs would become downcast and dejected. There were some extremely dry spells when only one out of ten projects was producing results, and that single success was usually a dull, bread-and-butter experiment that didn't excite anyone's imagination. But at the Palace, these were often the kinds of projects that satisfied your superiors and ended up fattening your publication score. A smart scientist tried to balance a few chancy personal long shots with a larger number of tried-and-true experiments, ones that would never produce any fireworks, but were solid guarantees.

One of my personal long shots was a project I designed to find the marijuana receptor. This was one I wanted to bag myself, to prove we had in our brains a natural marijuana that we could potentially access to create a natural high without smoking pot. I labored over this one for two years, trying hundreds of clever little things, all of which consumed an abysmal amount of my time and ultimately failed to produce results because I never managed to get hold of the right ligand. Without that ligand, all my efforts, hundreds upon hundreds of hours of hard labor, had less than a snowball's chance in hell of ever bringing me the marijuana receptor. So, eventually, I bowed to the inevitable and turned my attention to other things. Fortunately, there were lots of other ongoing projects to occupy me, far too many to explain, and lots of successes among them. In the end, and only recently, it was my friend Miles Herkenham, who, having procured the right ligand, was able to successfully visualize the cannabis receptors in rat brains.

I barely glimpsed it then, but the work we were doing in my lab during this time was laying the groundwork for a huge discovery, one that would lead us to formulate a radical theory that explained the link between mind and body, and how the emotions are directly involved with health and disease. My father's diagnosis of lung cancer in 1980, and the research I subsequently did in a desperate attempt to save his life, helped me begin to see this connection. But it wasn't until I was drawn into yet another, new level of personal involvement with my science that I was able to take a giant step out of the old paradigm and boldly follow what, deep in my heart, I knew to be the truth.

7

THE BIOCHEMICALS
OF EMOTION:
A CONTINUED LECTURE

I'M FEELING confident that my listeners have been properly introduced to the basic biochemicals of emotion, the neuropeptides and various other ligands and their receptors, and have learned a bit about how
we scientists came to understand them. They have also learned a little
about both the electrical and the chemical aspects of the brain, and how
the neurotransmitters making their electrically fired leaps across brain
synapses are just one part of a much more far-flung network of information carried by neuropeptides and their bodywide receptors. We are now
ready to explore my theory that these biochemicals are the physiological
substrates of emotion, the molecular underpinnings of what we experience as feelings, sensations, thoughts, drives, perhaps even spirit or soul.
As a result of the work done in my own lab, as well as many others, and
presented in a number of theoretical papers and lectures, there is now
much evidence to support this theory.

And so, with the house lights still dim, I gaze directly into the pool of
eager eyes, minds, and hearts before me and begin explaining the most
radical implications of my work—implications I could barely begin to
articulate at the time I was first describing the research in the professional journals. What I will describe to you today—why you feel the way
you feel—is my latest thinking on a subject that has consumed me for
over a decade. My ideas grew from a synthesis of many diverse sources,
ranging from my own work in the laboratory, to research done by today's
leading theorists of emotion, to the latest findings of a global community
of neuroscientists. My personal exposure to a number of mind-body

therapies emphasizing the importance of the emotions—particularly the power of their full expression to release us from the joy- and health-sapping patterns of the past—has gradually increased my confidence in the validity of these ideas.

WHAT DO I MEAN BY EMOTIONS?

I should say first that some scientists might describe the idea of a biochemical basis for the emotions as outrageous. It is not, in other words, part of the established wisdom even now. Indeed, coming from a tradition where experimental psychology textbooks (which focus on the observable and measurable) do not even contain the word *emotions* in the index, it was not without a little trepidation that I dared to start talking about their biochemistry! I grew bolder in 1984 when Paul Ekman, a highly respected psychologist who studies human emotion at the University of California at San Francisco, introduced me to Charles Darwin's book on the subject. If the great Darwin himself had thought it important, then surely I was on firm ground. In *Expression of the Emotions in Man and Animals,* Darwin explained how people everywhere have common emotional facial expressions, some of which are also shared by animals. For example, a wolf baring its fangs uses the same facial musculature as any human being does when angry or threatened. The same simple physiology of emotions has been preserved and used again and again over evolutionary eons and across species. On the basis of the universality of this phenomenon, Darwin speculated that the emotions must be key to the survival of the fittest.

To quote *The Selfish Gene,* by Richard Dawkins, on the subject of the relationship between evolution and survival mechanisms: "A duck is a robot vehicle for the propagation of duck genes." This is but another way of making Darwin's point that if emotions are that widespread across both human and animal kingdoms, they have been proved, evolutionarily, as crucial to the process of survival, and are inextricably linked to the origins of the species.

When I use the term *emotion,* I am speaking in the broadest of terms, to include not only the familiar human experiences of anger, fear, and sadness, as well as joy, contentment, and courage, but also basic sensations such as pleasure and pain, as well as the "drive states" studied by the experimental psychologists, such as hunger and thirst. In addition to measurable and observable emotions and states, I also refer to an assortment of other intangible, subjective experiences that are probably unique to humans, such as spiritual inspiration, awe, bliss, and other states of consciousness that we all have experienced but that have been, up until now, physiologically unexplained.

I must tell you that the experts—the emotional theorists who have their own scientific data to interpret—disagree about many things, including whether feelings are the same as emotions, how many basic or core emotions there really are, or even whether these are useful questions at all! They do agree, however, that there is now clear scientific experimental evidence that the facial expressions for anger, fear, sadness, enjoyment, and disgust are identical whether an Eskimo or an Italian is being studied. Facial expressions that register other emotions such as surprise, contempt, shame/guilt are probably also pancultural, meaning that these, too, are emotions with inborn genetic mechanisms for their expression. And there probably are more genetically based emotions to discover.

Robert Plutchik, a psychology professor at Hofstra University whose emotions research influenced me as an undergraduate, proposed a theory of eight primary emotions—sadness, disgust, anger, anticipation, joy, acceptance, fear, and surprise—which, much like primary colors, could be mixed to get other, secondary emotions. For example, fear + surprise = alarm, joy + fear = guilt, etc. Whether or not Plutchik's classification is borne out by more research, the idea of certain emotions being mixed to produce other emotions is interesting, and suggests that when other factors such as intensity and duration of emotion are considered, there may easily be hundreds of subtle emotional states.

The experts also distinguish among emotion, mood, and temperament, with emotion being the most transient and clearly identifiable in terms of what causes it; with mood lasting for hours or days and being less easily traced; and with temperament being genetically based, so that we're generally stuck with it (give or take certain modifications) for a lifetime. For example, Harvard psychology professor Jerome Kagan has proved that readily measurable traits like the tendency to be startled by novel stimuli can be shown most readily in those infants who go on to develop into shy children and adults.

THE LIMBIC SYSTEM OF THE BRAIN:
CLASSICAL SEAT OF THE EMOTIONS

For a long time, neuroscientists have agreed that emotions are controlled by certain parts of the brain. This is a big, "neurocentric" assumption—and I now think it is a wrong (or at least incomplete) one! Still, as a neuroscientist and a onetime believer in the brain as the most important organ in the body, I was led by this assumption to do the right analysis for the wrong reason. During the mid-1980s, with my NIH lab colleagues Joanna Hill and Birgit Zipser, I systematically analyzed the brain distribution patterns of twenty-two different neuropeptide receptors that our lab had mapped over the years, comparing them to the classical emotional brain areas of the limbic system—a hypothetical construct, known as the seat of the emotions, which has come to include more and more brain structures over the years. We confirmed for many other neuropeptide receptors what we'd first seen for the brain distribution of opiate receptors, the very first neuropeptide receptors to be mapped: Core limbic brain structures, such as the amygdala, hippocampus, and limbic cortex, believed by neuroscientists to be involved in emotional behavior contained a whopping 85 to 95 percent of the various neuropeptide receptors we had studied! This concordance fueled my conviction (which had first begun to develop as I mapped opiate receptors in the late 1970s and early 1980s) that there was such a thing as the molecules of emotion.

Human experiments showing the connection between emotions and those parts of the brain where we were now locating almost all the neuropeptide receptors had been done at McGill University in Montreal by Wilder Penfield in the 1920s. Working with conscious, awake individuals during open-brain surgery to stop severe and uncontrollable epilepsy, he found that when he electrically stimulated the limbic cortex over the amygdala (the two almond-shaped structures on either side of the forebrain, about an inch or so into your brain from your earlobes), he could elicit a whole gamut of emotional displays—powerful reactions of grief, anger, or joy as patients relived old memories, complete with the appropriate bodily accompaniments such as shaking with rage or laughter, weeping, and blood pressure and temperature changes.

Another indication that neuropeptides and their receptors might be plausible candidates for the locus of the emotions was that they met

Darwin's criterion: The physiological basis for the emotions, he predicted, would be found "conserved" throughout evolution. Given their important role in the survival of the species, they would appear again and again throughout the various evolutionary stages of the animal kingdom. In fact, the receptor-mapping experiments I'd done with radioactive opiates like morphine and naloxone had shown that identical opiate receptors could be found in the brains of all vertebrates, from the simple, hideous hagfish to the complex, exalted human. Even insects and other invertebrates could be shown to have opiate receptors. Darwin himself could write only about the physiology of emotions and not about their biochemistry or genetics because the concept of biochemistry, with its specific components, the proteins and peptides (direct products of genes), would not be discovered for almost a hundred years. But I think he would see in such work the confirmation of his brilliant hunch.

IT WAS NIMH researcher Paul MacLean who popularized the concept of the limbic system as the seat of the emotions. The limbic system was one constituent of his triune brain theory, which held that there are three layers to the human brain, representing different stages of humanity's evolution—the brainstem (hindbrain), or reptilian brain, which is responsible for breathing, excretion, blood flow, body temperature, and other autonomic functions; the limbic system, which encircles the top of the brainstem and is the seat of the emotions; and the cerebral cortex, in the forebrain, which is the seat of reason.

Back in 1974, I had an opportunity to visit with the prominent physician-scientist when I went to his NIMH laboratory to deliver a lecture about my then new opiate-receptor discovery. Afterward, Paul mischievously led me past cages of monkeys who shrieked and shook their genitals at me in an intense display evolved to frighten intruders off their turf. Even back then Paul was fielding questions about just how much the limbic-system concept was accurate science and how much was metaphor. But what really excited me that day was the discussion we had about the fact that opiate receptors are by far the densest in the frontal lobes of the cerebral cortex of the human brain, which share many interconnections with the amygdala, one of the so-called limbic structures. As Paul emphatically tapped his forehead in front of his frontal cortex—the most newly evolved of the brain structures, and the one that is most fully developed in human beings—I thought about the physiological and biochemical pathways that had had to be forged between that cortex and the rest of the brain to enable humans to learn to control their emotions and act unselfishly. Although the capacity for learning is to some extent present in even the simplest creatures, willpower is the uniquely human "ghost in the machine," and Paul was sure that it resided only in the frontal cortex.

DO EMOTIONS ORIGINATE IN
THE HEAD OR THE BODY?

Until 1984 I had assumed that Wilder Penfield's famous human experiments had proved beyond a shadow of a doubt that emotions originate in the brain. But that year, I went to give a lecture at the Second International Meeting of the Society for the Study of Emotion, which was held at Harvard, and there I met Eugene Taylor, the scientific historian in the psychology department. He was excited about the lecture I had just delivered, in which I presented the theory of peptides and other ligands as the biochemicals of emotion. Gene wanted to know where I stood on the famous James-Cannon debate, which, he reminded me, was about the ultimate source of emotions. Do they originate in the body and then get perceived in the head, where we invent a story to explain them, as William James said? Or do they originate in the head and trickle down to the body, as Walter Cannon posited?

In 1884, while an assistant professor of philosophy at Harvard, William James had published his essay "What Is an Emotion?" basing his theory on his own introspective observations and a general knowledge of physiology. He said he had concluded that the source of emotions is purely visceral, that is, originating in the body and not cognitive, originating in the mind, and that there is probably no brain center for emotional expression. We perceive events and have bodily feelings, and then after the perception, which joggles our memories and imagination, we label our physical sensations as one or another emotion. However, he believed that there was in fact no such entity as emotion. There is simply perception and bodily response. The immediate sensory and motor reverberations that occur in response to the perception—the pounding heart, the tight stomach, the tensed muscles, the sweaty palms—*are* the emotions. And the emotions are felt throughout the body as sensations, "each morsel of which contributes its pulsations of feeling, dim or sharp, pleasant or painful or dubious, to that sense of personality that every one of us unfailingly carries with him." Emotions consist of organic changes in the body, muscular and visceral, and are not a primary feeling directly aroused, but a secondary one, indirectly aroused by the body's workings.

Like many appealing armchair theories, James's seemed to collapse under the weight of real data, in this case, animal laboratory work performed by his student Walter Cannon, the experimental physiologist and author of *Wisdom of the Body*, who, by 1927, had explained the workings of the sympathetic autonomic nervous system. A single nerve called the vagus ("wandering") nerve exits the back of the brain through a hole in the bottom of the skull (the foramen magnum), then splits to run down the bundles of nerve cells, or ganglia, along either side of the spinal cord to send branches to many organs, including the pupils of the eye, the salivary glands, the heart, the bronchi of the lungs, the stomach, the intestines, the bladder, the sex organs, and the adrenal glands (from which the hormone adrenaline is released). When Cannon stimulated the vagus through electrodes implanted in the hypothalamus in the bottom of the brain just above the pituitary gland, he demonstrated physiological changes in all these organs consistent with what would be needed by the body in an emergency situation when resources had to be quickly, efficiently, and automatically managed without any time-wasting thought. As the result of this hypothalamic stimulation, for example, blood from the internal organs of digestion was quickly rerouted to the muscles for "fight or flight"—digestion could wait until the emergency was over—and an increased output of adrenaline stimulated the heart and caused the liver to release extra supplies of sugar for instant energy.

From Cannon's point of view, James's theory of visceral emotion was all wet. Cannon could accurately measure how much time it took from the moment the hypothalamus got an electrical jolt to the moment the bodily changes in blood flow, digestion, and heartbeat began to occur as a result. And his conclusion was that these changes were just too slow to be the cause of emotions rather than the effect. Also, artificial induction of visceral changes that were typical of strong emotions, such as the use of electrical current to produce a strong intestinal contraction like that which occurs in the first moment of panic, failed to produce the other signs of that emotion. Not only that, Cannon pointed out that animals whose vagal nerve had been cut, and presumably were incapable of sympathetic visceral bodily changes, still seemed to behave just as emotionally when placed in a threatening situation. According to Cannon, the hypothalamus of the brain was the seat of the emotions, which trickled down to the body through the hypothalamus's neuronal connections to the back of the brain, or brainstem, or through the secretions of the pituitary gland.

While Eugene Taylor waited expectantly for my late-twentieth-century spin on the somewhat arcane James-Cannon debate, I suddenly had a big aha!: "Why, it's both! It's not either/or; in fact, it's both *and* neither! It's *simultaneous* — a two-way street," I blurted out. I had just realized that the resolution of a debate whose origins went back over a century held the key to understanding a very modern conundrum: How can emotions transform the body, either creating disease or healing it, maintaining health or undermining it?

This also helped me to understand the reading I'd recently been doing on biofeedback, which is the technique of using monitoring devices to measure various bodily functions (for example, heart rate or blood flow) as a step toward gaining control of those functions. Biofeedback can enable ordinary folks (and not just ascended yogis) to attain a state of deep relaxation in which it is possible for them to take conscious control of physiological processes previously thought to be autonomic and not susceptible to voluntary intervention. For example, anyone can increase the temperature of his hand by 5 to 10 degrees, often on the very first try. Elmer Green, the Mayo Clinic physician who had pioneered in biofeedback for treatment of disease, had said, "Every change in the physiological state is accompanied by an appropriate change in the mental emotional state, conscious or unconscious, and conversely, every change in the mental emotional state, conscious or unconscious, is accompanied by an appropriate change in the physiological state." Taylor's question had led me to another insight into the meaning of the discoveries we'd been making about the location of peptides and their receptors, and about the theories we'd been formulating about these molecules of emotion.

BEYOND SYNAPSES: A NEW MODEL OF INFORMATION EXCHANGE

In the 1960s, the emerging science of neuropharmacology focused on neurotransmitters being released from nerve endings, traveling across synapses to ignite another electrical discharge, in a hardwired (neuron-to-neuron), point-to-point hookup of traveling neuronal impulses. All brain functions, even for the most complex levels of mental activity and behavior, were thought to be determined by the synaptic connections between billions of neurons. The synapses formed the networks and defined the neural circuits whose chatter was thought to dictate every aspect of perception, integration, and performance. At the synapse, the models for both the electrical and chemical brain seemed to merge. There were no discrepancies, just exciting concordances, as neurochemistry, the new field of mapping neurotransmitters, seemed to confirm the existence of the neuroanatomical brain circuits discovered previously, and to reveal new ones as well.

For example, Arvid Carlson and the Swedes (as we American neuroscientists collectively refer to the school of crackerjack neurohistochemists working in Stockholm) had invented a method for visualizing nerve endings in the brain that contained norepinephrine, also called noradrenaline. Using this new tool, they saw that a tiny cluster of previously unremarkable cell bodies in the hindbrain, called the locus coeruleus, projected its norepinephrine-containing nerve endings into the forebrain, and that all the norepinephrine in the forebrain comes from this one source. Then psychologist Larry Stein at Wyeth Labs and Bryn Mawr College showed that what previous research had called the "pleasure center" or "pleasure pathway"—an area of the brain that when electrically stimulated caused rats (and humans) to ignore the need for food and sleep in a frenzy of pleasure and excitement—contained within it this locus coeruleus. Unbeknownst to those earlier researchers, the electrical stimulus had worked by causing the release of norepinephrine from the nerve endings along the pathway. Amphetamines and cocaine were shown to work by amplifying this same "pleasure pathway," blocking re-uptake of the neurotransmitter norepinephrine and thus increasing the amount of it that came in contact with its receptors, all of which were believed to be located just across the synapse.

And so matters stood for a couple of decades, with neurochemists amplifying and elaborating on work that had been done in earlier decades by the neuroanatomists. But that work didn't go far enough.

Enter a new theory of information exchange outside the bounds of the hardwired nervous system, focused on a purely chemical, nonsynaptic communication between cells. My lab at the NIMH, having specialized in the neuropeptides, had not only mapped their receptor sites throughout the brain, but, by the early 1980s, with the help of my postdoctoral fellow Stafford Maclean, had also devised a new autoradiographic method for identifying where the neuropeptides are produced, a technique that enabled us to take a much broader perspective. Suddenly, we felt like we were flying above a forest rather than studying the bark on the trees.

While Miles Herkenham and I had previously confirmed the lovely concordance that we had hoped to find between certain electrical pathways and the chemical patterns of opiate receptors, the new method revealed a discrepancy. Studying the ton of new data on numerous neuropeptides and their receptors, data that had been produced in his lab, our own, and many others by the early 1980s, Miles was struck by a disturbing "mismatch" between what we thought we knew and what we were seeing. Something was wrong. If peptides and their receptors were communicating across the synapse from each other, they should be only minuscule distances apart, but their location was not conforming to this expectation. Many of the receptors were located in far-flung areas, inches away from the neuropeptides. So we had to wonder how they were communicating, if not across the synaptic gap. Miles concluded that the largest portion of information ricocheting around the brain is kept in order not by the synaptic connections of brain cells but by the specificity of the receptors—in other words, by the ability of the receptor to bind with only one kind of ligand. Miles has estimated that, counter to the collective wisdom of the neuropharmacologists and neuroscientists, less than 2 percent of neuronal communication actually occurs at the synapse. It was so radical an idea that for several years his observation of the mismatches was ignored and attributed to artifacts of the mapping techniques. In fact, the way in which peptides circulate through the body, finding their target receptors in regions far more distant than had ever previously been thought possible, made the brain communication system resemble the endocrine system, whose hormones can travel the length and breadth of our bodies. The brain is like a bag of hormones! Our view of the brain, and the metaphors we used to describe it, were permanently altered.

In 1984, at around the same time that Miles was teaching me the significance of the mismatch in the mapping studies, Francis Schmitt, an elder statesman of neuroscience from MIT who had originated the Neuroscience Research Program, introduced the terminology of "information substances" to describe a variety of transmitters, peptides, hormones, factors, and protein ligands. Alongside the conventional model of synaptic neuronal circuitry, Schmitt proposed a parasynaptic, or secondary, parallel system, where chemical information substances travel the extracellular fluids circulating throughout the body to reach their specific target-cell receptors. His idea was readily accepted, as was his vivid terminology.

THE MIND-BODY CONNECTION: EMOTION-CARRYING PEPTIDES

Suddenly, the number of possible lines of communication between the brain and the body seemed to explode for me. There were numerous alternatives to the synaptic nerve hookups that once seemed indispensable for mind-body communication, and we were beginning to learn what was getting communicated through those connections. For example, the receptors for sex hormones that had been unexpectedly identified in the brain and then ignored for many years were clearly the mechanism through which testosterone or estrogen, if released into the fetus during pregnancy, could determine neuronal connections in the brain and permanently affect the sexual identity of the child. John Money, the famous Johns Hopkins psychiatrist, had shown that female fetuses exposed to testosterone-like steroid hormones (aberrantly produced by their pregnant mothers' adrenal glands) were more likely to become tomboys and avoid dolls!

Also, numerous additional nerve hookups could now be discovered thanks to the invention of new biochemical tools with which to examine them. Scientists began to follow up on the pioneering mid-1980s work of Tomaas Hokfeldt (one of the Swedes), who had reported that the classical autonomic nervous system described in Cannon's work unexpectedly contained just about every neuropeptide that he had sought there. Neuropeptides could be found not only in the rows of nerve ganglia on either side of the spine, but in the end organs themselves. An era of discovery began that is still in full swing, as neuroscientists began tracing the precise connections among all the parts of the body. New peptide-containing groups of neuronal cell bodies in the brain called "nuclei," the sources of most brain-to-body and body-to-brain hookups, are now being elaborated upon every day.

To cite just one recent example, Rita Valentino of the University of Pennsylvania has shown that the nucleus of Barrington in the hindbrain, formerly believed to control merely micturition (bladder-emptying), sends axons containing the neuropeptide CRF down through the vagus nerve all the way to the most distant part of the large intestine, near the anus. Rita has proved that sensations of colonic distention (i.e., the feeling of needing to poop) as well as those of genital arousal are carried back to the nucleus of Barrington. From there, there is a short neuronal pathway (called a "projection") that hooks up to the locus coeruleus, the norepinephrine-containing source of the "pleasure pathway," which is also very high in opiate receptors. The pleasure pathway hooks up to the control area of these bathroom functions, which is located in the front of the brain. Goodness, is it any wonder, based upon Rita's neuroanatomical discoveries, that toilet training is loaded with emotional stuff! Or that people get into some unusual sexual practices involving bathroom behaviors! Clearly, the classical physiologists had grossly underestimated the complexity and scope of the neurochemistry and neuroanatomy of the autonomic nervous system. But the limitations of the past are now giving way before our newfound ability to track these fascinating connections.

If we accept the idea that peptides and other informational substances are the biochemicals of emotion, their distribution in the body's nerves has all kinds of significance, which Sigmund Freud, were he alive today, would gleefully point out as the molecular confirmation of his theories. The body is the unconscious mind! Repressed traumas caused by overwhelming emotion can be stored in a body part, thereafter affecting our ability to feel that part or even move it. The new work suggests there are almost infinite pathways for the conscious mind to access—and modify—the unconscious mind and the body, and also provides an explanation for a number of phenomena that the emotional theorists have been considering.

THE MIND IN THE BODY: FILTERING, STORING, LEARNING, REMEMBERING, REPRESSING

Because of the research I've been describing, we can no longer consider the emotional brain to be confined to the classical locations of the amygdala, hippocampus, and hypothalamus. For example, we have discovered other anatomical locations where high concentrations of almost every neuropeptide receptor exist, locations such as the dorsal horn, or back side of the spinal cord, which is the first synapse within the nervous system where all somatosensory information is processed. (The term *somatosensory* refers to any bodily sensations or feelings, whether it is the touch of another's hand on our skin or sensations arising from the movement of our own organs as they carry on our bodily processes.) Not just opiate receptors but almost every peptide receptor we looked for could be found in this spinal-cord site that filters all incoming bodily sensations. In fact, we have found that in virtually all locations where information from any of the five senses—sight, sound, taste, smell, and touch—enters the nervous system, we will find a high concentration of neuropeptide receptors. We have termed these regions "nodal points" (or, colloquially, "hot spots") to emphasize that they are places where a great deal of information converges. The information is carried by axons and dendrites from many nerve cell bodies that are passing near or making synaptic contact with each other.

These nodal points seem to be designed so that they can be accessed and modulated by almost all neuropeptides as they go about their job of processing information, prioritizing it, and biasing it to cause unique neurophysiological changes. For example, the nucleus of Barrington is one such nodal point, since it contains many neuropeptide receptors, and depending on what neuropeptide is occupying its receptors, feelings related to sexual arousal or bathroom functions can be switched or modified, made unconscious, or moved to the most urgent priority. Emotions and bodily sensations are thus intricately intertwined, in a bidirectional network in which each can alter the other. Usually this process takes place at an unconscious level, but it can also surface into consciousness under certain conditions, or be brought into consciousness by intention.

All sensory information undergoes a filtering process as it travels across one or more synapses, eventually (but not always) reaching the areas of higher processes, like the frontal lobes. There the sensory input—concerning the view, the odor, the caress—enters our conscious awareness. The efficiency of the filtering process, which chooses what stimuli we pay attention to at any given moment, is determined by the quantity and quality of the receptors at these nodal points. The relative quantities and qualities of these receptors are determined by many things, among them your experiences yesterday and as a child, even by what you ate for lunch today.

Think of the brain as a machine for not merely filtering and storing this sensory input, but for associating it with other events or stimuli occurring simultaneously at any synapse or receptor along the way—that is, learning. Let's look at how this occurs in the process of vision, which is very advanced and complex in humans. After a visual signal hits the retina, the light-sensitive part of the eye, it must make its way across five more synapses as it moves from the back of the brain, called the occipital cortex, to the frontal cortex. At each synapse, the neurophysiological patterns evoked by the visual image become progressively more complex, the simple lines and edges signaled at the first synapse accruing ever richer detail and associations as the visual image moves closer to the front of the brain. Do you ever think you recognize someone you miss in a place where they cannot be? When I'm traveling, for a few milliseconds I often think that blond teenager I glimpse in the airport is my son Brandon before I realize that's impossible.

By contrast, smell is an older, more primitive sense, with little potential for erroneous associations because it takes a quicker, unfiltered route into consciousness. It's only one synapse away from the nose to the amygdala, a nodal point that directly routes incoming sensory information in all forms to the higher centers of association in the cortex. This explains why our associations with odors are so strong and memorable. The other day, my husband suddenly realized why he had had an irrational hatred of bluejays all his life. As a seven-year-old, he had painted a bluejay model in a confined place with a foul-smelling paint that had made him vomit!

Using neuropeptides as the cue, our bodymind retrieves or represses emotions and behaviors. Dr. Eric Kandell and his associates at Columbia University College of Physicians and Surgeons have proved that biochemical change wrought at the receptor level is the molecular basis of memory. When a receptor is flooded with a ligand, it changes the cell membrane in such a way that the probability of an electrical impulse traveling across the membrane where the receptor resides is facilitated or inhibited, thereafter affecting the choice of neuronal circuitry that will be used. These recent discoveries are important for appreciating how memories are stored not only in the brain, but in a *psychosomatic network* extending into the body, particularly in the ubiquitous receptors between nerves and bundles of cell bodies called ganglia, which are distributed not just in and near the spinal cord, but all the way out along pathways to internal organs and the very surface of our skin. The decision about what becomes a thought rising to consciousness and what remains an undigested thought pattern buried at a deeper level in the body is mediated by the receptors. I'd say that the fact that memory is encoded or stored at the receptor level means that memory processes are emotion-driven and unconscious (but, like other receptor-mediated processes, can sometimes be made conscious).

STATE-DEPENDENT MEMORY AND ALTERED CONSCIOUSNESS: OUR PEPTIDES AT WORK

Back in my college days, at one of the graduate seminars held in the psychology department of Bryn Mawr, I heard the psychologist Donald Overton of Temple University, who had documented a widespread phenomenon in animals, which later proved to carry over to humans. A rat that learns a maze or receives a shock while under the influence of a drug (which you can now visualize as an external ligand that binds to receptors in the brain and body) will remember how to solve the maze or avoid the shock most efficiently if the rat is retested under the influence of the same drug. When we consider emotions as chemical ligands—that is to say, peptides—we can better understand the phenomenon known as *dissociated states of learning,* or *state-dependent recall.* Just as a drug facilitates recall of an earlier learning experience under the influence of that same drug for the rat, so the emotion-carrying peptide ligand facilitates memory in human beings. The emotion is the equivalent of the drug, both being ligands that bind to receptors in the body. What this translates into in everyday experience is that positive emotional experiences are much more likely to be recalled when we're in an upbeat mood, while negative emotional experiences are recalled more easily when we're already in a bad mood. Not only is memory affected by the mood we're in, but so is actual performance. We're more likely to be helpful to others and perform in altruistic ways when we are experiencing a good mood. Conversely, hurt the ones you love enough times, and they will learn to feel threatened in your presence and remember to act accordingly. It doesn't take an expert in emotional theory to recognize that there is a very close intertwining of emotions and memory. For most of us, our earliest and oldest memory is an extremely emotion-laden one.

One extremely important purpose of emotions from an evolutionary perspective is to help us decide what to remember and what to forget: The cavewoman who could remember which cave had the gentle guy who gave her food is more likely to be our foremother than the cavewoman who confused it with the cave that held the killer bear. The emotion of love (or something resembling it) and the emotion of fear would help secure her memories. Clearly, just as drugs can affect what we remember, neuropeptides can act as internal ligands to shape our memories as we are forming them, and put us back in the same frame of mind when we need to retrieve them. This is learning. In fact, we have shown that the hippocampus of the brain, without which we can not learn anything new, is a nodal point for neuropeptide receptors, containing virtually all of them.

Emotional states or moods are produced by the various neuropeptide ligands, and what we experience as an emotion or a feeling is also a mechanism for activating a particular neuronal circuit—*simultaneously throughout the brain and body*—which generates a behavior involving the whole creature, with all the necessary physiological changes that behavior would require. This fits nicely with Paul Ekman's elegant formulation that each emotion is experienced throughout the organism and not in just the head or the body, and has a corresponding facial expression. It's part of the constellation of bodily changes that occurs with each shift of subjective feeling.

Is there one kind of peptide that is specific to each emotion? Perhaps. I believe so, but we have a way to go to work this out. In the case of angiotensin, a classical hormone that is also a peptide, we have a good, simple example of the relationship between a neuropeptide and a mood state, and how that mood state can coordinate and integrate what happens in the body with what happens in the brain. It has long been known that angiotensin mediates thirst, so if one implants a tube in the area of a rat's brain that is rich with angiotensin receptors and drops a little angiotensin down the tube, within ten seconds the rat will start to drink water, even if it is totally sated with water. Chemically speaking, angiotensin translates as an altered state of consciousness, a mood state that makes humans and animals say, "I want a glass (or a trough) of water." In other words, neuropeptides bring us to states of consciousness and to alterations in those states. Similarly, angiotensin applied to its receptors in the lung or kidney will also cause bodily changes, all of them aimed at conserving water. For example, there will be less water vapor in each breath exhaled from the lung and less water in urine excreted by the kidneys. All systems are working together toward one goal—more water—which has been dictated by an emotion (or what the experimental psychologist would call a "drive state")—that of thirst.

Does the sum of the peptide secretions in our brains and bodies— our emotional state—bias our memory and behavior so we automatically get what we expect? Now, that is an interesting question that I will consider next.

CREATING OUR OWN REALITY, REALIZING
OUR OWN EXPECTATIONS

There is no objective reality! In order for the brain not to be over-whelmed by the constant deluge of sensory input, some sort of filtering system must enable us to pay attention to what our bodymind deems the most important pieces of information and to ignore the others. As discussed, our emotions (or the psychoactive drugs that take over their receptors) decide what is worth paying attention to. Aldous Huxley was on to this in *The Doors of Perception* when he referred to the brain as a "reducing valve." He was also on the right track when he assumed that what got through to command headquarters was just a tiny trickle of what could be absorbed at any given moment.

Since our sensing of the outer world is filtered along peptide-receptor-rich sensory way stations, each with a different emotional tone, how can we objectively define what's real and what's not real? If what we perceive as real is filtered along a gradient of past emotions and learning, then the answer is we cannot. Fortunately, however, receptors are not stagnant, and can change in both sensitivity and in the arrangement they have with other proteins in the cell membrane. This means that even when we are "stuck" emotionally, fixated on a version of reality that does not serve us well, there is always a biochemical potential for change and growth.

Most of our bodymind attentional shifts are subconscious. While neuropeptides are actually directing our attention by their activities, we are not consciously involved in deciding what gets processed, remembered, and learned. But we do have the possibility of bringing some of these decisions into consciousness, particularly with the help of various types of intentional training that have been developed with precisely this goal in mind—to increase our consciousness. Through visualization, for example, we can increase the blood flow into a body part and thereby increase the availability of oxygen and nutrients to carry away toxins and nourish the cells. As discussed, neuropeptides can alter blood flow from one part of the body to another—the rate of blood flow is an important aspect of prioritizing and distributing the finite resources available to our body.

Norman Cousins once told me that he had gotten over a broken elbow, which he had suffered while playing tennis, and got back on the court in record time simply by focusing for twenty minutes each day on increasing the blood flow through the injured joint, after his physician explained that poor blood supply to the elbow was why injuries to this joint healed slowly.

But I don't want to leave you with the impression that I am advocating that the unconscious must always be brought to consciousness in all successful therapies. In fact, the unconscious mind of the body seems all-knowing and all-powerful and in some therapies can be harnessed for healing or change without the conscious mind ever figuring out what happened. Hypnosis, yogic breathing, and many of the manipulative and energy-based therapies (ranging from bioenergetics and other psychotherapies centered on body work to chiropractic, massage, and therapeutic touch) are all examples of techniques that can be used to effect change at a level beneath consciousness. (Based on the drama and rapidity of some therapeutic transformations, I believe that repressed emotions are stored in the body—the unconscious mind—via the release of neuropeptide ligands, and that memories are held in their receptors.) Sometimes transformations occur through the emotional catharsis common to the many bodymind therapies that focus on freeing up emotions that have gotten lodged in the psychosomatic network, but not always.

For example, the famed psychiatrist and hypnotherapist Milton Erickson addressed the subconscious minds of several young women who, although having been subjected to all kinds of hormone injections, remained completely flat-chested. He suggested to them while they were in a deep trance that their breasts would become warm and tingly and would start to grow. Although none of them could remember anything that happened in his office, all grew breasts within two months, presumably because Erickson's suggestions caused the blood supply to their breasts to increase!

Emotions are constantly regulating what we experience as "reality." The decision about what sensory information travels to your brain and what gets filtered out depends on what signals the receptors are receiving from the peptides. There is a plethora of elegant neurophysiological data suggesting that the nervous system is not capable of taking in everything, but can only scan the outer world for material that it is prepared to find by virtue of its wiring hookups, its own internal patterns, and its past experience. The *superior colliculus* in the midbrain, another nodal point of neuropeptide receptors, controls the muscles that direct the eyeball, and affects which images are permitted to fall on the retina and hence to be seen. For example, when the tall European ships first approached the early Native Americans, it was such an "impossible" vision in their reality that their highly filtered perceptions couldn't register what was happening, and they literally failed to "see" the ships. Similarly, the cuckolded husband may fail to see what everyone else sees, because his emotional belief in his wife's faithfulness is so strong that his eyeballs are directed to look away from the incriminating behavior obvious to everyone else.

As investigations continue, it is becoming increasingly apparent that the role of peptides is not limited to eliciting simple and singular actions from individual cell and organ systems. Rather, peptides serve to weave the body's organs and systems into a single web that reacts to both internal and external environmental changes with complex, subtly orchestrated responses. Peptides are the sheet music containing the notes, phrases, and rhythms that allow the orchestra—your body—to play as an integrated entity. And the music that results is the tone or feeling that you experience subjectively as your emotions.

March 1981: During one of my many nightly visits to my father at the VA hospital during the course of his treatment for lung cancer, he looked up from his bed and asked sardonically, "Well, how's the cure coming?" Embarrassed and saddened, I couldn't meet his gaze. I had visited the lab earlier that day and found that his cancer, which had mutated and returned to his body since remission a year ago, was proving resistant in vitro to all known chemotherapeutic agents. My own attempts to understand his disease and find a cure were looking equally as hopeless.

"It's going well," I lied, trying to give him the hope I myself did not have, hope that seemed to me to be his last shot at somehow effecting a miracle cure of this deadly disease. "The results from the lab will show something is going to work, I'm sure of it." And then changing the subject, I said, "Look at what Vanessa made at school for you!" He seemed to brighten up as I brought out a rainbow mobile that my five-year-old daughter had sent along to decorate Grandpa's room.

He was dozing off as I hung this symbol of hope above his bed, and, with a heavy heart, I whispered a soft apology: "I'm sorry, Dad, science still doesn't have the answer."

I knew that even after decades of intense research there

essentially had been no treatment advances beyond the highly toxic drugs developed before 1965. What I didn't know was how fiercely the cancer establishment would resist the efforts of an outsider—me—to come up with new ideas for treatment. This would be my first personal exposure to the intransigence of old-paradigm thinking, and an experience so profoundly disillusioning that it made it almost easy for me to slip the traces of my own intellectual harness. I was about to enter a very long, dark tunnel of despair, and then, to my joy and astonishment, make my way toward a light that would illuminate a whole new world of thought.

8

TURNING POINT

SHIFT

It was inside the NIH cafeteria, on a winter afternoon in early 1982, that the last sparks of my faith in the Palace, the power boys, and the prevailing paradigm sputtered and began to die. There I sat at the lunch table, my salad wilting on the plate, halfheartedly engaged in word-to-word combat with an alpha male scientist over who would get the credit for the what, when, and how of the work we had done together. In my ninth month of pregnancy with Brandon, my marriage turning sour, and my father dead—I wanted to just get up and walk away.

Until the death of my father in 1981, in spite of the lessons I'd learned from the Lasker incident (and perhaps partially because of them), I had been willing to wrestle over rewards, to do whatever it took for the chance of more citations, and to grab turf from my competitors. I was as willing as the next guy to split one research project into as many self-contained papers as possible, selfishly disregarding the needs of researchers who might benefit from getting the news all at once. I had learned to play the game of science, and it had brought out a survivor's instinct in me. Stopping short only at sacrificing my personal integrity and honesty, and still holding strongly to my ideal of science as truth-seeking, I had gotten good at swimming with the sharks.

But when Dad died, I came up for air. For the first time in my career, I was forced to see a connection between the science I did and the people whose lives depended on it. Real people, not just statistical ones, were dying from diseases that lacked effective cures, and now one of these real people was my own flesh and blood. With this new perspective, the chicanery of politics, the sportsmanship, the ego battles all receded into the background, and a deeper sense of purpose emerged to guide me.

The diagnosis of my father's lung cancer was made in February 1980. It was the first time anyone in my family had been diagnosed with a life-threatening disease, and the news hit me hard. Even more shocking, however, was learning that he had a particular kind of cancer that I knew quite a bit about. Small-cell, or "oat cell," carcinoma, named from the cancer cells' resemblance to tiny oats under a microscope, is a nasty mutation of the body's natural processes. The tumors grow rapidly and spread quickly, metastasizing throughout the body and usually causing death within a very short time. Of the four major types of lung cancer a person can develop, about 25 percent are small-cell, and almost 100 percent of those who get it are, like my father, continual, heavy cigarette smokers.

As soon as I heard the diagnosis, I immediately started phoning around, asking for the name of the top small-cell cancer clinician at the Palace. The name I came up with was that of a major player who ran a titanic lab inside the NIH's National Cancer Institute. In one of the many instances of synchronicity that seem to mark my life on a regular basis, this same doctor had been trying to phone me for the past month, but I had never been able to find the time to answer his calls.

But now he went straight to the top of my list. Although it was a Sunday when we got the diagnosis, I called him at home to explain the situation and ask for his help. As a favor, he agreed to admit my father to his current trials, even though Dad's age exceeded the protocol's limit and his outcome wouldn't be counted in the trial data. But none of that mattered to me. Even if there was only a one-in-a-million chance that the latest experimental chemo-cocktail might offer a permanent cure, we now had hope where there had been none before, and I was grateful.

The doctor and I then got down to business. He had been phoning me because he wanted to follow up on research that showed his small-cell cancer cells were secreting neuropeptides. I, too, knew of this research, done in the 1960s by Dr. Rosalind Yalow, a woman who had won the Nobel Prize for her findings. Since then, however, many new neuropeptides had been identified, and what the cancer doctor and his lab wanted was an updated "peptide profile," showing exactly which of the newly discovered peptides were being secreted. He knew that if there was going to be a cure for small-cell carcinoma, it had to come from a better understanding of the nature of these cells and how they functioned. Only then could we define the precise molecular effects and devise a rational, effective treatment. He also knew my lab was state-of-the-art in the peptide field, and was probably the fastest route to getting the answers he needed.

It was a prototypical Palace transaction, this agreement of one lab to collaborate with and answer questions being raised by another—an example of the Palace in its prime, of crème-de-la-crème science. Had I been at a university when this request came up, I would have first needed to write a proposal for a grant, then submit it, and then wait around for the funding cycle to spin around in my favor. Even then, I would have had only one out of five chances of being allotted the money I needed. If I'd been conducting research in the business world, I would have needed to convince the shareholders or vulture capitalists of its profitability before anything could happen.

But at the Palace we had only to do the verbal equivalent of a hand-shake over the phone, and we had a deal. My father would be included in the cancer doctor's clinical trials, and I would help the doctor's lab identify the peptides being secreted by the cancer cells. Having served in World War II, Dad was given a bed in the Veteran's Administration Hospital in downtown Washington, D.C., one of the few beds reserved for the trial patients.

Within a few days, a hundred tiny test tubes arrived at my lab, each containing a minute, pellet-sized ball made from a different cell line, which is a sample of cells taken from a patient and arduously grown in a dish. These cell lines included several different kinds of lung cancer from scores of patients. I proceeded to prepare a peptide extract of each pellet, a precise job of adding a hot acid solution just so. Then I transferred the contents of each test tube into ten other test tubes, giving me ten samples for each cell line of lung cancer, for a total of one thousand test tubes. I planned to look for ten different peptides, handling the endorphins myself and giving the bombesin to my former postdoc, Terry Moody, who was now at George Washington University on the other side of town, but who had done the receptor-mapping research on bombesin while he was still in my lab. The other eight projects I farmed out to pep-tide buddies who I knew were set up to make quick searches for spe-cific peptides. What better way, I reasoned, to accomplish the goal of understanding these cancer cells than to measure everything possible, a tactic not uncommon among researchers hot on the trail, often referred to as a "fishing expedition." Time was short, and my dad's life was depending on my being able to make a quick run. I prayed for a swift turnaround.

Certainly, a new approach to the treatment of cancer was—and still is—desperately needed. Although the cancer establishment has been trying to crack this disease for years, it continues to kill more people every year, often a slow and painful death made even more excruciating by toxic treatments. The highly toxic chemotherapeutic drugs developed in the 1950s kill all rapidly dividing cells in the body, which means not only cancer cells but many kinds of healthy cells as well. Unfortunately, the immune system, the body's natural defense system against cancer, is itself composed of cells that are rapidly dividing. So both the disease *and* the protection against the disease get zapped.

In the Cancer Institute's trials for small-cell lung cancer, there was only one patient who was still alive five years later. Most chemotherapy patients were dead within two years. State-of-the-art chemotherapy in those days was nothing more than different combinations of the same old toxic drugs given on different schedules. If my father was going to survive, I knew that a new approach, a major breakthrough in understanding and treating this disease had to happen. But I was hoping the chemotherapy would buy him some time—enough time for me to do the necessary work.

Dad rallied right after the chemotherapy, and within weeks came back from the near-dead to looking nearly normal. He had had a remission, as expected, but would soon have a relapse, which was also expected. I knew this, but I couldn't bring myself to tell him or my mother. I believed, intuitively, that he needed all the hope he could get, in spite of his natural tendency toward skepticism. For this reason, I chose to emphasize only the "good news" during my daily visits to fill him in on how the race to find a cure for his cancer was progressing.

In my lab, however, I was not feeling so optimistic. As part of my quest to understand his disease, I dove into the oncology literature with a million questions. Why were these strange small cells, which divided rapidly, so full of peptides? Why were the cells so unlike those typically found in lung tissue? I thought that if only I could answer these questions, I'd be able to save my father's life.

Dad was far from convinced. As his condition worsened, and the relapses followed the remissions, he stoically watched my frantic intellectual activity with detached amusement. My father was a man of the world, an artist, a big-band jazz arranger, a sophisticate, a skeptic—in short, no easy believer in miracle cures. What he wanted most of all was to be made as comfortable as possible while enduring the intense nausea brought on by the chemotherapy, and I did my best to see that his doctors and nurses were providing him with everything available.

When the data came in from my fellow peptide researchers, I entered it on a big spreadsheet and raced over to the cancer lab. None of us knew which cancer had been in what test tube, since ignorance (known as "being blind to the protocol") was a necessary part of the scientific etiquette. Now, bending over the numbers with one of the lab director's postdocs, I watched intently as the name slots were filled. It didn't take long to see what was happening: Every one of the tubes containing small-cell cancer cell lines was characterized by a detectable—and sometimes extremely high—level of the peptide bombesin.

Bombesin! I got a chill remembering that Terry Moody and I had taken this very peptide from obscurity to molecular neuropeptide stardom. First we had located the bombesin receptor, and then we'd gone on to use the bombesin antibodies to find the peptide itself inside neurons in the brain.

Until this breakthrough, I'd been running on wild romanticism, fueled with hope and fear for my father's fate. But the new finding brought us much closer to actually figuring out what makes these cells replicate so fast. If we could understand the mechanism of the rampant out-of-control growth, perhaps we could find the key to stopping it. Identifying the substance that stimulated the growth would put us in a good position to find an antagonist to block that action. I now seriously began to think we had a chance to find a treatment before it was too late.

I welcomed the chance to work with Terry again, so enjoyable and productive had our work been before, and we quickly set about trying to answer the question, Was bombesin's presence the key to the cancer cells' wild, rapid proliferation? This was pure conjecture, but most cell "growth factors" had finally turned out to be peptides when they had been purified and chemically identified. Peptide growth factors acted at receptors in the cellular membrane, causing cells to divide and then multiply as part of their normal, healthy development. Insulin was one of these peptides; epidermal growth factor (EGF) was another. For obvious reasons, growth factors had become an area of intense interest in cancer research. If bombesin was a growth factor secreted by the tumor cells to promote their own growth, it could explain the mechanism by which the lung cancer cells were proliferating so rapidly. When our research showed that the cancer cells were not only secreting bombesin but were presumably acted on by it as well, since they had bombesin receptors on their surfaces, we thought we had identified the mechanism. Bombesin, it turned out, was not only a growth factor but an autocrine growth factor, a substance secreted by the very cell upon which it acts.

We dutifully communicated our findings to the cancer lab director. Two weeks later, one of his researchers called me and, with an audible tremble in his voice, explained how bombesin had made his cell lines grow faster! He was confirming our hunch that the presence of bombesin and bombesin receptors on these cancer cells must be why they were growing out of control!

I was thrilled to include his confirmation of our findings at the end of our paper describing the bombesin–small-cell cancer connection, which I rapidly wrote up for *Science*. I cited it as a "personal communication" from Adi Gazdar, a researcher at the cancer lab—the standard way of acknowledging research that has not yet been published. I submitted the paper, putting Terry as first author and the Cancer Institute lab chief as last author, with the rest of us distributed in between. When it appeared, the cancer lab director felt that he had been scooped by my use of his lab researcher's data, which could conceivably enable his competitors to use the information and move in on his territory. My own feeling was that time was of the essence. The lung cancer community urgently needed this information, and the sooner it got out, the better. In my mind, playing the political game was secondary to making our findings available to as wide an audience as possible, and I didn't care at all who got the credit.

Meanwhile, my father's condition had seriously worsened. He'd undergone a bone marrow transplant, a very painful surgical procedure used when patients are not responding to normal doses of chemo. Immune cells are extracted from their sources in the bone marrow and refrigerated for later reinfusion after the patient undergoes massive doses of killer chemotherapy. The assumption is that the bone marrow has no cancerous cells and so can be reintroduced after the system has been "cleaned" by chemotherapeutic drugs to act as seeds to grow a new immune system. What's left of the patient after the chemotherapy is expected to become fertile ground for the new immune cells to take root and flourish.

In my father's case, the strategy hadn't worked. Possibly the chemotherapy hadn't killed all the cancer cells, and those remaining continued to grow. Or, as we were to explore in later research, perhaps there were precursor immune cells from the bone marrow that were themselves implicated in the cancerous growth, and so continued the disease process once they were returned to the body after chemotherapy.

Even though this latest round of chemo had caused him to lose all his hair and appear haggard, he was still recognizable as his jaunty, joke-cracking self. I remember plying him daily with megadoses of vitamin C in hope of countering some of the toxicity from the drugs. At one point, I even suggested bringing in a controversial cancer doctor with an alternative treatment I'd read about. But in spite of my upbeat efforts, Dad was rapidly losing interest in exploring new avenues that might lead to his recovery.

Gradually, he was being moved closer and closer to the nurses' station, not a good sign since it meant they wanted to keep a more constant vigil. Then came a heavy blow, the news that the cancer had spread to his brain and would require radiation treatment. Until then, Dad had kept his spirits relatively high, even endearing himself to the nurses by constantly playing jazz riffs on his guitar. Now with this latest diagnosis, he took a swift and sure emotional dive. Dad was an intellectual, an artist, and the news that the cancer was destroying his brain demoralized him, destroying what little hope he had left. Nonetheless, he proceeded with the radiation.

On the seventh day of the radiation treatment, I noticed a distinct shift in my own feelings, from hopefulness to a numbing emptiness. Although my brain was lagging behind in acceptance, I knew intuitively that my father wasn't going to make it.

That night when he said good-bye to my mother, he made an unusual request. I hadn't gone with her to the hospital that day, because once I'd given up hope, I couldn't face looking him in the eye.

"Go to Candy's," he repeatedly told her, as she held his hands and assured him she would. Dad knew he was going to die that night, and he wanted my mother to be with me, which she was, when the final phone call came around 2 A.M.

My father died almost a year to the day after being diagnosed. When I went to the VA hospital to pick up his effects, I noted a half pack of cigarettes in the drawer next to his hospital bed, not a surprising find considering how heavily addicted he had been. As I was leaving, an administrator gave me an American flag to drape over his coffin at the funeral. But remembering his frequently expressed opinion, "War is for idiots," we had no intention of using it. Instead, it went into storage until some years later when, remembering his fun-loving nature and still missing him terribly, we brought it out to cover our bounteous Fourth of July picnic table.

As for my research, I hadn't moved quickly enough. Although we now had some further understanding of what was going on in this disease, we hadn't had time to devise a treatment for it. Instead of a cure, I had another paper in *Science*—and another enemy. The Cancer Institute lab chief was furious at how I'd squandered the hot finding about the growth of the cancer cells by releasing it, in his eyes, prematurely. I'd done the unspeakable, packing our paper with data that any self-respecting scientist would have spun out over three or four papers, thereby increasing the number of publications, to say nothing of all the chances to get cited.

IT WAS AT this juncture in the episode that I found myself at an NIH lunch table, sitting opposite my onetime collaborator, verbally wrestling with him over control of the research. As the power boy sat across from me making demands, I was in no position to mount an effective defense, nor did I want to. In tough tones, he told me that he'd prefer to cut me out altogether, and deal only with Terry. This was his arena, he told me in no uncertain terms, and I needed to understand that. After all, he was the chief, and if I was a good girl, *maybe* I would be considered for the position of research associate on future projects.

It was déjà vu, a replay of my final conversation with Sol before trading Hopkins for the Palace. Once again, I was being told by a powerful male scientist not to work on a part of the research that we had begun jointly. Some things, it seemed, were too important, too prestigious for acknowledgment to be shared, and from his viewpoint, obviously, the bombesin–small-cell cancer connection was one of them.

Of course, he could justify his reasons for playing the cards this way. Evidently, I had added insult to injury when, with the help of his associates, I had checked out an obvious prediction that bombesin levels should be elevated in the blood of patients with small-cell lung cancer. We had cut our teeth on peptide blood assays and could do them in our sleep, so it was a natural and easy confirmation. I communicated our results to the cancer lab director, letting him know that I intended to get off a brief paper to the *Lancet*, a prestigious British medical journal, suggesting that the symptoms of patients with this type of lung cancer, such as itching, low temperature, and loss of appetite, were due to the surplus of bombesin in the bloodstream.

It was this move that was the final straw, causing him to reconsider any further collaboration with me. To this day, I don't know if the real source of his problem with me was that he didn't trust my work, or that he saw me as invading his turf, messing with his plan of slowly spinning off our discovery into as many articles as possible. Later, I suspected that he was partly motivated by a feeling that it was politically unwise to appear with me on too many papers, a speculation that was supported when he ordered his name removed from the *Lancet* paper. When it was finally published, the paper had my name and the name of my technician on it, with an acknowledgment to the Cancer Institute lab.

I remember thinking about all this as I sat across from my lunch table combatant, barely listening as his booming voice harangued me for well over an hour. I also remember feeling the late-term stirrings of my unborn child, which made me oddly peaceful in the midst of this ordeal. Perhaps the message that new life was on its way gave me strength and afforded me some distance from the attack I was now being subjected to.

The next morning, I received a four-page, single-spaced letter from the cancer lab director that read like a formal contract, spelling out exactly who would do what, when and where, in regard to further research. I read it with a dulled responsiveness, having no intention of ever bothering to reply. Clearly, my ex-collaborator believed I had been poaching on his turf, and I believed, just as clearly, that his territorial maneuvering, driven by a self-aggrandizing motive to get as much credit for the research as he could, was the stuff that kept medical science from finding desperately needed treatments. My father was dead, and I no longer had a shred of a reason to stay in the cancer doctor's good graces.

I returned to my brain maps, my receptors, my peptides. What had been a thrilling and meaningful search for answers to the questions of why these lung cancer cells were full of peptides, why they didn't fit the profile of other lung cells, was now being shelved. I fantasized that someday it would be resurrected, perhaps in a time when cold-hearted ambition mattered less than a genuinely motivated search for the truth.

But for now, I let it go, thinking of it only in those moments when I remembered, with an ache, my father and how I had tried in vain to save his life.

CONNECTION

My weariness and disillusionment would soon fade, however, as I began a new intellectual journey, one that started with a casual social encounter, picked up steam as it involved me in yet another major quest for a disease cure, and had such momentum that it sustained me through criticism and adversity that would make what I'd already endured look minor. I met Dr. Michael Ruff in the fall of 1982 at the now-defunct Palace bar, a clubhouse and watering hole for brain-weary scientists, located in a donated stone house right off the campus grounds. Next to the cafeteria, it was the most interdisciplinary spot in the Palace, a fruitful place where the usual boundaries dropped away, and the talk flowed freely.

I rarely went there, but late one afternoon, several months after my separation from Agu had been formalized, I stopped by on a psychic hunch. With my new baby tucked securely into a carrier strapped on my chest, I certainly didn't feel very glamorous or sexy. But I had a premonition as I climbed the stairway to the main room that I was about to meet someone very interesting.

I found myself gravitating toward the end of the bar, where two young, good-looking postdocs were holding forth. A few flashes of friendly eye contact passed between us, and I could tell that they had recognized me. "That's Candace Pert," I could see one mouth to the other. Soon we were all talking.

Michael Ruff and Rick Weber, I found out, had studied immunology in graduate school together, and were now getting some seasoning as postdoctoral students at the Palace. Mike told me later that he'd remembered seeing me on a TV science documentary, talking about how endorphins from the testes caused the orgasmic spasms of the vas deferens. I have to admit that my feminine ego soared as I contemplated playing the alpha, older, wise-woman scientist to these betas. But what really excited me was the fact that both guys were immunologists. Because I'd had an idea in the back of my mind for a long time—a notion that schizophrenia might be an autoimmune phenomenon—I'd been hankering for a congenial immunologist to grill.

Rick's specialty was the study of the chemistry of antibody molecules, those spongelike substances made by certain immune system cells to recognize and eliminate invading pathogens (disease-causing agents) that threatened the organism. Rick was eloquent in his description of how these antibodies vibrated and changed shape as they encountered the bacteria, virus, or tumor cells, latching on to them and escorting them out of the system. We all had a good laugh when he described the scene in the movie *Fantastic Voyage* where Raquel Welch enters the bloodstream and is engulfed by a swarm of antibodies, each perfectly shaped to cup on to her amazing breasts.

Michael, who seemed quieter, more reserved, was interested in the cellular portion of the immune system, in particular the highly mobile scavenger cells known as macrophages, whose function is to keep the blood clean of debris left over from the battle to kill invaders. He talked about how "eating dirt" was but one function of these cells. They also played a key role in repairing the body fabric, manufacturing tissue when needed and orchestrating a chemical and cellular cascade that resulted in a healing response. Michael was beginning to question how they did all of this without some ability to communicate with each other or with the rest of the body—a concern that didn't bother other immunologists in the least.

Like Michael, I, too, had some radical ideas stewing on the back burner of my imagination. One of these involved my hunch that schizophrenia might be the result of an autoimmune response, which is what we call the phenomenon in which the immune cells go awry and attack part of the organism itself rather than the invaders they are supposed to attack. In schizophrenia, I theorized, the immune cells were secreting antibodies that targeted brain cells via their receptors. Throwing caution to the winds, and trusting my sense that Mike and Rick were young and open enough to follow me, I put it right out: "I want to find a real cure for schizophrenia," I announced, noticing their attention shift rather dramatically. "My idea is that the disease is caused by antibodies being made against the brain-cell receptors."

There was a thoughtful silence as they pondered this possibility.

"Do you guys know if there's a way to test this?" I asked more tentatively.

We agreed on the spot to explore the hypothesis. A first step would be for me to teach them brain-receptor science and for them to teach me immunology, a project we began that afternoon in the bar! I didn't know it then, but the deal we had just struck marked the beginning of a collaboration that would bear much fruit by merging our two fields to bring about a new approach to healing and disease. The research we would do would circle out from schizophrenia, leaving it temporarily behind as we explored the connection between nervous and immune systems, mind and body, as it pertains to cancer and AIDS, returning only years later to the question of schizophrenia.

Soon after our meeting at the clubhouse, the three of us took to hanging out together. One afternoon, we were bouncing up and down in Rick's open Jeep when he pushed a journal reprint toward me.

"Look at this, Candace," he said. "It's written by my buddy Ed Blalock from the University of Texas."

"What is it?" I asked, since I was jiggling around too much to be able to read the title.

"He's found immune cells that make endorphins."

"Really?" I said, hesitating while I let this astounding piece of information settle into my consciousness. "Does this guy know what he's doing?"

"It looks rock solid to me," Rick responded. "Read it yourself."

Rick pulled the Jeep over, and with both him and Mike leaning over my shoulder, I read the paper. Blalock, an immunologist who had been in graduate school with both Mike and Rick only a few years earlier, had been studying interferons, peptides that are made by certain white blood cells known as lymphocytes. Like antibodies, interferons have the job of fighting off invading pathogens and thus help to preserve the integrity of the body. In his studies, Blalock noticed that interferons sometimes mimicked the activity of hormones, an observation that prompted him to put the lymphocytes in a dish and stimulate them to produce interferons, so that he could see if they produced anything else at the same time. To his shock and surprise, he found the lymphocytes were also secreting the mood-altering brain peptide endorphin, as well as ACTH, a stress hormone thought to be made exclusively by the pituitary gland, the main gland in the endocrine system.

"My God!" I exclaimed. "If this guy's right, it makes the immune system look like a floating endocrine system, a bunch of tiny pituitary glands!"

In our excitement, we jumped rapidly to a possible conclusion: The immune system was communicating not only with the endocrine system but with the nervous system and the brain, as well, by using a chemical mechanism that consisted of the neuropeptide endorphins and their receptors to code for information. But there would be many steps—and about two years—between formulating this hunch and feeling confident enough about it to put it in print.

As the news of Blalock's discovery spread, very few of my colleagues shared my enthusiasm. They either ignored Blalock or dismissed him as wrong. This was to be expected. Whenever something doesn't fit the reigning paradigm, the initial response in the mainstream is to deny the facts. To suggest that systems historically defined as separate were actually interrelated was paradigm-busting at its best! For a while after his discovery, wherever Blalock went, he heard the whispered charges: "Sloppy work! Artifact! Dirty test tubes!" And they kept up until the number of labs that had repeated Blalock's observation grew too great to ignore. What he had seen was no "artifact," that is, something that was a product of the experiment itself. Finally, in 1983, an editorial in *Nature* admitted the presence of brain peptides in the immune system, but warned the scientific community against those "radical psychoimmunologists" who might prematurely interpret this work to mean that "no state of mind exists that is not reflected by a state of the immune system." Rick, Mike, and I embraced the moniker with pride, proudly referring to ourselves as radical psychoimmunologists from then on.

But even while the weight of the evidence compelled Blalock's critics to accept the data, those critics still had no intention of addressing how that data challenged the ordained view of the body.

As a budding radical psychoimmunologist who didn't think twice about disregarding traditional boundaries to get at the truth, I immediately plunged into extending and exploring the implications of Blalock's work.

Michael, whose home base was the Dental Institute over at the NIH, moved into my lab temporarily so we could do the work together. Rick joined us frequently. The first idea we investigated was that if the immune cells were secreting endorphins, there were probably opiate receptors on the immune cell surfaces. I knew there had been numerous papers published claiming to show opiate receptors on immune cells; one was even written by Pedro Cuatrecasas, my former teacher at Hopkins. He had found opiate receptors on immune cells using the traditional grind-and-bind method to isolate the receptors, but his paper, as well as several others, had been ignored because of various anomalies. Again, this kind of finding was far out in front of what the current paradigm accepted. Receptors for brain peptides on immune cells? What could that possibly mean? You'd have to have been some kind of radical psychoimmunologist—a scientific category that didn't even exist when Pedro wrote his paper—to be interested!

We decided to take a more compelling route, one that would be harder for our colleagues to ignore. We would prove our hypothesis through what is called a "functional assay," one that would elicit a specific and measurable activity, rather than one that simply showed the receptor. The main question we asked was: What function of the cell changes as the result of binding?

As part of his work on tissue inflammation at the Dental Institute, Michael had studied a process known as *chemotaxis*, the ability of a cell to pick up the "scent" of a peptide by the receptors on its surface, literally getting on its track and traveling toward where the peptide was most concentrated until it could bind with the peptide, whereupon the peptide commences its job of ordering the cell's activities. We decided to use chemotaxis as a tool to demonstrate the action of opiates and their receptors on immune cells.

To do this, we chose ten different opiate drugs, including the various endorphins, and showed that immune cells chemotaxed to them in the same order of relative potency that they bound to the receptor. Later, we extended this work and showed, using the same method, that there were receptors on immune cells for virtually every peptide or drug we had identified in the brain, such as Valium, Substance P, and many others.

We published our findings and moved to the next logical question, which was the inverse of the one we'd just answered: If there were neuropeptides in the immune system, would we find immunopeptides in the nervous system? Finding brain correlates of peptides that had originally showed up in other parts of the body was what our lab had been doing for years, and so we decided to take a look. This time, working with Mike and another immunologist, Bill Farrar, a collaborator from the Cancer Institute, I chose interleukin-1 as our first target immunopeptide.

IL-1, as it's known in the jargon, is a polypeptide hormone produced mainly by macrophages in the immune system, and is one of fifty or so identified peptides that mediate the inflammatory reactions caused by injury, trauma, or an activated immune system. In a molecular cascade, IL-1 causes fever, activates the T cells, induces sleep, and puts the body in a generally healing state of being, allowing it to mobilize its energy reserves to fight pathogenic intruders with maximum efficiency.

Sure enough, there were interleukin-1 receptors in many areas of the brain, the second immunopeptide receptor found there. (The first to be discovered was Thy-1, short for thymus, and Rick Weber and Joanna Hill and I had done the autoradiographic mapping that showed its distribution pattern in the brain.) We weren't at all surprised, but the immunologists, who had previously known only about the presence of Il-1 receptors in the hypothalamus, where they had long been recognized as a cause of fever, were quite puzzled by the finding that Il-1 receptors were also in the cortex and higher brain centers (mainly on glial cells and the tough membranes around the brain). Today we know that numerous—perhaps all—of the peptides discovered by immunologists can be made in the brain under some circumstances, and can act on receptors in the brain.

What we were now seeing was astounding and very revolutionary. The immune system was potentially capable of both sending information to the brain via immunopeptides and of receiving information from the brain via neuropeptides (which hooked up with receptors on the immune cell surfaces). Our work confirmed Blalock by pointing irrefutably to the existence of a chemical mechanism through which the immune system could communicate not only with the endocrine system but with the nervous system and the brain, as well. Previous work my colleagues and I had done demonstrated quite convincingly that the brain communicated with many other bodily systems. But the immune system had always been considered separate from the other systems. Now we had definitive proof that this was not the case.

POTENTIAL

It was in the spring of 1983 that Michael and I decided we were in love. The many hours we spent working closely together in the lab had certainly contributed to this happy state, but ours was more than an intellectual merger. We had recognized something in each other that promised a new direction, a common quest based on a desire to step outside what was ordinarily accepted and bring forth something greater, both personally and scientifically. This became our bond, which was to serve us well in the coming years when nothing we would do would be possible without the support and strength we gave each other.

I remember the moment I realized that Michael and I had a future, although never in my wildest imagination could I have guessed what that future would hold. I had taken to hanging out with some frequency at the clubhouse, often meeting on a casual basis with Mike, Rick, and other friends. Mike had been away from the lab for a week, and on his first day back had arranged to meet me at our usual corner for an informal update on the work we'd been doing together. I looked up as he slid into the booth and noticed his eyes were wider, more clear and deep, and he seemed present in a way I hadn't notice before.

"Wow, Mike, where have you been?" I couldn't help myself from asking. "You look like a different person!"

He smiled and in his low-key way proceeded to tell me that he'd been at a week-long bioenergetic workshop.

"It was great," he responded enthusiastically. "We did all these exercises and made all these noises. I can't believe how much better I feel!"

Bioenergetics, he went on to tell me, was a kind of alternative therapy that had been created by Alexander Lowen and inspired by Wilhelm Reich. Now, this was hardly a subject you'd expect to hear two Palace denizens deep in conversation about, especially since Reich had been banned from the realm of "real science" for his wild and crazy experiments with human sexual energy—but I was fascinated and pressed for more details. Bioenergetic therapy, Mike continued, made use of various physical postures and exercises to access deeply stored emotional traumas or blockages. The theory was that these emotions are trapped in the physical body and can only be released physically, through movement accompanied by loud, emotive expressions. The result was a freer, more abundantly flowing sense of energy, something I was certainly seeing manifested before me in Mike's transformed appearance.

As we talked, I shared with him some of my own formal and informal personal excursions into the mind-body experience. One of these had occurred in 1977, when I took Est training at the recommendation of a colleague at NIH. Est personal-growth seminars were popular in the seventies, and were presented as a two-weekend affair with as many as two hundred people crammed into a hotel banquet room, literally sequestered for long stretches of time with few breaks. Although I had a lot of skepticism at first, I decided to throw myself into the training, intending to get the full experience and then make up my mind afterward. A team of sensitive if domineering trainers led us through the paces, from guided visualizations to confrontational dialogues, and on to mind-boggling explorations into the nature of reality. At one point, I witnessed a woman's physical body change. As she re-experienced an incest trauma buried for years, the hunched shoulder she'd had since child-hood appeared to spontaneously transform itself, healing before our eyes.

I emerged from the training with one conclusion: "God is in the frontal cortex!" As the part of the brain that gives us the ability to decide and plan for the future, to make changes, to exert control over our lives, the frontal cortex seemed to me to be the only way I could explain what I had seen and experienced. It seemed to me to be the God within each of us. I struggled over the next few weeks to integrate that remarkable experience, trying to transplant the sheer poetry of it into my scientific mind-set, while Agu watched with suspicion and alarm. In retrospect, I realized that what had happened to me during the training was that, for the first time in my life, I had directly experienced my own raw emotions. The sleep and food deprivation had broken down my defenses, putting me in touch with how I really felt—my sadness, loneliness, anger, as well as my joy and love for people. I was left with a new freedom to feel and a new faith in the future, both of which empowered me during the days of the Lasker flap and the period of ostracism that followed.

I was beginning to realize that Mike, like me, was willing to allow his life to be broadened and enriched by the science he did, exploring in real life what he was seeing under the microscope, a rare quality in a scientist, especially at the Palace. The idea that the mind and body could be treated as a whole, that the emotions could be accessed through the body, not just the mind, for healing, and that the result of this could vastly improve the health of the organism appealed to our deepest instincts.

That afternoon, it dawned on me that I'd found a true companion, a soul mate, and maybe even a co-conspirator with whom I could explore a whole new and exciting frontier. The feeling, it turned out, was entirely reciprocal, and soon afterward, we began to date. It wasn't long before Michael and I became an "item," demonstrating by our relationship a union of two separate disciplines that would soon evolve into a whole new field of science, one that would dramatically bridge and heal the mind-body split that had been entrenched in Western medicine for over two hundred years.

My own naturally evolving awareness made me ripe for the next radical shift in my consciousness. This occurred one day as I was help-ing Michael clean out the trunk of his car, and I came across a copy of Norman Cousins's *Anatomy of an Illness*. I took the book home and, in one sitting, practically inhaled it, so compelling was the thesis and so closely did it resonate with my own nascent thinking at the time. Cousins, the editor of a major literary magazine, *The Saturday Review,* had been diagnosed with a life-threatening illness, an experience that had led him to question the whole foundation of Western medicine. Although not a doctor himself, he'd had a few brushes with the medical establishment as a patient, and had come to some rather sensible conclusions about its shortcomings. Rejecting what little help was offered by his doctors, Cousins had checked out of the hospital and checked into a hotel, where he holed up with an assortment of Charlie Chaplin videos and literally laughed himself back to health. He had felt, intuitively, that what the body needed was the life-affirming, joyous experience of laughter. What he was suggesting from this experience was that state of mind, thoughts, and feelings, all of which were completely ignored by the medical model, did in fact play a major role in his recovery. He even postulated that the laughter had triggered a release of endorphins, which, by elevating his mood, had somehow brought about a total remission of his disease.

I hung on every word. Truly, he was speaking my language, and I could relate to his experience directly from my own. Like him, I had chosen to do it my own way when, after one high-tech, heavily drugged hospital delivery, and a second natural childbirth, again in the environment of a hospital (which required me to fight off a constant barrage of unnatural interventions), I'd decided to have my third child at home. Instead of laughter, my magic bullet had been breathing, which is a surefire, proven strategy for releasing endorphins and quelling pain. Obviously, this is what previous generations of women, in the days before IV drips and synthetic painkillers, had relied on. Both they and their babies must have been better off for the experience, as I certainly felt myself to be.

Even though I questioned Cousins's notion that endorphins are also the key to the healing power of laughter, there was no doubt in my mind that he was on to something wonderful. What it was hit me in a sudden flash of awakening: Here were the direct implications of all our efforts to understand the neuropeptides, the brain chemicals of mood and behavior, to trace the chemical pathways by which they communicated with the immune system and every other bodily system as well! Cousins was saying that the work I'd been doing all along pointed to a new avenue for healing. With this new insight, something in me woke up, leading me to see clearly what I'd barely glimpsed back when my father was dying and

I was racing to understand his condition—that receptor science could lead to a whole new way of understanding and treating cancer and other diseases as well. On an intellectual level, I felt like I was shedding an old skin, the last vestiges of old-paradigm thinking.

CONCEPTION

One unusually warm spring afternoon, Mike and I were cruising around in my Fiat convertible through Rock Creek Park, seeking the perfect spot to enjoy the grass, a bottle of beer, and each other. Our conversation had turned to the research I'd done in an effort to save my father from lung cancer. The mystery of the small-cell–bombesin connection was always in the back of my mind, and now, with Michael listening, I could wonder out loud why in the world a cancerous lung cell would be secreting peptides. Suddenly, Michael blurted out, "Maybe it's because the cancer cells are really macrophages!"

As soon as the words were out of his mouth, I instantly had the feeling I get when I know a wild idea is right. Now, more than ever, I was willing to trust my intuitions and design experiments around them with an enthusiasm my colleagues often dismissed as unscientific.

In fact, I had accumulated some background knowledge about macrophages by this time, because these white blood cells were Michael's passion. The commonly held view was that macrophages were designed to do very basic functions. If you get a splinter in your finger, for instance, hordes of macrophages descend upon the invading bacteria to gobble them up, releasing enzymes to digest the debris and then cart it away. The lungs contain macrophages whose function it is to suck up all the dirt—pollen, dust, carbon particles, and other chemicals—we take in with each breath. Theoretically, if you were to fill a normal lung with water, shake it, and turn it upside down, billions of macrophages would flood out. If you did this with a lung from a cigarette smoker, there would be ten times as many macrophages.

But what Michael was considering was brilliantly radical. He was suggesting that small-cell carcinoma was not a matter of normal lung cells turned cancerous, the traditional view, but, instead, stemmed from the macrophages that had been drawn into the lung to clean up the dirty residue left by smoking. Somehow, the healing work of the macrophages had gone terribly awry, and the macrophages had mutated, becoming cancerous. It was the kind of idea only someone who saw beyond the reigning paradigm would have dared to suggest. Although Michael was well versed in the literature and even in the unpublished musings of immunology, there was still room in his thinking for the word *somehow*.

Not being an immunologist myself, I could afford to be equally fresh in my perspective. Moreover, being in love, I always found Michael's "somehows" terribly plausible. But this was a somehow we could test, a wild proposition that would be borne out—or not—by a simple experiment. Long chances like this were just what I'd been trained to grab at, regardless of what the literature claimed.

We designed the experimental approach that very afternoon in Rock Creek Park. The instant Michael had said the word "macrophage," I'd pounced on it and pulled the Fiat over to the side of the road. Happily, we leapt out, six-pack in hand, to sprawl on the grass with a pad and a pen, and started brainstorming the science. We became so engrossed in our discussion that we failed to notice an approaching policeman who abruptly ordered us to hand over the beer and wrote us a ticket. To his extreme annoyance, we barely acknowledged him, so intoxicated were we with the new idea that small-cell lung cancer might be a case of mutated macrophages.

LABOR

It took a year to finish the work. Our assumption was that if these lung-cancer cells were actually macrophages, then they ought to look and act like macrophages. They certainly had very little in common with normal lung cells, which had been our first clue. If, indeed, they really were macrophages, it would explain how they replicated so quickly and traveled, or metastasized, so widely, both activities commonly associated with macrophages, and much more typical of small-cell compared to other types of lung carcinoma. Our research strategy involved using antibodies that typically bound to macrophages to see if they also bound to the cancer cells. We devised a simple method of detecting antibody-binding to receptors on the cancer cells. If the small cells had the same surface receptors as macrophages, then they probably were macrophages.

We picked up the cancer cells from a tissue bank facility up the road in Rockville, Maryland, where two different small-cell tumor lines had been left on deposit by, of all people, my old collaborator Adi Gazdar, who had grown them while he was working in the lab of my dad's doctor. I thought it ironic, even poetic justice, that if the samples hadn't been left there, the testing of our hypothesis would have ground to a halt. But from this prominent researcher's perspective, I suppose, it was his worst nightmare: outsiders, led by the unstoppable Pert, stealing into his field and using cell lines started in his lab to prove an idea he hadn't thought of. We proceeded in utmost secrecy, partly to keep low and out of the line of fire, and partly because we didn't yet know whether our idea was crazy or spectacular.

Michael took off on the project. Using a modified version of the original opiate-receptor assay, he assayed the cancer cells to identify their receptors. He did most of the work after hours, using an old, discarded Triple M machine he found in the basement at the Dental Institute, and brought the data back to my house in the evening. Even though the machine had a big crack in it, which Michael had to continually shoot full of silicone goop to maintain the vacuum seal, the Triple M provided a quick and dirty method for getting our results.

We stayed up late, poring over the data, trying to reduce the long sheets of numbers to a few facts we could be sure of. Although we kept mum about what we were finding, not wanting the cancer guys to get wind of our activities and accuse us of poaching on their turf, they eventually heard rumors about the work we were doing. And we, in turn, heard rumors that they might be doing work and coming up with data similar to our own. But neither could communicate with the other to aid in the process, for all the old familiar reasons of power, ego, and turf.

In contrast to our lab, theirs was huge, a gigantic machine churning out data, measuring anything and everything possible to measure on the small cells without a specific hypothesis to guide them. We, on the other hand, were two people with a definite hypothesis: These cancer cells were somehow related to macrophages. More focused than the cancer researchers, we chose to scan for only those markers relevant to our theory.

DELIVERY

Eureka! It worked.

Michael's assay clearly demonstrated that the macrophage antibodies had bound to the cancer cells, and we concluded that these cancerous lung cells looked and acted too much like macrophages for this to be just a coincidence. The macrophage antibodies had bound to the cancer cells because those cells *were* macrophages—or, to be more precise, mutated macrophages. We further concluded that the cancerous cells had originally arisen not from lung cells but from macrophages that had migrated to the lung from the bone marrow to participate in the cleanup and repair of damaged tissue. Somewhere on their journey between the bone marrow and the lung, they had mutated and turned into cancers that metastasized and spread everywhere, eventually causing death.

The startling implication of our research, so radical it frightened even us, was that there was a clear connection between cancer, the immune system, and toxicity in the body. Small-cell lung cancer, it appeared, was one disease that was entirely related to a toxic condition in the body. The "dirt" in the lung from cigarette smoking (and presumably from other forms of pollution as well) had caused the immune system to go into hyper-response, in the form of sending in more and more macrophages to try to repair the damage, a situation that could not go on forever without some kind of mutation or "mistake" occurring in the DNA of these cells. Eventually, the mutated cells lost their ability to do the job, and grew wildly in response to peptide hormones like bombesin, metastasizing all over the body, following peptide signals. The paper describing this research, "Origin of human small-cell lung cancer," was accepted by the highly visible journal *Science*, and appeared in September of 1984—the first of many Ruff and Pert papers to come.

While we didn't dare to include this speculation in the paper, we wondered privately: Had we found the underlying mechanism that explains how cancer is a response to toxicity from environmental pollutants in the air and chemical additives in the diet? This connection had long been suspected but little investigated by the cancer establishment, and now we were seeing a mechanism by which it could be explained.

Preferring to explore that particular speculation more thoroughly before committing it to print, we used our closing paragraphs to open up an equally controversial possibility concerning the intercommunication among three key systems of the body. We summarized the data suggesting that the same peptides found in the brain were also found in the immune system, and that the nervous, endocrine, and immune systems are functionally integrated in what looks like a *psychoimmunoendocrine* network. This was a key realization, which now appeared in print for the first time. We boldly postulated that this network should be seriously considered in explaining the pathology of not only cancer but other disease states as well—a theory soon to become the basis for the emerging field of psychoneuroimmunology (now often referred to as PNI).

What we had seen in our research was that the brain, the glands, the immune system, indeed the entire organism, were joined together in a wonderful system coordinated by the actions of discrete and specific messenger molecules. These findings had prompted us to ask some interesting questions: Did the endocrine system communicate with the immune system? Yes, Blalock had already shown this, so we figured there was nothing too alarming in saying it again, despite the fact that Blalock was still viewed as a virtual heretic in his field. Did the immune system, via these peptidergic messengers, communicate with the nervous system or the brain? Yes, there were many ways that peptides from immune cells could affect the brain through their action on peptide receptors in the brain's blood vessels, on surrounding membranes, or even on neurons (brain cells) themselves! But we also had to consider a slightly more troubling question as a result of our research, and that was: Did the brain communicate with the immune system? And did this have implications for cancer-growth spread or for antitumor immune responses? Now, it was barely acceptable to suggest that the body influenced the mind, but to even hint that the mind might influence the body—well, that reeked too much of mind-over-matter, and only wild-eyed Californians and out-of-print Russians dared do that, at least in 1984!

But Michael and I knew that what we were looking at was incredibly paradigm-smashing and revolutionary—the part played by the mind in the growth and development of cancer tumors. At the time we organized our findings into a paper, we could cite only a handful of papers that dealt with these ideas, and then only in a rudimentary and fragmented fashion. We weren't alone in questioning the old static model, but no coherent synthesis was possible; indeed, much of the key data hadn't existed until then. This was our contribution—and it all had to do with what we had discovered about neuropeptides, building on work I'd been doing for ten years.

Neuropeptides, those chemicals secreted by the brain and known to mediate mood and behavior, were clearly signaling the cancer cells via their receptors and causing them to grow and travel, or metastasize, to different parts of the body. In the case of small-cell lung cancer, the peptide mechanism seemed to be bombesin (rather than the endorphins), which could pull over the cancer cells through chemotaxis, latch onto their receptors, and then program their growth and division.

In a 1985 follow-up paper, we speculated: Could excess or inappropriate production of neuropeptides released by the immune system, or by the brain, or by any other organ system in the body, promote other forms of cancer also? Was the cancerous tumor really part of a network, receiving and sending information that linked it to the brain and the immune system? (The "link" provides a mechanism by which these body systems may regulate, control, promote, or retard the actions of one another.) In later papers, we showed that besides the immune cells, many different kinds of cancerous cells were also chemotaxing according to neuropeptide signals. This process became a key to our thinking about the mind-body basis for cancer and other diseases, especially those that were a part of the psychoimmunoendocrine system. Because so many peptides were active, we could put forth a new precept: Cancer cells have neuropeptide receptors. This was antidogma and therefore profound, with rich implications that we and others were eager to explore.

COLLISION

In the euphoria of our romance, neither of us had given much thought to how vulnerable Michael was in all this. As a lowly postdoc, he had stuck his neck out, working with his lover on a paper to appear in one of the most prominent journals in the scientific arena, while at the same time jousting with a major player whose institute had God knows what connections with Michael's own institute. He had been "on loan" to my lab thanks to his wonderful boss, a brilliant section chief named Sharon Wahl, who generously lent her own energy and resources to aid our studies at the time. If Michael humiliated his superiors or caused a downpour of bad publicity on his branch, he'd be scrubbing test tubes in the Dental Institute basement for the rest of his career.

The day our paper was accepted, and prior to its publication, Michael went alone to the Cancer Institute to show the director our data. I certainly had no desire to see him again; the mere memory of the cafeteria harangue over a year and a half before was upsetting enough. I almost expected Michael to come staggering back with a knife in his chest. Instead, he reported, the cancer researchers had pulled out a ton of recent data showing results similar to ours. The situation was a classic: Their work had also revealed the connection between macrophages and small-cell cancers, but since they didn't have a hypothesis to make any sense out of it, they had overlooked it, and had run off chasing ten other ideas.

In my view, an old-paradigm insistence on the separateness and autonomy of the individual disciplines had blinded them to the significance of their data, preventing them from understanding that it all added up to the fact that the nervous and immune systems were clearly in communication with each other. Their own interpretation was quite different. In a letter that appeared in *Science* shortly after our article, they wrote: "We noticed the similarity, but *we* believe it was insignificant." The obvious implication was that since they were the cancer specialists, they knew what they were talking about.

When the editors of *Science* gave us a chance to answer, which is how these disagreements are handled formally, we were happy see our "wild" ideas in print once again, exactly a year after the initial report. We had drafted the response together, seeing it as an opportunity to reiterate our theory and discuss our conclusions further. In fact, we were elated to find that in our opponents' remarks there was information we could turn around and use to bolster our own theory.

Our approach was threateningly interdisciplinary, in complete violation of all kinds of traditional boundaries between scientific fields, bureaucratic departments, and medical specialties. We were investigating the origins of cancer—even illness itself—in an area far away from the cancer-genes-gone-awry-in-a-vacuum mentality that was fashionable and highly funded. But for scientists at NIH to follow up on our work, they would have to talk openly with trust, respect, and even mutual admiration to folks from other branches and institutes, a not very likely scenario given the ongoing interdepartmental funding competition.

In the ensuing *Science* debate, our opponents tried to destroy our argument but, in the end, only succeeded in confusing themselves. The paper was cited and mentioned for about a year before the field moved on, essentially ignoring our points and giving the game to the cancer lab. Years later, the field would finally return to the role of peptides in connection with cancer and the immune system, and Terry Moody would be lured away from his full professorship at George Washington University to the NIH's Cancer Institute to conduct research on the effects of bombesin on the growth of cancer cells. Slowly, the field would come around to accepting the possibility that if bombesin made these cells grow, then finding an antagonist to bombesin—a receptor blocker—might provide a useful therapeutic, a magic bullet. But over ten years would elapse before that possibility would be seriously investigated again.

It seems shameful that the doorway we stood before more than a decade ago is just now being reopened, and still only at the level of basic research, not in clinical trials where treatments are tested for use by the public. Michael and I later went on to explore the possibility that if the cancer cells were indeed macrophages, then perhaps they would act like macrophages if given macrophage growth hormones. Adult macrophages do not replicate, so growth hormones might cause the immature macrophages to grow up and stop dividing.

Both of these ideas, that of blocking a growth factor like bombesin by using a receptor antagonist, and that of providing a growth factor in the hopes of causing these tumors to differentiate and stop dividing, are examples of the new field of peptide pharmacology, as opposed to the old toxic treatments. One example of our newfound ability to use receptor antagonists to retard or stop a cancer involves the peptide LHRH (luteinizing hormone releasing hormone). Since LHRH plays a role in the development of the male prostate gland, and seems to be necessary for continued growth of the youthful cells that line the tube leading away from the gland into the penis, it's not surprising that doctors have been able to use LHRH antagonists successfully to treat tumors containing receptors for LHRH.

I do not want to give the impression that peptides are the only substances important in understanding cancer. Other information substances that are not peptide-based, such as the sex hormones, also play a part in the network, acting to promote growth that may lead to cancer. One of these, estrogen, has been shown experimentally to increase the growth of certain breast tumors. And, once again, the use of an antagonist to block the receptors has shown great promise as a cancer treatment. The antagonist, a drug known as tamoxifen, has been extraordinarily successful in treating women who have estrogen-dependent breast cancers. (Since not all breast cancers are estrogen-dependent, a sample of the tumor can be readily tested to determine whether or not it contains estrogen receptors before treatment is begun.)

The Cancer Institute appears to be slowly shifting gears, but old paradigms die hard, and the resistance to new ideas continues to delay progress, so that the promise of ideas we had proposed so many years ago still remains unfulfilled.

PNI

While our paper had relatively little impact on the cancer establishment at the time, it did make an impression on certain other researchers who were moving, mostly silently, behind the scenes to establish the new field of psychoneuroimmunology, or PNI. By providing PNI with a clear scientific language, that of neuropeptides and their receptors, we helped legitimize it.

It's quite amazing that PNI was able to arrive on the scene relatively unscathed, given the climate of scientific conservatism that generally ensures that newcomers will be vigorously and sometimes brutally hazed before being admitted to the club. The first time cellular and molecular PNI was seriously presented in a public forum was in 1984, when I was invited to a conference in Rome that went by the name "Endorphins and Opiate Receptors in the Periphery." There I assembled a panel of speakers, including doctors Michael Ruff, Ed Blalock, and several others, to speak specifically to our finding of a psychoimmunoendocrine system, the term Michael and I had used in our paper for the linkup of the three systems. For my talk, I'd prepared a slide that showed a triangle to graphically represent the three systems that used peptides to communicate with one another. It was a pleasant and synchronous surprise to see that two other presenters had created the exact same slide, which meant that we'd all arrived at the same understanding more or less simultaneously. Shortly after that, Herb Spector, an NIH psychologist, organized a private, but higher profile, event at the Palace, one that had three Nobel laureates in the audience, and put PNI more solidly on the map, a separate field with its own secure funding niche.

In the beginning, several names for the new science were proposed. One of these was "psychoimmunology," originally coined by psychiatrist George Solomon back in the fifties. This granddaddy term grew from Solomon's observations on how profoundly personality affects disease. A competing name was proposed by Herb Spector, who was one of the few Americans who bothered to keep track of what the Russian behaviorists, the heirs of Pavlov, had been up to. He knew they were light-years ahead of us in understanding the holistic balance of the body, and had been proving that the immune system could be shown to respond to classical conditioning for decades, thus implicating the nervous system as part of the process by which the body maintains health or lapses into a disease state. Spector's proposed term was "neuroimmunomodulation."

The term that triumphed, psychoneuroimmunology, was championed by Dr. Robert Ader, an experimental psychologist who had coined the word for both a conference and a book that appeared under that title in 1981. Inspired by the Russians, Ader had done some interesting experiments with rats that showed the immune system could be conditioned, and therefore did not operate autonomously, as immunologists had always believed, but was under the influence of the brain. Within the field, Ader had taken up a rather right-wing banner, fighting hard against what he considered a left-wing drift into pseudoscientific thinking, the kind of thing that he thought Californians were expressing by including PNI in their "New Age" modalities. *His* PNI was scientifically solid, he insisted, grounded in meticulously designed rat experiments, and in the resolutely unflaky tenets of behavioralism.

Psychoneuroimmunology: I personally find it a misnomer, a term that is not only inaccurate, because it reveals only part of the picture, but also redundant. Of course, it's necessary to concede that I speak as an also-ran in the naming game. The term Michael and I proposed, "psychoimmunoendocrinology," made a point of including the endocrine system, to make it clear that we were looking at a network hookup of multisystems, not just the brain and the immune system. To us, *psycho* meant the same thing as *neuro,* and did not need double emphasis, as if *psycho* weren't really good enough and needed *neuro* to make it legitimate. However, our proposal was met with a deafening silence and has since gone the way of the dinosaur.

A BODYWIDE SYSTEM

In the early days of our professional relationship, Michael had asked me what I meant by the term *neuropeptide.* Why add the prefix *neuro,* he argued, if the same peptide is found in the gut and immune system, as well as in the brain? And why call it a *neuro*receptor if it is also found in the gut, in the immune system, alongside the spinal cord, and who knows where else? By tossing out these linguistic distinctions and simply using the term *peptides* or *information substances* to refer to all peptides regardless of where they occurred, it became more obvious that we were describing a bodywide communication system, one we suspected was ancient in origin, representing the organism's first try at sharing information across cellular barriers. The brain, or *neuro,* component was only one part of the organism's nonhierarchical system to gather, process, and share information (albeit the most complicated and sophisticated component by far).

But what was this bodywide system? How did it translate into our experience, our behavior as human beings? These were some of the questions we were asking. I knew from my brain mapping over the years that the communicating chemicals were most dense in certain areas of the brain and along the sensory pathways. We also knew how the mind drugs heroin, opium, PCP, lithium, and Valium entered the network and worked on the receptors, and how the endogenous, or "in-house," substance, endorphin, communicated over a wide range. If we were to characterize exactly what these chemicals were·doing, we would have to say they affect the emotional state of the person who takes them, making him or her happy, sad, anxious, or relaxed, or something somewhere along the continuum of these emotions. And when we focus on emotions, it suddenly becomes very interesting that the parts of the brain where peptides and receptors are richest are also the parts of the brain that have been implicated in the expression of emotion. I don't remember whether it was Michael or I who said the words first, but both of us had the gut feeling that we were right: "Maybe these peptides and their receptors are the biochemical basis of emotion." Finally, we were looking at the implications of the fact that the limbic system had the densest concentration of these receptors.

Could it be that what we were seeing were the molecules of emotion?

Unfortunately, *emotion* is another of those words mainstream science likes to spit out at the very first taste. I was enormously bolstered when UCSF psychologist Paul Ekman taught me that Charles Darwin had been a theoretician of emotions as well as of the origins of species. Still, I was nervous the first time I stood up in front of my peers and suggested that this bodywide network of peptides and receptors might be the molecular basis of emotion. I'd hoped that, as strict materialists, they might find some satisfaction in hearing that emotions could now be understood as a basic, molecular, biological process. But, no, too many borders had been crossed, too many taboo words spoken. People did what they do to ideas that won't fit the reigning paradigm—ignore them. The pop journals, however, picked up the theory several years later with no attribution.

We published the key paper on our theory in the *Journal of Immunology* in 1985. To quote from the abstract:

A major conceptual shift in neuroscience has been wrought by the realization that brain function is modulated by numerous chemicals in addition to classical neurotransmitters. Many of these informational substances are neuropeptides, originally studied in other contexts as hormones, gut peptides, or growth factors. Their number presently exceeds 50, and most, if not all, alter behavior and mood states, although only endogenous analogs of psychoactive drugs like morphine, Valium, and phencyclidine have been well appreciated in this context. We now realize that their signal specificity resides in receptors rather than the close juxtaposition occurring at classical synapses. Precise brain distribution patterns for many neuropeptide receptors have been determined. A number of brain loci, many within emotion-mediating brain areas, are enriched with many types of neuropeptide receptors, suggesting a convergence of information processing at these nodes. Additionally, neuropeptide receptors occur on mobile cells of the immune system: monocytes can chemotax to numerous neuropeptides via processes shown by structure-activity analysis to be mediated by distinct receptors indistinguishable from those found in the brain. *Neuropeptides and their receptors thus join the brain, glands, and immune system in a network of communication between brain and body, probably representing the biochemical substrate of emotion.*

The molecules of emotion. This was our new paradigm, newborn and a bit shaky about its place in the universe, but lusty in its cries for attention, its insistence on life. Meanwhile Michael and I would be rocked by the death throes of the old paradigm in one of our next adventures—the race to develop a treatment for AIDS.

9

THE PSYCHOSOMATIC NETWORK:
A CONCLUDING LECTURE

IT IS USUALLY at this point in my lecture, when I've leaned hard on the science and still have more to present, that I try to lighten things up a bit and throw in a slide that will amuse my audience. One slide that suits this purpose well is that of a brightly colored MRI scan of the human brain, a visual delight, almost as beautiful as the rainbow butterfly pattern Miles and I saw when we first began to autoradiograph animal brain slices. But this isn't just any brain, I tell my audience, looking forward to the reaction I'll get when I announce that the brain they are looking at is my very own. And then I proceed to explain how one day we'll be able to tell from the variety and density of receptors in certain areas what kind of life I have led, which substances I have abused, and, in general, what the biochemicals of emotion are doing in my brain.

After that bit of fun, I move on to a slide that introduces the subject of the next segment of my lecture. It reproduces an editorial that appeared in *Nature* magazine commenting on Ed Blalock's shocking discovery in 1982 that our immune system cells are secreting peptides, most notably endorphins. The idea that there were brain peptides in the immune system was so unsettling to immunologists that Blalock's work was not believed at first—a virtual replay of the disbelief that had greeted Jesse Roth's work showing insulin in the brain. The establishment was still committed to the separation of body and brain. But, finally, *Nature* had printed this editorial in which it grudgingly acknowledged the validity of Blalock's research, while disputing its implications. *Nature* warned the scientific community to beware of those "radical psychoimmunologists" who would dare to use Blalock's work to propose that body and mind were in communication with each other, in fact that the body mirrored the mind. Which is exactly the path I chose to pursue in my subsequent research at the NIH. As I've mentioned elsewhere, my colleagues and I loved calling ourselves the radical psychoimmunologists.

THE IMMUNE CONNECTION

We have seen how the neuropeptides and their receptors, the biochemicals of emotion, act to orchestrate many key bodily processes, linking behavior and biology to effect a smooth functioning of the organism. At this point, taking my lead from Ed Blalock's work, I would like to introduce a new layer to this dynamic, the role played by the immune system, which forms a vital link in the overall hookup of the biochemistry of emotions.

I have talked about how the endocrine system, which historically had always been studied as completely separate from the brain, conceptually resembles the nervous system. The brain is a big hormone bag! Pockets of peptide juices are released from both glands and brain cells, after which they bind with specific receptors that enable them to act at sites far from where the juices originated. (This is what endocrinologists call "action at a distance." Looked at in this way, endocrinology and neuroscience are really exploring two aspects of the same process.)

Now I want to show you how the immune system is part of the same network as the endocrine and nervous systems, even though most immunologists still consider it a separate and autonomous field of study.

The immune system is made up of the spleen, the bone marrow, the lymph nodes, and various kinds of white blood cells, some of which circulate throughout the body, while others reside in the various tissues of the body, including the skin. Its overall purpose is to defend against pathological invaders that threaten the health of the organism and to repair any damage they cause. To do this, the immune system must define the boundaries of the organism, distinguishing between what is self and what is not self, that is, determining what is part of the organism and needs to be repaired and restored versus what is part of a tumor and needs to be killed.

A key property of the immune system is that its cells move. Unlike brain cells, which, for the most part, do not move, the cells of the immune system do their job by traveling throughout the organism to wherever they are needed to mount a defense or repair damage. Certain white blood cells known as *monocytes* (called *macrophages* in later stages of growth), for example, whose function is to ingest foreign organisms in the bloodstream, start life in your bone marrow and then diffuse out to travel through your veins and arteries, deciding where to go by following chemical cues. Monocytes and other white blood cells, such as *lymphocytes,* travel along in the blood and at some point come within "scenting" distance of a given neuropeptide, and because these cells have receptors for that particular neuropeptide on their surface, they begin literally to *chemotax,* or crawl toward it. This action is very well documented, and there are excellent ways of studying it in the laboratory.

Now, monocytes are responsible not just for recognizing and digesting foreign bodies, but also for wound healing and tissue-repair mechanisms. For example, we have enzymes that produce and degrade collagen, an important structural material out of which the body's very fabric is made. What we are talking about, then, are cells with vital health-sustaining and healing functions.

When Michael Ruff and I read Ed Blalock's astonishing paper on brain peptides in the immune system, we went looking for neuropeptide receptors there. And the radical discovery we made was that every neuropeptide receptor we could find in the brain is also on the surface of the human monocyte. Human monocytes have receptors for opiates, PCP, and other peptides such as bombesin. These emotion-affecting peptides, then, actually appear to control the routing and migration of monocytes, which are very pivotal to the overall health of the organism. They communicate with the other lymphocytes, called B cells and T cells, by interacting through peptides called cytokines, lymphokines, chemokines, and interleukins and their receptors, thus enabling the immune system to launch a well-coordinated attack against disease. The action looks something like this: A health-sustaining cell like the monocyte circulates through the blood until it is pulled over by the chemical attraction of a peptide—for example, an endorphin, the body's endogenous opiate. It can then connect with that opiate because it has the receptor to do so.

But immune cells don't just have receptors on their surfaces for the various neuropeptides. As demonstrated by the paradigm-shaking research of Ed Blalock at the University of Texas in the early eighties—and confirmed by research done by Michael Ruff, Sharon and Larry Wahl, and me—immune cells also make, store, and secrete the neuropeptides themselves. In other words, the immune cells are making the same chemicals that we conceive of as controlling mood in the brain. So, immune cells not only control the tissue integrity of the body, but they also manufacture information chemicals that can regulate mood or emotion. This is yet another instance of the two-way communication between brain and body.

Or that's how we see it. But such an idea is so astounding to both neuroscientists and immunologists that many will still maintain, as the *Nature* editorial did, that just because these communication molecules are there doesn't mean they're actually used to communicate. After all, their education was based on the idea of an impenetrable "blood-brain barrier," the existence of which had been "proved" by turn-of-the-century experiments in which huge dye molecules injected into the body could not get into the brain. And, certainly, it is true that many drugs are absorbed slowly, if at all, into the brain. But very recent evidence shows many ways that cytokines, chemokines, lymphokines, interleukins, and other immunopeptides can breach the barrier. One well-documented route of access involves their binding with receptors on the surface of the brain in such a way as to affect the permeability of the brain's surface membranes. From there they can propagate a signal that gets picked up by other peptides and receptors deep within the brain. In fact, they probably do this routinely.

The question remains: What is the purpose of such communications? To answer, let's look at an example of a neuropeptide that has receptors in several different bodily systems—not just the brain and the immune system but the gastrointestinal system as well. Consider CCK, a neuropeptide governing hunger and satiety, which was first discovered and then sequenced by chemists who were exploring its action on the gut. As discussed earlier, if you were given doses of CCK, you would not want to eat, regardless of how long it had been since your last meal. Only recently have we been able to show that both the brain and the spleen—which can be described as the brain of the immune system—also contain receptors for CCK. So brain, gut, and immune system are all being integrated by the action of the CCK. Why would this be so?

There are nerves that contain CCK all along the digestive tract and in and around the gallbladder. After a meal, when the fat content is moving through the digestive system to your gallbladder, you experience a feeling of satisfaction, or satiety—thanks to the signal CCK sends to your brain. CCK also signals your gallbladder to go to work on the fat in the meal, which enhances the feeling of fullness. This much we know. As to what the CCK receptors are doing on cells in your immune system at this point, I can only conjecture. It certainly would not be a good idea to have your immune system revved up shortly after eating, when the food is still undigested, because you don't want your immune system mounting an attack response on the undigested meal! For this reason, it would make sense if the same CCK system that creates a sense of satiety in your brain and makes your gallbladder go to work would also be able to signal the immune system to slow down.

NETWORK

Let me summarize the basic idea I have been developing. The three classically separated areas of neuroscience, endocrinology, and immunology, with their various organs—the brain; the glands; and the spleen, bone marrow, and lymph nodes—are actually joined to each other in a multidirectional network of communication, linked by information carriers known as neuropeptides. There are many well-studied physiological substrates showing that communication exists in both directions for every single one of these areas and their organs. Some of the research is old, some of it is new. For example, we've known for over a century that the pituitary gland spews out peptides throughout the body. But it's only been a few years that we've known that peptide-producing cells like those in the brain also inhabit the bone marrow, the place where immune cells are "born."

The word I want to emphasize in regard to this integrated system is *network*, which comes from the relatively new field of information theory. In a network, there is a constant exchange and processing and storage of information, which is exactly what happens, as we have seen, as neuropeptides and their receptors bind across systems. The informational nature of these biochemicals led Francis Schmitt of MIT to introduce, in 1984, the term *information substances*, a wonderfully descriptive way of referring to all of the messenger molecules and their receptors as they go about their job of linking brain, body, and behavior. Schmitt did us a great favor by giving us a metaphor to explain the purpose of the complex overlapping of these multiple-functioning substances as they move from one system to another, one job to another. He included in his new generic category both long-familiar substances such as the classical neurotransmitters and the steroid hormones, and newly discovered ones such as peptide hormones, neuropeptides, and growth factors—all ligands that trigger receptors and initiate a cascade of cellular processes and changes.

So what we have been talking about all along is information. In thinking about these matters, then, it might make more sense to emphasize the perspective of psychology rather than of neuroscience, for the term *psycho* clearly conveys the study of mind, which encompasses but also goes beyond the study of the brain. I like to speculate that what the mind is is the flow of information as it moves among the cells, organs, and systems of the body. And since one of the qualities of information flow is that it can be unconscious, occurring below the level of awareness, we see it in operation at the autonomic, or involuntary, level of our physiology. The mind as we experience it is immaterial, yet it has a physical substrate, which is both the body and the brain. It may also be said to have a nonmaterial, nonphysical substrate that has to do with the flow of that information. The mind, then, is that which holds the network together, often acting below our consciousness, linking and coordinating the major systems and their organs and cells in an intelligently orchestrated symphony of life. Thus, we might refer to the whole system as a psychosomatic information network, linking *psyche*, which comprises all that is of an ostensibly nonmaterial nature, such as mind, emotion, and soul, to *soma*, which is the material world of molecules, cells, and organs. Mind and body, psyche and soma.

This view of the organism as an information network departs radically from the old Newtonian, mechanistic view. In the old paradigm, we saw the body in terms of energy and matter. Hardwired reflexes, caused by electrical stimulation across the synapse, ran the body in a more or less mechanical, reactive fashion, with little room for flexibility, change, or intelligence. With information added to the process, we see that there is an intelligence running things. It's not a matter of energy acting on matter to create behavior, but of intelligence in the form of information running all the systems and creating behavior. Walter B. Cannon, William James's debater, was onto this when he referred to the "wisdom of the body," and today certain manipulative healers such as chiropractors refer to it as the body's "innate intelligence." But, classically, there is no such thing as an intelligent organism, and to say so is heresy to the old guard who cling to a concept of the body as unintelligent, a bundle of mass and matter stimulated by electrical impulses in a predictable way. Theirs is the ultimate godless, mechanical universe, peopled by clocklike organisms, as conceived by Cartesian and Newtonian models.

While much of the activity of the body, according to the new information model, does take place at the autonomic, unconscious level, what makes this model so different is that it can explain how it is also possible for our conscious mind to enter the network and play a deliberate part. Let's look, for example, at the role of opiate receptors and endorphins in modulating pain. Pain researchers all agree that the area called the periaqueductal gray, located around the aqueduct between the third and fourth ventricles of the midbrain, is filled with opiate receptors, making it a control area for pain. (It is also loaded with receptors for virtually all the neuropeptides that have been studied.)

Now, we've all heard of the yogis of the East and practitioners of certain mystical disciplines who have been able, through breath training, to alter their perceptions of physical pain. (Other people, known as mothers, demonstrate mastery equal to that of the yogis, when, with proper training such as Lamaze, they use breathing techniques to control pain in childbirth.) What seems to be going on here is that these people are able to plug into their PAG (their periaqueductal gray), gaining access to it with their conscious intention, and then, I believe, are able to reset their pain thresholds. Reframed by conscious expectations and beliefs, the pain is abolished, reinterpreted as either a neutral experience or even pleasure. The question is: How can the mind mediate and modulate an experience of pain? What role does consciousness play in such matters?

To answer, I must return to the idea of a network. A network is different from a hierarchical structure that has a ruling "station" at the top and a descending series of positions that play increasingly subsidiary roles. In a network, theoretically, you can enter at any nodal point and quickly get to any other point; all locations are equal as far as the potential to "rule" or direct the flow of information. Let's see how a concept like this explains the process by which a conscious intention can reach the PAG and use it to control pain.

Conscious breathing, the technique employed by both the yogi and the woman in labor, is extremely powerful. There is a wealth of data showing that changes in the rate and depth of breathing produce changes in the quantity and kind of peptides that are released from the brain stem. And vice versa! By bringing this process into consciousness and doing something to alter it—either holding your breath or breathing extra fast—you cause the peptides to diffuse rapidly throughout the cerebrospinal fluid, in an attempt to restore homeostasis, the body's feedback mechanism for restoring and maintaining balance. And since many of these peptides are endorphins, the body's natural opiates, as well as other kinds of pain-relieving substances, you soon achieve a diminution of your pain. So it's no wonder that so many modalities, both ancient and New Age, have discovered the power of controlled breathing. The peptide-respiratory link is well documented: Virtually any peptide found anywhere else can be found in the respiratory center. This peptide substrate may provide the scientific rationale for the powerful healing effects of consciously controlled breath patterns.

We are all aware of the bias built into the Western idea that the mind is totally in the head, a function of the brain. But your body is not there just to carry around your head. I believe the research findings I have described indicate that we need to start thinking about how the mind manifests itself in various parts of the body and, beyond that, how we can bring that process into consciousness.

MIND IN BODY

The concept of a network, stressing the interconnectedness of all systems of the organism, has a variety of paradigm-breaking implications. In the popular lexicon, these kinds of connections between body and brain have long been referred to as "the power of the mind over the body." But in light of my research, that phrase does not describe accurately what is happening. Mind doesn't dominate body, it *becomes* body—body and mind are one. I see the process of communication we have demonstrated, the flow of information throughout the whole organism, as evidence that the body is the actual outward manifestation, in physical space, of the mind. *Bodymind,* a term first proposed by Dianne Connelly, reflects the understanding, derived from Chinese medicine, that the body is inseparable from the mind. And when we explore the role that emotions play in the body, as expressed through the neuropeptide molecules, it will become clear how emotions can be seen as a key to the understanding of disease.

We know that the immune system, like the central nervous system, has memory and the capacity to learn. Thus, it could be said that intelligence is located not only in the brain but in cells that are distributed throughout the body, and that the traditional separation of mental processes, including emotions, from the body is no longer valid.

If the mind is defined by brain-cell communication, as it has been in contemporary science, then this model of the mind can now be seen as extending naturally to the entire body. Since neuropeptides and their receptors are in the body as well, we may conclude that the *mind* is in the body, in the same sense that the mind is in the brain, with all that that implies.

To see what this means in practice, let's return for a moment to the example of the gut. The entire lining of the intestines, from the esophagus through the large intestine, and including each of the seven sphincters, is lined with cells—nerve cells and other kinds of cells—that contain neuropeptides and receptors. It seems entirely possible to me that the density of receptors in the intestines may be why we feel our emotions in that part of the anatomy, often referring to them as "gut feelings." Studies have shown that excitement and anger increase gut motility, while contentment decreases it. And then, because this is a two-way network, it's also the case that the movement of the gut as it digests food and excretes impurities can alter your emotional state. "Dyspeptic" means grouchy and irritable, but originally it referred to having poor digestion. Or let's look again at the autonomic nervous system, which runs all the unconscious aspects of your body, such as breathing, digestion, and elimination. You would think that if any part of the body functioned independently of the mind, it would most surely be the autonomic nervous system. There, the ability to make your heart beat, your intestines digest, and your cells replicate is carried on below conscious awareness. And yet, surprisingly, as we discussed in the example of yogis and women in labor, consciousness *can* intervene at this level. This is the radical lesson of biofeedback, which many doctors now teach their patients so that they can control pain, heart rate, blood circulation, tension and relaxation, etc.—all processes previously thought to be unconscious. Up until the early sixties, we thought that the autonomic nervous system was run by two neurotransmitters, acetylcholine and norepinephrine. But it turns out that in addition to the classical neurotransmitters, all of the known peptides, the information molecules, can be found abundantly in the autonomic nervous system, distributed in subtly different intricate patterns all the way down both sides of your spine. It is these peptides and their receptors that make the dialogue between conscious and unconscious processes possible.

IN SUMMARY, the point I am making is that your brain is extremely well integrated with the rest of your body at a molecular level, so much so that the term *mobile brain* is an apt description of the psychosomatic network through which intelligent information travels from one system to another. Every one of the zones, or systems, of the network—the neural, the hormonal, the gastrointestinal, and the immune—is set up to communicate with one another, via peptides and messenger-specific peptide receptors. Every second, a massive information exchange is occurring in your body. Imagine each of these messenger systems possessing a specific tone, humming a signature tune, rising and falling, waxing and waning, binding and unbinding, and if we could hear this body music with our ears, then the sum of these sounds would be the music that we call the emotions.

Emotions. The neuropeptides and receptors, the biochemicals of emotion, are, as I have said, the messengers carrying information to link the major systems of the body into one unit that we can call the bodymind. We can no longer think of the emotions as having less validity than physical, material substance, but instead must see them as cellular signals that are involved in the process of translating information into physical reality, literally transforming mind into matter. Emotions are at the nexus between matter and mind, going back and forth between the two and influencing both.

HEALTH AND EMOTIONS

What, then, is the relationship of mind and emotions to an individual's state of health?

As we have seen, the neuropeptides and their receptors are the substrates of the emotions, and they are in constant communication with the immune system, the mechanism through which health and disease are created. One of the ways we now know the immune system affects our health is through plaque formation in the arteries. Immune cells squirt out peptides that either increase or decrease the buildup of plaque in coronary blood vessels—a key factor in heart attacks. And although we don't know what the role of the emotions is in all this, epidemiological evidence suggests there is a link. It's well documented, for example, that people have more heart attacks on Monday mornings (when the work week begins) than any other day of the week, and that death rates peak during the days after Christmas for Christians and after Chinese New Year for the Chinese. Since these are all days with high emotional valence, one way or another, it seems clear that the emotions in some way correlate with the state of people's hearts.

Another possible immune system connection has to do with viruses.

Viruses use the same receptors as neuropeptides to enter into a cell, and depending on how much of the natural peptide for a particular receptor is around and available to bind, the virus that fits that receptor will have an easier or harder time getting into the cell. Because the molecules of emotion are involved in the process of a virus entering the cell, it seems logical to assume that the state of our emotions will affect whether or not we succumb to viral infection. This might explain why one person will get sicker from the same loading dose of a virus than another person. I don't know about you, but I never get sick when I'm about to go skiing! Could an elevated mood, one of happy expectation and hope for an exciting possibility or adventure, protect against certain viruses? One possible explanation for how this might work is that the rheovirus, shown to be a cause of the viral cold, uses the receptor for norepinephrine—an informational substance thought to flow in happy states of mind, according to the main psychopharmacological theories—to enter the cell. Presumably what happens is that when you're happy, the rheovirus can't enter the cell because the norepinephrine blocks all the potential virus receptors.

Over the centuries, much attention has been paid to the influence of the mental and emotional processes on health and disease. Aristotle was among the first to suggest the connection between mood and health: "Soul and body, I suggest, react sympathetically upon each other," he is credited with saying. But it is only since the early twentieth century that researchers have had tools powerful enough to discern the links and to demonstrate that one of those links, the immune system, was trainable. In the 1920s and 1930s, pioneering Russian scientists showed that classical Pavlovian conditioning could both suppress and enhance the immune response. Working with guinea pigs and rabbits, for example, they paired cues such as a trumpet blast with injections of bacteria to stimulate the immune system. After repeated trials, the animals "learned" to activate their immune systems without the stimulus of the bacteria injections whenever they heard the sound of the horn.

An American picked up this thread and did more research on the communication links between brain and immune system. Psychologist Robert Ader of the University of Rochester School of Medicine (who was later to coin the term *psychoneuroimmunology*) and his colleague Nicholas Cohen did a series of groundbreaking experiments in the 1970s. They trained lab rats to associate certain stimuli with an event, much as Pavlov trained his dogs to associate the sound of a bell ringing with the approach of food. In Ader and Cohen's studies, rats were given an immune-suppressing drug flavored with sweet-tasting saccharin. Eventually, they became so conditioned to the effects of this drug that the saccharin taste alone, divorced from the drug, caused a suppression of their immune system—another demonstration of mental cues altering physiology.

While these studies showed that the immune system could be conditioned at the subconscious, or autonomic, level, it remained for Howard Hall to show us in 1990 that the immune system could also be consciously controlled. In the pivotal experiments Hall conducted at Case Western Reserve University in Ohio, he instructed his human subjects in cyberphysiologic strategies. The word *cyber* derives from the Greek "kybernetes," meaning "that which steers" or "the helmsman," and in this context refers specifically to self-regulatory practices such as relaxation and guided imagery, self-hypnosis, biofeedback training, and autogenic training. Using several control groups, Hall showed that those with cyberphysiologic preparation could use these techniques to consciously increase the stickiness of their white blood cells, as measured by saliva and blood tests. Up until his work, there were anecdotal reports of an association between hypnotherapy and clinical improvements in warts and asthma, both of which may be mediated by immune changes under subconscious control. But there were no measurements of change at the cellular level, and no work demonstrating the potential for conscious control. Hall was the first to show that psychological factors, that is, conscious intervention, could directly affect cellular function in the immune system.

If the immune system can be altered by conscious intervention, what does this mean for the treatment of major diseases such as cancer? The idea that emotions are linked to cancer has been around for a while. In the 1940s, Wilhelm Reich proposed the then heretical idea that cancer is a result of the failure to express emotions, especially sexual emotions. Reich was not only ridiculed by the medical and scientific establishment, he was actually persecuted. It was perhaps the only time in history that the government of the United States held an official book burning, calling for all available copies of Reich's life's work to be rounded up by the FDA and incinerated. However, the heretical idea did not die in that bonfire. The German psychoanalyst Claus Bahnson, among others, continued this line of work in the interim until, today, it links up with much of contemporary biology. More recent, 1980s, studies by Lydia Temoshok, a psychologist then at UCSF, showed that cancer patients who kept emotions such as anger under the surface, remaining ignorant of their existence, had slower recovery rates than those who were more expressive. Another trait common to these patients was self-denial, stemming from an unawareness of their own basic emotional needs. The immune systems were stronger and tumors smaller for those in touch with their emotions.

Can suppressed anger or other "negative" emotions cause cancer? In addition to the recent studies by various researchers like David Spiegel of Stanford who have convincingly shown that being able to express emotions like anger and grief can improve survival rates in cancer patients, we now have a theoretical model to explain why this might be so. Since emotional expression is always tied to a specific flow of peptides in the body, the chronic suppression of emotions results in a massive disturbance of the psychosomatic network. Many psychologists have interpreted depression as suppressed anger; Freud, tellingly, described depression as *anger redirected against oneself*. Now we know something about what this looks like at a cellular level.

Take cancer, for example. It's a fact that every one of us has a number of tiny cancerous tumors growing in our bodies at every moment. The part of the immune system that is responsible for the destruction of these errant cells consists of natural killer cells whose job it is to attack these tumors, destroy them, and rid the body of any cancerous growth. In most of us, most of the time, these cells do their job well—a job coordinated by various brain and body peptides and their receptors—and these tiny tumors never grow large enough to cause us to become ill. But what happens if the flow of peptides is disrupted? Is it possible we could learn to consciously intervene to make sure our natural killer cells keep doing their job? Could being in touch with our emotions facilitate the flow of the peptides that direct these killer cells at any given moment? Is emotional health important to physical health? And, if so, what is emotional health? These are the sort of questions we have to start addressing if we take the links between body and mind seriously.

Let me begin to answer by saying that I believe *all* emotions are healthy, because emotions are what unite the mind and the body. Anger, fear, and sadness, the so-called negative emotions, are as healthy as peace, courage, and joy. To repress these emotions and not let them flow freely is to set up a dis-integrity in the system, causing it to act at cross-purposes rather than as a unified whole. The stress this creates, which takes the form of blockages and insufficient flow of peptide signals to maintain function at the cellular level, is what sets up the weakened conditions that can lead to disease. All honest emotions are positive emotions.

Health is not just a matter of thinking "happy thoughts." Sometimes the biggest impetus to healing can come from jump-starting the immune system with a burst of long-suppressed anger. How and where it's expressed is up to you—in a room by yourself, in a group therapy situation where the group dynamic can often facilitate the expression of long-buried feelings, or in a spontaneous exchange with a family member or friend. The key is to express it and then let it go, so that it doesn't fester, or build, or escalate out of control.

THE UNITY OF LIFE

I'd like to conclude my lecture for today with my final slide, that of a single-celled animal, the tetrahymena. This is a critter so widely studied in basic science laboratories that it has earned the title of the "workhorse of biology." What is truly amazing is that this primitive unicellular animal makes many of the same peptides, including insulin and the endorphins, that we humans do. On its single-cell surface, Blanche O'Neil found opiate receptors just like the ones in our brains. These same basic building blocks, then, are found in the earliest and simplest forms of life as well as in the most complex ones. And just as there are four basic molecules that code for all DNA in living organisms, there is some given number, not yet finally determined, of informational molecules that code for communication, for the information exchange that runs all systems in all living things, whether that communication is inter- or intra-cellular, organ to organ, brain to body, or individual to individual.

I like to bring the tetrahymena to your attention because it both illustrates an important biological fact and gives me a chance to end my lecture on a philosophical note (after which I'll go on to discuss some of the more practical implications of these ideas—i.e., how you can bring more consciousness into your life and use it for achieving better physical and emotional health). Think about what it means that the same basic informational network found in the tetrahymena is still to be found in us. If these peptides and their receptors—the molecules of emotion—have not only been conserved since their origins in the earliest and simplest forms of life but have continued to grow into the incredibly elaborate psychosomatic network we have discovered in the human body, we have to conclude that their role in evolution has been a powerful and critical one. To me, this is a stunning demonstration of the unity of all life. We humans share a common heritage, the molecules of emotion, with the most modest of microscopic creatures, a one-celled being, even though evolution has caused us to develop into trillion-celled creatures of astonishing magnificence.

I leave you with that thought, and thank you for your attendance here at my lecture today.

THE HOUSELIGHTS come up as the image on my slide screen fades, and I am once again aware of the real, live people who have been sitting out in the audience, the ears and eyes and hearts and minds my talk has been directed to—the trillion-celled creatures themselves.

～♨～

It was December 1987 in Puerto Rico. The piña coladas were flowing as we American neuropsychopharmacologists greeted one another at our annual conference. Fellow NIMHer Peter Bridge and I spotted each other across the packed floor.

Normally reserved, even sardonic, Peter seemed really excited as he began to brief me on the first two Americans to receive the experimental AIDS drug Michael and I had just invented.

"Did anything . . . happen?" I asked, knowing already from the way my heart was pounding that something had.

"Two of them—both of them—had terrible neuropathy, one could barely walk. The other couldn't."

"And now?"

"They're both walking normally. Their neuropathies went away. I've talked to three neurologists who've seen a lot of AIDS patients. They said it never happens."

"What do you mean, 'it never happens'?"

"When their neuropathy's as bad as these guys' were, they don't usually ever get any better. These guys got better a few days after starting your drug." Peter shrugged while we hugged, both of us thrilled but trying to stay skeptical, or at least unemotional.

Just then, the crush of neuropsychopharmacologists suddenly moved to the window to see the rainbow that had appeared along the horizon where the black storm clouds of the Caribbean rainy season were starting to move off at last. It was a huge double rainbow that lasted almost an hour, practically filling the small slice of blue sky. Later, I watched it from my balcony with my sister Wynne, marveling at how a rainbow at one end of the horizon could share the sky with lightning bolts tearing through the blackness at the other end.

10

CHILD OF THE NEW PARADIGM

· ·

PROMISE

It was in Maui, the navel of the earth, in 1985, that the promise of our theory about mind-body cellular communication as the foundation for understanding health and disease was most spectacularly revealed.

That year my lab was as large as it had ever been. I had a team in place, twelve people under my immediate direction and a larger group within my circle of influence, all supporting the work I was doing in an informal, collaborative association. Joining my people in brain biochemistry were Michael, courtesy of the cellular immunology section of the Dental Institute, Frank Ruscetti from the Laboratory of Molecular Immunoregulation, and Bill Farrar from the National Cancer Institute. By this time, I had earned tenure and was secure in my position as a senior scientist. I was pleased that people were working well together and our projects were benefiting from the interdisciplinary efforts.

The work consisted mainly of finding confirming evidence for our theory of an organismwide information system linking the brain and glands with the immune, digestive, and autonomic nervous systems. It was becoming clear to us that any receptor on an immune cell would also be found on cells in the brain, and that at this molecular level there really was no distinction between the mind and the body. We were just beginning to ask the questions this knowledge raised: What implications might this system of shared information have for our understanding of disease? And how does it help us to develop approaches to their treatment? It was this kind of inquiry that led our lab to uncover a very significant finding, something only we were staffed and equipped to pursue, and that catapulted us smack into the middle of a race to find a cure for AIDS.

That Thanksgiving, Michael and I had announced to our families that we were planning to get married the following summer. Shortly after Thanksgiving, we were on our way to Maui to present our latest findings at the annual American College of Psychoneuropharmacology Conference. We arrived a week early with plans to do some camping and hiking in the Haleakala Crater, a dormant volcano whose long-ago eruptions had formed the island. Michael had mapped out a route from Hanna, a remote ranching town, up the difficult rear slope to the crest of the crater, where we would find a trail that would take us down into the interior. After stopping to camp for a night in the crater, we would complete our journey by climbing back up to the crest and then hiking back down the rear slope, a three-day journey—one day for the hike up, one day to hike around the inside of the crater after the ascent, and the final day for the descent back to Hanna. Ambitious, yes, but doable, Michael was certain of it. Earlier in our romance, he had introduced me to nature hikes, and while I thoroughly enjoyed them, I didn't have enough experience to judge my own limits. But I was in love, and no challenge seemed insurmountable. We packed our equipment and supplies and drove to the trailhead.

The ascent was strenuous and took twice as long as we had planned, for halfway up what we had thought to be a four-mile ascent we discovered was actually eight miles, with a 4,000-foot vertical gain. So a hike that had begun at dawn did not end until we pitched our tent at 7 P.M. To this day that hike remains my most grueling physical feat.

The headset I had brought received only one station from nearby Hawaii, the Big Island, but I managed to enjoy the rock tunes and kept my spirits high. Halfway up we turned a bend and suddenly came upon a spectacular rainbow, as vivid and complete as any I'd ever seen. As we oooh-ed and ahhh-ed, I remember thinking we were being given a sign that meant in spite of our misplanning, we were on the right trail, and that this could apply also to the direction our research was taking back in the lab. The rainbow, long a symbol to me of the promise of science to eventually reveal ultimate truths, now graced our way, beckoning us on.

Once over the top, we descended into the crater, and a truly magical landscape unfolded before us. The interplay of light and shadow over a constantly changing terrain revealed reddish hues of cinder cones, black lava flows dotted by stunning silver swords jutting up out of nowhere, a moonscape of vast emptiness. I remember feeling a sense of sacredness as we hiked along the solitary trail, a certainty that this was a special place and we were walking on holy ground. Haleakala, the House of the Sun, where myth had it that the demigod Maui had captured the Sun and made it do his bidding! All of this—the gorgeous nature, the mystical aspect, and the sheer physical challenge—was having a profound effect on me, and I experienced an expansion of my heart and consciousness that left me in a state of deep awe and humility.

When we returned from our unexpectedly heroic journey, we were exhausted and dehydrated, yet exhilarated in our triumph. Looking back, I see how our ordeal was a harbinger of things to come, of the labyrinth that lay ahead and would consist of infinitely more twists and turns than our hike into the crater, and of an infinitely more strenuous journey than our hike up the rear slope. In our work back at the lab, we stood on the brink of an abyss we would soon descend, completely oblivious to the events that would lead us through the war-torn land of the AIDS establishment and toss us from our cozy Palace nest.

Michael drove the rental car to our rented condo, where we collapsed after soaking our aching muscles in a hot Jacuzzi. That night, our first spent indoors since we'd arrived on the island, I slept deeply, lulled by the sweet smells and gentle sounds of the ocean lapping at our lanai edge. I was rested and refreshed the next day when I arrived at the conference and took my place among the speakers for the opening session, entitled "AIDS and the Brain."

OVERLAP

An exciting series of events, precipitated by our investigation into the link between the immune system and the brain, had brought us to the Maui conference and had placed us at the doorway of the then emerging arena of AIDS research. It had all begun when Michael and I discovered that many peptide receptors thought to be confined to the brain were also found on immune cells. Once we knew that, we began to wonder if receptors that were found on immune cells might be in the brain. A fortuitous phone call from an immunologist proved to be crucial to the work we were about to begin. Knowing of my interest in neuroimmune connections because of the papers I had published, Bill Farrar called me one day to discuss his own work in that area. When I told him we were trying to map immune receptors in the brain, he offered to supply me with the antibodies we would need to help us find them. The next morning a tall blond bodybuilder type, dressed in shorts and sandals, showed up in my office, carrying ice buckets of antibodies—the delivery boy from Bill's office, I assumed (since scientists don't tend to look either so casual or so athletic). But in fact it was Bill himself. Once I saw him in action, it was hard for me to believe I'd ever mistaken him for a delivery boy, for Bill had a decisive and masterful way of getting things done, thanks to the years he'd spent as a navy fighter pilot launching fighter planes off aircraft carriers. For all my feminist leanings, I was intrigued by the idea of working with someone whose style and presence were so quintessentially masculine.

Some weeks after we'd begun our immune-receptor mapping, I was even more intrigued when Bill called to tell me that three different research teams had more or less simultaneously discovered the receptor that the AIDS virus used to enter and infect cells—the T4 receptor. The T4 was found on key lymphocytes in the immune system, called T4 or CD4 lymphocytes. A severe depletion of T4 lymphocytes is one of the signals of the presence of the AIDS virus, and also one of its deadliest effects, for it is the lack of these lymphocytes that makes AIDS victims susceptible to the normally benign microbes that cause their numerous and sometimes fatal opportunistic infections.

No sooner did Bill deliver the news about the T4 receptor being the entry point for the AIDS virus, than he began grilling me excitedly. "T4," he said, the words spilling out of his mouth. "I know I gave you an antibody that would bind with the T4 receptor. Have you used it yet? Did you find anything with it?"

"You bet we did!" I answered triumphantly. "And it went right for those receptors and lit up the brain like a Christmas tree."

Subconsciously, the significance of our T4 mapping began to dawn on me. If this receptor was the entry point for the virus in the body, then it must also be the point of entry in the brain as well. And if this was so, our expertise in receptor mechanisms could lend itself to a deeper understanding of how this process actually happened, and maybe even of how it could be stopped.

We also began to suspect that we might be able to use our knowledge of the virus receptors in the brain to help explain "neuro-AIDS," the dementia, memory loss, neuropathies (nerve degeneration), and depressions that were just starting to be recognized by neurologists and psychiatrists, who were now seeing more and more AIDS patients with these symptoms. Very little research attention had been given to this aspect of the disease. Since virologists and immunologists had no contact with psychiatrists, much less neuroscientists, their awareness of the growing phenomenon of neurological complications was limited, and what little they did know about it they tended to attribute to the understandable emotional depression of patients who were critically ill.

Now that we knew the T4 immune receptor was the entry point for the virus, we would focus our brain-mapping effort on it. We knew that no one else would be looking for immune receptors in the brain, because almost no one else even believed they were there—as was clear from events then taking place right across the way.

ENTER HIV

Within a few hundred yards of where we were working, in the part of the Palace where people looked only at the body and not the mind, a team of NIH immunologists and virologists at the Institute for Allergies and Infectious Diseases (NIAID) was following the newly discovered human immunodeficiency virus, HIV. Earlier, Dr. Robert Gallo of the Cancer Institute at NIH had made headlines when he announced that HIV was the cause of AIDS, a disease that had first been identified when it broke out quite suddenly in the male homosexual population. Gallo showed that the HIV was infecting the cells of the immune system, using their DNA to replicate and spread. As a result, the immune system was severely compromised, allowing opportunistic diseases to proliferate and eventually kill the host. So the NIH scientists, like us, were focused on how to prevent the HIV virus from doing its deadly work.

But their approach would, of necessity, be quite different from ours. With a few notable exceptions, among them the knowledge that the rabies virus used the acetylcholine receptor, virologists had never gained much understanding of how a virus gets into a cell. The process they most favored, *viroplexis,* was frequently described as the virus somehow glomming on to the cell surface and then fusing with the outer membrane to gain entry. The "glomming" step was a big unknown and not considered too important. Up until this point, virologists had been interested mostly in the molecular processes that regulated the reproduction of viruses—in other words, how did a virus replicate itself? And the answer, insofar as it was known, was that viruses replicated autonomously inside cells, where they could not be attacked without drugs that also destroyed the cell. Because of this, any cure that attempted to interfere with the replication of a virus after it had entered and "infected" a cell would be extremely toxic. Nonetheless, this became the focus of their research effort.

We could, however, go after the virus in a different way, for the question of how the virus could find and enter an immune cell was no stumper for a neuroscientist. We could easily understand how viruses might operate like exogenous ligands, binding, just like peptides, to specific receptors. Viruses were known to contain various proteins on their surfaces that were important in determining which cells they could infect. Thus, different viruses exhibit what we term a "tropism" for different cells, so that we would say, for example, that the HIV virus is T4-tropic. To a neuroscientist it made perfect sense that some of these viral invader proteins might resonate with some of the body's own molecular vibrations. In other words, we believed that there must be viral keys that could unlock receptor keyholes, and thereby enter the cell.

To see it under a microscope, the HIV virus looks like something out of *Star Wars,* a sphere whose surface is covered by hundreds of sharp protein spikes. It is this part of the virus, the surface protein envelope gp_{120}, that has a particular molecular sequence that allows it to latch on and bind to immune, brain, and other cells, initiating infection and, as we and a few others were to discover, many other *receptor-mediated* processes important in causing the signs and symptoms, indeed the disease, of AIDS.

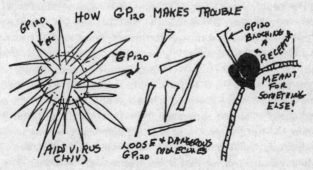

Once we had zeroed in on the T4 receptor, Joanna Hill, the skilled neuroanatomist of our group, generated gorgeous images of its autoradiographic distribution pattern in rat and monkey brains. And then one day shortly afterward, I received a chance call from Dr. T. Peter Bridge, an NIMH psychiatrist with an interest in psychoneuroimmunology, who was organizing a symposium on his new interest area, neuro-AIDS. "Do you have anything on AIDS and the brain?" he asked me. Slightly stunned by his prescience, I told him what we were up to. And it was because of that conversation that we received an invitation to present our work at the symposium on AIDS being offered at the 1985 American College of Psychoneuropharmacology Conference in Maui, Hawaii.

INTERVENTION

The descent from Haleakala had, if anything, been harder than the ascent. I hadn't expected the fatigue of the descent, thinking, "Ah, it's all downhill from here," but as anyone who has done this sort of thing knows, the faster pace down quickly puts the burn to the quads. So when I showed up for the meeting on the first day of the conference, my body ached exquisitely with the pleasure of a hard job well done. My mind, however, was unusually quiet as I listened to my colleagues present their findings about AIDS, realizing for the first time that the word *pandemic*, or global plague, was no exaggeration when applied to this rapidly spreading disease.

My knowledge of the disease had been limited to what I'd read in the newspapers, and, of course, I was familiar with the announcement by Bob Gallo's office that the cause of AIDS, the HIV virus, had been found by researchers in his lab. It had made quite a stir when Margaret Heckler of Health and Human Services announced that huge amounts of federal money were being allocated to the NIH for the battle against AIDS now that there was a known virus they could target. And, occasionally, news and gossip wafted across the lines, from NIH to NIMH, via Bill Farrar, who had one foot planted in each camp. But mostly I knew nothing, for reasons best left to the account in Randy Shilts's book *And the Band Played On*, where it was explained how little information the public had access to at that early stage. Now I was watching a slide show depicting the terrible agonies of AIDS sufferers and hearing how the virus destroyed the immune system and ravaged the body, exposing its host to any number of rare but lethal opportunistic infections. For the first time, I began to think about the human cost of this disease, and a sense of urgency came over me, a strong desire to make some kind of contribution to the efforts of researchers to understand and treat it.

When it finally became my turn to speak—I was the last one on the program—I stepped up to the podium and presented our findings. I described how we had found a T4-like molecule in our brain mapping, with high densities in the hippocampus and cortex. The slide Joanna had made of the monkey brain came up, projecting the colorful pattern T4 made in the brain for all to see. As I gazed at it admiringly, I suddenly became aware of a curiously altered state of awareness. I began to speak, but my voice sounded strange, as if it were coming from a distance.

"Our data clearly suggests the T4 receptor could be a neuropeptide receptor, since its pattern is reminiscent of the patterns of known brain peptide receptors," I reported, the words echoing as they left my mouth. And then, following that, seemingly from out of the blue: *"If we could find the body's natural peptide ligand that fits the T4 receptor, it could yield a simple, nontoxic therapeutic to stop the virus from entering the cell."*

There was a hushed silence as both my audience and I let these remarkable words sink in. Had I just proposed a discovery path for the cure for AIDS? It was the very first time such an idea had occurred to me.

And then I heard a louder voice, this time not recognizable as my own and not spoken out loud, but echoing inside my own head! It was a strong male voice that commanded: *"You should do this!"*

I certainly wasn't used to hearing voices comment on my lectures, and at first I assumed the command was some kind of hallucinatory aftereffect of the exhausting hike up the volcano. But the logic of the approach was so compelling that I trusted that voice. Even the fact that it was distinctly male didn't rankle my feminist sensibilities, because whatever that voice was—hallucination, voice from God, my own higher wisdom—I knew exactly what it was telling me to do! My whole scientific career, it seemed, had been a preparation for answering the question I had just proposed to myself: What natural peptide fits the HIV receptor in the brain and in the immune system, and how can we make a synthetic version of it to block the receptor and thereby stop the entry of the HIV virus?

Hadn't this been the route followed once before, when we all went looking for the enkephalin/endorphin discovery? The CD4 receptor wasn't there exclusively to bind to the virus, any more than the opiate receptor existed to bind to morphine! It made perfect sense, and I was mystified that it had never occurred to me before that moment. My thinking was now traveling down what was for me a well-trodden path of theoretical trailblazing, which had begun with the discoveries of the opiate receptor and its endogenous ligand, endorphin. Just as before, we had a receptor (T_4), and now we would go looking for its unknown ligand. But this time we would use a computer to help us find it.

I could barely wait until the next morning to call the mainland and have my lab set in motion the process needed to begin the quest. Bill Farrar was on hand to direct the computer that would help us search a worldwide peptide database. We were looking for a molecular sequence identical to that of the external viral envelope of the HIV virus, the gp_{120}, which was the part of the virus that fit into the receptor.

The identification of a receptor entry path for the virus, and the subsequent description of the T_4 receptor, had suddenly opened up many new avenues for AIDS research. Thus, there were soon a number of government and industry teams doing the same thing we were doing—looking for pieces of gp_{120} that bound to the T_4 receptor—but doing it much more indiscriminately. Since their method involved making nearly every possible peptide permutation, and since there were over 600 positions for amino acid candidates to be filled in the sequence, their chance of success was analogous to throwing 100 monkeys and 100 typewriters up into the air and waiting to see if, eventually, the complete works of Shakespeare would appear, printed out perfectly to the letter.

Not that our path was as direct as we had hoped it would be. We had thought that if we designed some well-thought-out computer searches, we would quickly arrive at the sequence that would enable us to identify the one natural ligand; but what we got was not nearly so clear-cut. The computer identified multiple sequences of other proteins that closely matched the gp_{120} sequence, none of which, alas, were underlined in red with a note that said, "This is the one!" We were going to have to put our seat-of-the-pants feeling for the material into action so that we could narrow the search down to just a few of the numerous candidates for synthesis and further testing. We'd simply have to hope that one of our hunches would pay off in the discovery of a substance that would act in place of the still-unidentified natural ligand, and that it would be a good enough mimetic to successfully displace the gp_{120} from the receptor, bumping it off at the point of entry.

So one night, I brought home the printout of all the possible sequences our computer had generated and spread it out on the dining room table—but not without a sense of foreboding about what would happen if we chose the wrong one. Pharmacology is an exacting science, and there are no "almosts" or "close tos." If one letter of the sequence was wrong, or we missed one, then our synthesized substance would be useless to us, failing in any tests or assays we would do to prove its abilities to displace the virus. We could be very close but never know it.

Together Michael and I pored over the pages of letters for a week. It was Michael who finally made the decision to focus on an octapeptide contained in the Epstein-Barr virus, another "lymphotropic" virus, albeit for B cells, not T cells, which we guessed, rightly or wrongly, might use the same receptor as HIV. It didn't, but as we were eventually to discover, we had picked a winner, having gotten to the right place for the wrong reason.

I placed a call to my old friend Dr. Jaw-Kang Chang at Peninsula Labs in San Francisco. It was New Year's Eve, and, as I suspected, he was working late in his lab. In a near-replay of what had happened ten years earlier with enkephalin, I carefully read him the sequence of the eight-amino-acid peptide we had picked and asked him to synthesize it for me. Since the first amino acid in the sequence was alanine (just like the one I had asked Chang to change to make the long-acting version of enkephalin), I instructed Chang to make three D-alanine analogs for a total of four closely related octapeptides. Again, I swore him to secrecy and told him he'd have to do it without asking too many questions. Chang accepted the challenge, and two months and $10,000 of the taxpayers' money later, we had the four synthesized peptides in our hands, enough to begin our tests.

Bill Farrar was the broker for a deal with a lab in Frederick, Maryland, that had made the HIV virus and separated out its various protein components, essential ingredients that we would need. He got us the viral protein we needed, the gp_{120}, which we then had labeled with radioactive iodine. Now we had both of the crucial elements on hand, the synthetic peptide and the radioactive viral protein, and we were ready to begin our receptor-binding assay to see what these peptides would do.

I remember feeling excited to finally begin the experiments, but also apprehensive, as if I were about to dive into a swimming pool drained of water. Everything about this project, it seemed, had depended on some hard-to-fathom combination of intuition and/or mystical intervention and/or sheer good luck, all of which were somewhat suspect to my scientific mind at the time. The direction had been dictated by a voice in my head while I stood at a podium in Maui, and the choice of sequences we'd made, while based on a firm rationale, had involved a heavy dose of intuition, a fact that other researchers would either marvel at for decades to come if we were successful, or ridicule mercilessly if we were not. Now the moment of truth was upon us. Would our magic peptides do the deed and prove us right?

Michael and I distributed half of our synthetic peptides to collaborators Frank Ruscetti and Bill Farrar, who were working at one of the Cancer Institute's many labs, and we set to work using the remainder ourselves. The goal of Frank and Bill's assay was to show that the peptides blocked the virus from growing in human cells. Frank was the only person other than Bob Gallo himself who had access to the actual virus from fresh patient isolates, not the old stuff passed around in lab cultures for many years. This would prove to be a crucial advantage to us, because while other researchers were enticed by the ease of using these lab-adapted strains, they often ended up wasting much effort, resources, and precious time studying what would only years later be revealed as artifacts—that is, phenomena that were not natural to the virus, that were not the way the virus behaved in humans.

The receptor-binding assay our lab was developing was aimed at determining whether the test peptides could actually block the attachment of the viral protein envelope gp_{120}, even displace it from the receptor on the T cell. If it did, we had a potential therapeutic, a new drug that we could begin to develop as a possible treatment for AIDS.

The hands for the assay belonged to Robbie Berman, a newly graduated Yale man spending the summer at the Palace, before going on to medical school. Robbie came into the lab every day and set up the test tubes, pipetted the many different ingredients into them, ran the experiment, and then brought me the numbers. He was brilliant and secure, performing every tiny step in the assay with the precision I required, enduring many hours of my thorough cross-examination about the daily progress of the experiment. He was as young and energetic as a typical graduate student but as smart as a postdoc, and best of all, from my point of view, he lacked the kind of oversensitive male ego that required an excess of diplomacy. We were able to work very closely together as he cheerfully tolerated my leaning hard on his shoulder barking out minuscule technical changes, something that would have caused just about any postdoc to get up and exit in a huff.

Before Presidents' Day weekend of 1986, Robbie and I did the key experiment. Together we dissolved the peptides into a solution of radioactive gp_{120}, using a number of different concentrations, and allowed them to react with the T4-laden membranes. Since we had a three-day weekend before us, and we were apprehensive that the counts would be so low that we would have trouble getting a meaningful signal off them, we decided to set the counter to counts of twenty minutes for each filtered sample—much longer than usual. It was a luxury to be able to do such a patient, careful measure, and, as it turned out, one of several instances of good luck.

On Tuesday morning, I arrived early and eagerly pulled the tapes from the counter, scanning the numbers. It took only minutes to realize that we had something here. The counts showed that our peptides had knocked the binding of gp_{120} down to half, competing with the HIV for the receptors, just as we had theorized they would. What was particularly exciting was that the data showed that while three of our four peptide analogs worked, the fourth was almost inactive. This was a crucial test, because it demonstrated some specificity and selectivity in the binding inhibition, the hallmark of a receptor-mediated event.

Just hours after our Eurekas had died down, news came from Frank and Bill's lab that the peptides had also inhibited the virus from growing in human cells in test tubes. There was about 80 to 90 percent effectiveness. But Frank, who didn't get excited too easily, was quick to point out the apparent downside of his results. "Well, Candace, three worked and one didn't." Even the fact that one hadn't worked turned out to be good news for us, however. When we compared the data from the two labs, it was clear not only that our most active peptide was also his most active peptide, but that a different peptide was also inactive in both his and our assays. This type of result, showing comparable relative potency in two very different labs using two completely different methods, was the old standard for a receptor effect, and it clinched it for us. We knew we really had something.

We were ecstatic that our prediction had seemingly been confirmed! Perhaps we had found the substance that would prevent HIV from entering the cell and replicating. And a bonus surprise was that the concentration of peptides it took to occupy the receptor and accomplish these results was astoundingly low. In fact, this computer-generated synthetic peptide compared in sensitivity with the most potent neuropeptides themselves. Later, when we did the calculations, we found that an effective concentration was equivalent to an aspirin tablet dissolved in the amount of water contained in a railway tanker car. We named our child of the new paradigm Peptide T, the "T" referring to the presence of threonine, the dominant amino acid in the synthesized substance's molecular sequence.

We were so flush with the success of our experiments, we'd completely forgotten what our original goal had been—the identification of the endogenous ligand, the body's own natural substance that binds to the HIV receptor on brain and immune cells. We had found our mimetic, Peptide T, and our direction seemed clear: We should publish our findings and test our new therapeutic through human trials. However, the search for the endogenous ligand was not over, even though we ourselves had gotten distracted from it. A few months later, our colleague Ed Ginns, a molecular biologist whose help we had sought, found the ligand in a Peninsula Labs catalog. He'd been flipping through the pages that listed peptides available from the manufacturer when he came across what he recognized as a sequence identical to that for Peptide T. Right there, contained within the printed sequence for one of their catalog peptides, was what we had been looking for—VIP, our vasoactiveintestinal peptide.

As it happened, we knew quite a bit about VIP. It's found in the frontal cortex of the brain, in the thymus gland, the gut, the lungs, some immune cells, and parts of the autonomic nervous system. Eventually, we would come to understand how the HIV virus competes with VIP for receptors on the surface of immune and brain cells, and some other cells as well, latching on when the VIP molecule is not occupying them. Just how much VIP the receptor is "dripping with" at any given moment will affect the system's susceptibility to infection at that time.

Much later, I was to speculate about what emotional tone VIP was associated with. Could a particular emotion generate or suppress quantities of VIP in the body, affecting how much of it was available to block or leave open a pathway for the HIV to enter the cell? Clinicians have the impression that increasing self-esteem seems to slow the progress of the disease. This leads me to speculate that VIP might be the hormonal manifestation of self-love, just as endorphins are the underlying mechanisms for bliss and bonding.

OBSTACLE COURSE

The next challenge was to describe our Peptide T findings in a short, concise paper and have it reviewed for publication in a scientific journal. We hoped that this would be easy, a swift fait accompli, to allow us to go on to our next major step, developing the drug through human trials. But we were also aware that our approach was radically cross-disciplinary and would be difficult for reviewers to understand. Also, we had based our work on concepts not altogether acceptable to most immunologists and virologists at the time: that the brain and the immune systems have many cell-surface receptors, and that viruses use these receptors to enter the cell.

Believing we had a piece of an enormous discovery on our hands, we decided to try to position our paper in the most prestigious and widely read journal, *Proceedings of the National Academy of Science*. PNAS is the journal published by the National Academy of Sciences, a federally funded organization founded in Abraham Lincoln's time that still retains many customs dating from that era (including the opportunity to blackball potential new members, a practice many believe encourages the cronyism that is its hallmark). One of the academy's more antediluvian traits at the time was that only about 2 percent of its members were women.

PNAS is a slick and beautifully produced journal that has a very high impact, as witnessed by the number of times its articles are cited in other journals. Through an elaborately cautious procedure that supposedly discourages favoritism, members of the academy can submit a limited number of papers to the journal each year, their own or those they think important and worthy by others. But despite the apparent safeguards against favoritism, having a few academy members as your friends is the only way to guarantee quick, prestigious publication—provided one of them can be persuaded to relinquish one of his own precious slots.

What we needed was an academy member who would agree to evaluate our paper and then find two reviewers to give it the nod, at which point it could be submitted for publication. We had tried to do this once before with our earlier paper showing how we'd found the T4 receptor in the monkey brain. But that paper was still circulating unpublished, having made the rounds with no success. In fact, it had been rejected in a particularly humiliating fashion by someone for whom I had had great respect. Frank Ruscetti had suggested we give it to the virologist Albert Sabin, a visiting emeritus scientist at the NIH, who had years ago created an oral polio vaccine that became more popular than Jonas Salk's injections. I was eager to meet Dr. Sabin and naively expected that he would be more than glad to read our paper and wave us on into the journal. I sent the paper over to him by messenger and, two days later, with

Michael along as coauthor, visited the famous doctor in his tiny office located in the basement of the NIH library.

Remembering how my children had been inoculated with the Sabin vaccine, I gushed about how honored I was to meet someone who had a human vaccine named after him. Sabin accepted the praise but then, without warning, began to tear into our paper. In a mounting harangue, he proceeded to critique it, frequently referring to his handwritten notes, in terms that made no scientific sense to either Michael or to me.

Finally, his rant drawing to a close, he said, "And what's this about virus receptors in the brain? We cured polio without needing to invoke virus receptors in the brain—or anywhere else!"

To Sabin, this new idea was flatly unacceptable. He slid the paper across his desk and, with unconcealed scorn, announced he positively would not recommend it for submission to·PNAS. At this point I lost my ability to fight back tears, so huge was the gulf between what was happening and what I'd expected. I signaled Michael to leave and got up to make my way to the door, when suddenly Sabin's mood shifted. Seeing a tear roll down my cheek, his face lit up, and he even chuckled as he escorted us out. "I can't believe it, I made Candace Pert cry!"

Eventually, I got over the shock of Sabin's personal hostility to me, and I was able to forgive him for the extremity of his reaction. This happened when I realized what an affront our paper must have been to him, a slap in the face to his somewhat tentative status as the man who had bested Jonas Salk. But back then, this bizarre event had only left me hurt and confused.

In looking for a champion for our new paper, we decided to ask for help from Fred Goodwin, my boss at the Palace after Biff left. Fred had been closely following my work and generously supporting my lab for several years, and he instantly understood the concepts proposed in the paper; but he also recognized that our more specialized peers probably would not be able to do the same. To the greater scientific community, our paper would seem like it came from the Tower of Babel, a hopelessly multilingual report, and its significance would be missed unless it was guided through the process by someone who Fred had under his direct influence. With that in mind, he suggested we seek the endorsement of one of the few NIMH scientists who was a member of the academy, a well-known neuroscientist who had made his mark with the first functional brain scans. ·

We sent the paper to our new potential champion, but after many weeks we had still heard nothing from him. However, it wasn't unusual for the chosen reviewers to take up to eight weeks before responding to a paper, and since this was the case, well, we'd just have to be patient, Fred reminded us. Meanwhile, Michael and I prepared for our rapidly approaching wedding in Lyme, Connecticut. Our hope was that we'd get word about our paper before departing, just in case we needed to be on call for any changes or requests to fix parts of it.

The silence continued. We pleaded with Fred to prod the foot-dragging scientist, to find out if the reviews would be in before our June 7 wedding, something Fred did with reluctance. Our plan had been to leave D.C. and drive to Lyme in plenty of time to get our marriage license at the Town Hall and supervise details of the elegant wedding on Saturday. But we didn't want to leave until we got the long-awaited response to our paper.

The wedding was to be a dream come true, the wedding I'd never had, since my marriage to Agu had been arranged rather hastily, a no-frills event. Michael and I had organized an extravaganza, complete with engraved invitations for over a hundred people, fancy tents on the lawn, and elegant catered eats. Many hours of planning had gone into it, and we looked forward to thoroughly enjoying every minute of the result.

At the last possible moment, two days before the wedding, the much-anticipated phone call came. Could we come over immediately? We went to meet with our potential champion at his office, feeling hopeful and certain that this long delay could only mean our paper had passed muster with the two required reviewers. But only minutes into the meeting, the doctor began to sputter and things took a dive.

"Virus receptors, virus receptors," he bellowed, his face dark red, spittle flying. "No one I know has ever heard of a virus receptor!"

In what was an incredible instant replay of our earlier encounter with Sabin, he made it perfectly clear, as he shoved the paper across his desk at us, that he wasn't a virologist and couldn't possibly submit the paper to the journal. This time I shed no tears over the hostility directed at our efforts.

Later that same morning we left to drive north, fuming that we'd put everything on hold only to discover that our paper had simply been gathering dust all those weeks. We stopped to shop for my wedding dress and trousseau at the White Flint Mall, arriving in Lyme with just one day to get our marriage license. When we got to the Town Hall, a clerk abruptly informed us that in this township there was a four-day wait between application and issuance of the license, a fact that threw us into a panic. Should we just go ahead, have the wedding illegally and fake it? This kind of sham didn't appeal, especially since I had wanted things to go so perfectly. In my pleading, I mentioned that my uncle Bill Beebe was the town treasurer, a fortuitous connection that enabled our by now totally sympathetic clerk to help us out with the higher-ups. In the end, a judge wrote a special edict to enable us to skip the wait, and Uncle Bill, the church choir director, who was a talented musician as well as our savior, played "Somewhere Over the Rainbow" on the organ for our walk down the aisle.

We spent our honeymoon in Provincetown, on the tip of Cape Cod, biking in the rain. It was blissful, a much-needed retreat, but as we passed through the streets of this community, long-favored by gays, I couldn't ignore the many ravaged faces and bodies I was seeing. The knowledge that we had a contribution, one that could possibly lead to a treatment for their disease, if only we could somehow get out the gate with it, was immensely frustrating. During one long ride along the sand dunes, we saw a rainbow, an image that had also been taken up by the gay community as a symbol of pride and unity, and again I felt affirmed in my quest. It seemed that the rainbow had been following me since the beginning of my career, and now it was even more visible, a symbol for others as well as myself, a symbol for those who could benefit from the work I was doing.

On our return, Fred was apologetic about not having been able to persuade his man to fall in line with our plan to pull off a quick and clean publication. But, beyond that, he had no other suggestions for us. As noted earlier, the NIMH has very few members of the academy, which may seem odd for such a prestigious institution, but scientists have a longstanding prejudice against psychiatry and psychology, a reluctance to acknowledge the behavioral sciences as authentic science, and they admit only a small number from those disciplines to their pantheon. Hence, Fred had few personal connections within the academy.

We moped for days, but our spirits took a turn for the better when we found that awaiting us in our pile of mail was an invitation to the NIMH's fortieth birthday party, an event that was to take place on June 26, 1986, the exact date of my own fortieth birthday! The fact that the NIMH had been created through an act of Congress on the very day I was born made me feel that the birth of the neuroscientific approach was closely aligned to my own birth, which gave me a renewed sense of hope. Furthermore, this event would bring me together with my old mentor, Dr. Sol Snyder, a longtime academy member and, perhaps, if he had thawed sufficiently from the Lasker incident, a potential benefactor—the key to getting our paper published in the academy journal.

The NIMH party was an elaborate bash with a sumptuous spread of food and plenty of awards being handed out. I spotted Sol almost immediately and approached him at the very first chance. We made the usual small talk, tense but cordial, at least on the surface. Then I decided to come right to the point, and I told him of the trouble we'd been having with our paper. He listened politely, but when I asked him to take a copy of it, he put both hands up and backed away, shaking his head, pleading ignorance about virology and saying he couldn't possibly do an evaluation. I stood there alone, feeling embarrassed and hurt, and contemplated the significance of this latest rejection.

Had it not been for the Lasker incident, I believe Sol would have leapt at the chance to help his former student advance a new drug using receptor theory, especially one discovered using an adaptation of the radioreceptor assay we had developed together. It was hard to swallow, but I had to accept the fact that my earlier actions had cost me Sol's support at a time when it might have made all the difference. Yet, in spite of this regret, I knew that had I never been handed the challenge of struggling back from ostracism and disrepute, I might never have gone in the direction that led me to the discovery of Peptide T.

Clearly, this was turning into more than the usual effort to get a controversial paper published.

Michael and I were now completely out of ideas about potential sponsors for our paper. One night, to divert ourselves from what we hoped was a temporary impasse, we rented the video of *Amadeus*. In the film, the genius Mozart is given a review by his peer, the jealous musical expert Salieri, who pronounces his latest composition as having "too many notes." It struck us that the problem with the Peptide T paper was that it also had too many notes, causing the "experts" to find it too unfocused to comprehend. Most papers reported on only one or two facts, allowing the writers to stretch out the data and publish two or three more papers down the line. But in our case, thinking we might have just this one shot, we had wanted to get the whole story succinctly and efficiently told in the five-page limit, and so had packed the paper full of details. In the paper, we had included the color illustration of the distribution of T_4 receptors in the monkey brain, a brief description of

214 ≈ MOLECULES OF EMOTION

how we arrived at Peptide T's molecular structure, and figures showing the drug's ability to block access of the virus to the receptor, as revealed by the binding experiments Michael and I had done and by the infectivity experiments Frank and Bill had done. Finally, and possibly most annoying of all to our peers, we had included a brief discussion of how the synthesized peptide might be used as a potent antiviral therapeutic to prevent the HIV virus from entering cells. What we needed in the way of a reviewer was a Mozart, for whom too many notes were not a problem!

It was Carleton Gajdusek, a Nobel laureate from the National Institute of Stroke and Neurological Disease, an NIH, not NIMH, academy member, who finally did the deed for us and got our paper into *PNAS*.

Even though he resided in the "body" section of the Palace, Gajdusek was a pediatric neurologist who specialized in diseases of the brain, especially the viral kind. Michael knew the renowned professor from his undergraduate days, when, during a visiting lecture, Gajdusek had regaled him and his classmates with tales of virus hunting in the South Pacific. I'd never met him myself, but from what I knew about him, he was a genius who had already ascended to science's highest ranks, someone with no particular interest in the AIDS arena and no political agenda, and therefore he would probably at least agree to give it a look.

I took a deep breath and picked up the phone. When he answered, I told him who I was and that I had a paper I'd like him to evaluate for possible submission to *PNAS*. He asked a few intense but brilliant questions about the content of the paper, and then, after a brief silence, responded.

"Yes, definitely," he said firmly. "Get me the names of scientists you know who can understand it and review it for errors. I'm flying off tomorrow but I'll be back in a couple of weeks."

I exhaled. The whole transaction had taken less than a half hour.

As promised, he had the paper reviewed, and when the science of it was confirmed as acceptable, he submitted it. We received notification of its official acceptance less than two weeks after my phone call to Gajdusek. By September, one month after it was submitted, the paper went to press, slated for publication in December 1986. We had found our Mozart and solved the dilemma of the paper that had too many notes.

TRIALS

It's to the credit of the Palace to say that Peptide T could only have been invented there. Only at the Palace was there a critical mass of free-flowing money, brilliant minds, and state-of-the-art equipment all assembled in one location. The irony was that the Palace, having given birth to a drug like Peptide T, would never give it the support it needed for full testing and development. The reasons for this were many, some related to my own tactical errors and past history, others the result of the harsh realities of Palace politics and government funding choices. But underlying all of it was a very fundamental but less visible drama—the shift from old to new paradigm. Conceived by believers in the mind-body connection, Peptide T was truly a child of the new, more holistic paradigm. And that was a big problem for a large establishment institution.

The reigning paradigm held firmly to the denial of any meaningful connection between mind and body as they pertained to health and disease. Itself a product of this old-paradigm thinking, the Palace in its institutional structure mirrored the Cartesian split: The NIMH attended to everything above the neck, while the larger, better-funded NIH took care of everything below. And although there were occasions when the twain did meet, those were the exceptions, not the rule. AIDS was a disease of the body, and it was the body boys at the NIH who would be trusted to come up with a treatment.

My sojourn into the AIDS arena in some ways paralleled my earlier foray into the world of cancer research, when we failed to convince the narrowly specialized field that neuroscience had something to offer in developing treatments for cancer. Now I was again confronted with the deeply ingrained theoretical division, not only between body and mind but between separately studied systems of the organism. Only this time I was joined by other scientists in my efforts, an interdisciplinary team of some of the brightest and most forward-thinking researchers at the Palace, many of whom were willing to venture out and cross lines. Still, I was playing in a much bigger league now, dealing with political funding issues that we hadn't come up against before. Huge amounts of money were coming down the pipeline for AIDS research, and to get a piece of the pie, we needed the goodwill of powerful people in high places, something we soon found out we did not have.

•   •   •

FOR MONTHS I had been banging on Fred's door, trying to get in to see him and discuss moving Peptide T to the next level, the phase I clinical trials. Fred's support of Peptide T had been unwavering until a moment of truth, which occurred early one Saturday morning during a government budget meeting when the director of the Institute of Allergy and Infectious Diseases took some $11 million he had put into NIMH back out of Fred's pockets. The rationale for this sudden withdrawal, the director informed him, was that development of antiviral AIDS therapies was not the business of the NIMH, and Fred certainly didn't need that much money to pursue a nonstarter like Peptide T.

And Fred was avoiding me because he already knew what I learned only later, which was that the NIH and the Cancer Institute had their own candidate for an AIDS treatment, a highly toxic, very conventional, but promising, drug called AZT. They'd already reserved the beds in the clinical center for the trials they planned to conduct, and there wasn't an inch of space left over for anyone else, even in a matter of life or death.

AZT was a chemotherapeutic drug, originally used for cancer in the early sixties, which worked by terminating virus replication. But the price to be paid for this benefit was a big one: AZT undermined a person's health by destroying not just the virus but also healthy cells, particularly those of the immune system. The resulting side effects, or toxicities, were sometimes severe, although some patients could handle them better than others. AZT showed some promise in reversing symptoms of AIDS and giving patients valuable time, but it could not be considered a "cure," since, in a process similar to what happens when chemotherapy is used for cancer, it causes the virus it attacks to eventually develop resistance so that the disease usually recurs.

Unaware that the Palace had decided to concentrate totally on AZT, I continued to spin my wheels, trying to find out whom I needed to talk to and what forms I was supposed to fill out to enable us to proceed to the next logical level. I was getting nowhere fast.

In retrospect, I see that in my haste to effect a swift and clean entry onto the turf of the body boys, I had failed to court favor. Perhaps I should have gone begging, hat in hand, to an NIH AIDS power boy for help in getting our paper published. But my thinking was that the sooner we got our paper into print, the sooner testing could begin on people with our new drug, and I didn't want to waste time by doing what I thought of as massaging male egos.

How incredibly naive I was! And the hubris I had! Who was I but a fairly accomplished bench researcher oblivious of the fact that testing a new drug was the big time—*showtime!*—for most scientists, and I hadn't even read the script. Stumbling into a scene where major players were about to sit down and carve up the funding bonanza AIDS research promised to be, I was clearly an intruder. It is understandable, in retrospect, why the arrival of Peptide T, coming out of left field, was not greeted with enthusiasm.

Up until that point, some twelve years into my career at the Palace, I had been protected, first by Biff and then by Fred, from the nastiness of the political process that led to the distribution of the government's money among the competing agencies of NIH and NIMH. On occasion, Fred would trot me out during visits by congressional aides to perform the perfunctory dog-and-pony show with my rainbow slides of brain receptors. But other than that, I'd been left to do my science, becoming a totally impractical, head-in-the-clouds, full-time scientist, unbothered by monetary concerns and good only for making breakthrough discoveries. I had wandered freely through the corridors and buildings of the Palace, becoming multilingual, absorbing all the science, talking, listening, observing, becoming totally lost in a fantasy-come-true of scientific heaven.

Now I was trying to enter a whole new playing field. Clinical trials involve millions of dollars, the futures of entire companies, the clash of many, usually male, egos, the necessity to thread your way through the intricate maze of the FDA, and a political sense that was entirely foreign to my direct, honest, in-your-face style. I wasn't even an M.D., and it's almost always M.D.s who are at least nominally in charge of clinical trials. In sum, I was completely unprepared to deal with the kind of real-world, big-business wheeling and dealing that was necessary if I was going to have any hope of having a direct impact on people's health. Just as before, when I had been seen as invading the cancer lab director's turf, I was trying to stray beyond the traditional confines of my world without understanding how rigid the turf boundaries were, and how threatened people became when anybody made a move to cross them.

RESIGNATION

The final blow came in June 1987 when a Harvard researcher with a private biotech company affiliation announced during his talk at a major conference that Peptide T could not be effective as an AIDS therapeutic and should not be tested. In a perfunctory manner, he closed his talk by showing three quick slides on Peptide T, and then explaining how he and several of his colleagues at the NIH had failed to replicate its antiviral effect in vitro.

The effect was like a bucket of ice water being flung in your face just as you're waking up from a long cozy nap. Michael and I both snapped to attention. *Failure to replicate*—the scientific kiss of death! All it took was a few brief words by an "expert" who claimed he and his colleagues couldn't replicate our experiment, and Peptide T was dead, killed before it ever got out the gate. The press was at this particular conference and grabbed on to the news for their evening headlines: "Novel AIDS Drug Bites the Dust," read one. "Experts Say Promising AIDS Drug Doesn't Work," said another.

Initially, putting our shock and confusion aside, we decided to treat it as a collegial disagreement. First we addressed the obvious questions: Why had no one called us up or walked the few hundred yards across the quad to tell us about their results? And what exactly were they finding that caused them to pronounce "a failure to replicate"? The objection, we found out, was against our claim that Peptide T stopped the HIV virus from growing in the test tube. But the harder we pressed, the less we understood, until it dawned on us that maybe the experiment hadn't been performed correctly.

When we finally got to examine the data, we discovered that the other labs had not closely followed our steps. They had increased the concentration of the virus by a factor of 100,000, while keeping the concentration of Peptide T at the level we had used. They had also used "lab-adapted" cultures of the virus, not the fresh isolate that Frank had gotten us from blood taken directly from AIDS patients' veins.

How could this have happened? Was the failure-to-replicate pronouncement the result of a simple mistake rooted in an old-paradigm blindness that made an unbiased approach impossible? Or, a more disturbing possibility, was it a crude ploy at effectively trumping a pesky competitor out of the race and thus eliminating a potential funding threat? Or perhaps the Lasker controversy was still dogging my heels. I am not prepared to say. But one thing I know for sure, the hoopla amounted to a death sentence for the further development of Peptide T within the Palace walls. Only after certain of the experts' favored approaches bombed and technology had advanced enough to expose the limitations of many of the methods used in 1985 would the search for virus receptors and their natural ligands be reinitiated—some ten years later. For the time being, I was going to have to confront the hard fact that in order to continue with my new-paradigm child, I needed to look elsewhere than the United States government.

It was in the depths of our discouragement, disgust, and disgrace that we received a very intriguing phone call. I listened to the voice of a fast-track biotech lawyer, who had learned of our predicament from his NIH scientist wife and had taken it into his own hands to step in.

"Dr. Pert?" he began. "We hear you have a miracle cure for AIDS that the government won't develop. Is that true?"

He went on to give us details of a proposition that he believed would make us all wealthy. In one hand he held a multibillionaire private investor and his group of venture capitalists. In the other hand was the Second Biggest Drug Company on the Planet, which just happened to be shopping around for some new products to add to its AIDS line. The drug company, of course, had tested Peptide T in its own labs, replicating our experiments and finding our claims to be 100 percent accurate. If we would just say the word, our savior promised, he would clasp these two hands together in a business handshake that would seal the deal and give us whatever we needed to develop our drug.

IN AUGUST OF 1987, one year from the date Gajdusek had submitted our controversial paper to *PNAS*, I handed in my resignation. Following the established protocol, I met my lawyer at the entrance to Building 10 and we rode the elevator to Fred's office suite, where Form 52 would be finalized. In a brisk and businesslike manner, Fred's secretary handed the papers to Fred, who passed them on to my lawyer, who gave them to me. By signing on the line, I was about to end my tenure and stroll away from the best deal in science that exists anywhere, the chance to work at the NIH. But I didn't hesitate for a second. I was so determined to carry Peptide T forward that had my dead father appeared in a vision and pleaded with me to reconsider, I would have ignored his wishes and walked away without a single glance backward.

11

CROSSING OVER, COMING TOGETHER

SPRINGTIME IN Washington, D.C.! The place is afloat on pink and white cherry blossoms, the air filled with elation as the East Coast emerges from the winter of 1995–1996—the worst of the century. For weeks we were all buried in mountains of snow, unable to leave home, go to offices, carry on any semblance of normal everyday life. But this morning, the day after Easter Sunday, I noticed the daffodils in my front yard finally starting to open, weeks later than usual. I am heartened, even though the early-morning TV weather report had cast a shadow by forecasting the outrageous possibility of more white stuff coming our way.

From my office at Georgetown University Medical School, where I am currently a professor of research, I place a call to the office of Dr. Wayne Jonas, newly appointed director of the Office of Alternative Medicine (OAM) at the NIH. My purpose is to see if I can drop by for a few moments and, in the spirit of hope and new beginnings, pay a long-overdue visit to him. The OAM was established four years ago by the NIH to investigate and evaluate many of the alternative therapies and practices—including acupuncture, homeopathy, manipulative healing techniques such as chiropractic and massage, visualization, and biofeedback—that have become so visible over the past ten years that the mainstream can no longer ignore them.

The creation of the OAM is a sign that the NIH is finally catching up with the American public, which is well aware of the benefits of alternative medicine, as shown by a study done by David Eisenberg at Harvard. In an article he published in the January 28, 1996, issue of *The New En-*

gland Journal of Medicine, he showed that one out of three Americans had used at least one unconventional therapy in the previous year, for which they spent $13.7 billion, three-quarters of which came out of their own pockets, not their insurers'. His study prompted a few of the smaller insurance companies to include alternative therapies in their benefits, but the majority of insurers offer no such coverage.

The media has also "discovered" alternative medicine, it seems, as evidenced by a rash of articles and TV shows on the topic, especially in the last year. An article in the *Washingtonian*, featuring my friend Dr. Jim Gordon, a D.C. psychiatrist and adjunct professor at Georgetown, who emphasizes prayer, yoga, and juice fasting, showcased the increasing number of local mainstream physicians who are combining alternative treatments with more Western, allopathic approaches. The article got the attention of quite a few people on Capitol Hill, a hopeful sign, since they hold the purse strings to the funds for all the research done at the NIH. Still, I wonder if all the publicity hasn't contributed to the backlash I've sensed from talking with some of my former colleagues at the Palace. Lately, people seem afraid to discuss any possible implications their research might have in helping to understand the mechanisms of alternative medicine. It seems, at least from an insider's view, that while the public is fed glowing reports of the advances of alternative medicine, it only grows harder for the OAM to gain acceptance in the highly resistant mainstream environment at the NIH.

Thus, I'm quite surprised when the receptionist answering my call gives me the address: Building 31 on the NIH campus. Less than a year ago the OAM office was located off-campus, purposefully kept outside the Palace walls for fear that its very presence might somehow taint the purity of the "real" science being done by establishment scientists on their hallowed grounds. This new move seems a promising sign of the fledgling office's progress and acceptance in some quarters, and perhaps its tiny budget, which currently constitutes less than a tenth of one percent of the NIH's funding, could someday be increased if acceptance grows.

There's a familiar chill in the air as I drive through the cherry blossom–lined streets of Bethesda on my way to the NIH, intent on my mission, which is to do whatever I can to help the OAM get a stronger toehold at the NIH. Specifically, I have some ideas about how to bridge the research gap—a phrase I use to refer to the lack of basic, data-gathering laboratory research being done on alternative therapies. This is the missing piece that I believe is necessary to legitimize what is now still severely marginalized.

In my capacity as chairperson of an OAM study section on mind-body medicine, a position I had undertaken at the behest of the previous director of the OAM, Joe Jacobs, I had had the opportunity to review many studies in the field of alternative medicine, finding good data, some as solid as anything in the mainstream, to show that the mind and emotions can influence immunity. The experience made me ask the serious question: If we know thoughts and feelings can influence disease, why aren't we doing more expanded hard research to determine which diseases these modalities are most applicable to, and performing the experiments that could lead to answers and possible cures? Guided imagery was one of the modalities on which, to my great surprise, I had found studies proving beyond a doubt that it could influence recovery rates for patients with cancer. Why, then, weren't these studies being followed up?

Acupuncture, too, looks very promising despite its having been dismissed because knowledge about the points and meridians, evolved over five thousand years of empirical medicine, do not correspond to any existing Western concepts of anatomy. But absence of proof is not proof of absence. In my mind, meridians may be the pathways that are followed by immune cells as they move up and down an anatomical highway, a discovery that could be just one experiment away. The peptide-containing skin cells, called Langerhans cells, could provide the clue, but no one has ever looked at their distribution.

Funding preferences determine the areas that get researched, and acupuncture simply isn't high on the list for research dollars, and never has been. Many mainstream researchers still refuse to believe that acupuncture has any validity, very much in the same way they didn't believe the opiate receptor existed before a simple lab method allowed us to measure it.

I wait in a partitioned area, early for my 1 P.M. appointment, and read the handsome OAM brochure sitting on the coffee table. I notice that the new phraseology is "complementary *and* alternative medicine," and that there's been some consideration of renaming the office OCAM in the future. I like the change. The term "alternative" is too confrontational, implying an "us or them" mentality, as if only one can survive and the other must die. This sort of positioning does not do well in mainstream science, as evidenced by a long history of resistance to new ideas.

In my lectures, I sometimes drive home the point of how hard it is for new ideas to gain a hold in medicine by recounting what happened to the Hungarian doctor Ignaz Semmelweis in the 1840s. Practicing in an obstetrics ward in Vienna, he noticed that the impoverished women, who were under the care of hospital midwives, were not nearly as susceptible to fatal childbed fever as were the wealthier women, who were cared for by doctors, and he figured out that the discrepancy could be due to the fact that the doctors were not washing their hands before examining the women. Since the doctors were on a daily schedule that took them straight from the morgue, where they did research, to the obstetrics ward, where they performed their examinations, their hands were often still covered with the blood and germs of the corpses when they saw the patients—but no one knew of the existence of germs then. In fact, it was considered a status symbol to have your white coat decorated with gore, showing that you had been doing research and therefore were worthy of much respect! As an experiment, Semmelweis tried washing his hands before seeing patients, with the result that his patients no longer contracted the dreaded fever. But when he implored his colleagues to do the same, they scoffed and laughed, paying no attention to his seemingly outrageous idea. Finally, in 1862, in a desperate attempt to make his point, he cut off one of his fingers and plunged his hand into the open belly of one of the corpses, only to develop a fever and die within a few days—or so one version of his untimely demise would have it.

Still, nothing changed. The world was not ready to act on Semmelweis's observations, despite ample evidence of their accuracy, because, without any knowledge of the existence of germs, those observations seemed to make no sense. It wasn't until the advent of germ theory, based on the research of Louis Pasteur and the urgings of Joseph Lister, that finally, in the 1880s, the reluctant doctors were forced to comply with new rules of cleanliness and antiseptic conditions. When you consider that such blind resistance actually costs human lives, it seems unforgivable, yet the track record shows that such ignorance is quite the norm. As late as the 1950s, there were still some professors teaching medical students that syphilis could be cured by giving patients the poison arsenic, an archaic turn-of-the-century belief that had long since gone the way of blood-letting. But old ideas die hard, and even in the face of something as miraculous as penicillin to treat venereal disease or as simple as washing the hands before touching patients, new ideas meet with resistance that goes beyond all rational, logical boundaries. The parallel in modern times could be made for alternative medicine and its dominant theme that mind and emotions directly affect health and disease. The embracing of new concepts need not threaten the establishment so much as move it along, to make modern medicine better, more capable of carrying out its mission of curing disease. Using the term "complementary" instead of "alternative" would be both more accurate and more politic.

SNOW AND CHERRY BLOSSOMS

Interrupting my reverie, Wayne Jonas peers over the partition wall that sections off the waiting area in the Office of Alternative Medicine. As he takes me into his office, I'm instantly aware of how spacious a room it is, and how expansive the view from its generous bank of windows—both signs that some blessings are flowing toward the OAM these days. Outside, the snow has started, white flurries blanketing the cherry trees that dot the NIH campus, their blossoms barely visible. Wayne offers me the politically correct chamomile tea, and we begin to talk just as the snow crescendoes to a near whiteout. I share with him that it has long been my wish to see a program started inside the NIH that would do the basic research to put new-paradigm science on a secure footing. Only at the level of the bench, where the important, legitimate discoveries get made, where the belief systems get set and the paradigm gets forged, can there be a real interface between the traditional and basic scientist and the alternative approaches.

For example, all kinds of practitioners, from the nurse healers who practice therapeutic touch to the holistic chiropractors who do manipulations, tell us that they "feel" subtle energies moving through the bodies of their patients. My own hunch is that these energy emanations get created as ligands bind with receptors in the body, doing their intricate humming dance with each other. Now, these energies have not been convincingly measured by objective devices, although a few physicists have worked at devising more sensitive means of measuring quantal events. Why couldn't the NIH fund some research on this subject?— either on measuring the body's energy field, or on studying how energy healers can influence this flow, perhaps by using their own energy to trigger receptors in a manner analogous to the way electromagnetic energy fires neurons.

Wayne Jonas agrees with me on the necessity for basic research and, in a great example of synchronicity, relates how just the day before he had presented his idea for an intramural research program to a gathering of NIH institute directors. This is wonderful news, and if he can make it happen, it will be a terrific breakthrough, because up until now, the OAM has concentrated its minuscule budget on *extra*mural activities, programs set up at universities to do research in alternative medicine—which means lots of people scrambling for very few funds at such places as the University of Maryland. An *intra*mural program would shift the focus radically. Scientists hired by the OAM would work within the labs on the NIH campus, the attraction being that they would come with funded positions, which are rare at the NIH these days.

This would also be a big step toward putting alternative medicine on a serious scientific path. Traditionally, research is divided into two types, clinical science and basic science. Clinical science deals mostly with human beings who take part in clinical trials, where specific, very practical questions are asked: Does this drug work? Does that manipulation work? In other words, the people are the rats. These are not easy experiments to do, costing millions and often involving serious ethical issues that can seem to get in the way of progress. By contrast, in basic research, you're asking different kinds of questions, ones that don't necessarily lead to practical results. Basic scientists never know where their experiments are gong to lead, and they get nervous when asked what it all means, for although they certainly hope to make it possible for major medical breakthroughs to occur, they see their role as the assemblers of myriads of little tiny pieces of a huge jigsaw puzzle, which will at some vague time in the future enable the big picture to become clear. Wayne and I share an appreciation of basic science, and we would both like to see Congress funding the NIH to do more of that kind of work, some of which would focus on alternative medicine.

We end on this note of agreement, and I offer to help in any way I can to get an intramural program up and running this year.

The snow has stopped. As I make my way to my car in the unusually chilly air, I am feeling heartened by the new possibilities for science at the NIH. Snow and cherry blossoms, I think, as unlikely a pair as the Palace and the alternative-medicine movement, yet here we are, and it is a synthesis that seems to be coming full circle, affirming that the direction I've been going in all along is the right one, the one that will lead us to a more expanded, more inclusive, more truthful science.

RISE AND FALL

In 1987 I left the Palace, and, along with my husband Dr. Michael Ruff, embarked on a business venture to develop our AIDS drug, Peptide T. On the day I signed the release forms and handed in my resignation to the NIH, I took my very first limo ride to the lawyer's office. Outside Building 10, a limo was waiting. And champagne. As we drove to his office, our lawyer got on his car phone and called the Second Biggest Drug Company on the Planet to let them know that the deal was going down. A little later, he notified our private investors, who were wealthy third-generation industrialists. The key player, a genteel multibillionaire, was one of the major biomedical venture capitalists in the world.

Our investors offered us a six-million-dollar deal to support the research and development of new peptide drugs for the marketplace, research we were ideally suited to do because of our advanced receptor technology. Peptide T itself would be in the bailiwick of a nonprofit medical research institution we called Integra. The Second Biggest Drug Company was responsible for setting up the phase II human trials for Peptide T, a task we expected them to do while we developed other peptide drugs. Or at least that's what we were told would be happening.

I knew Peptide T was going to be vindicated in the clinical trials. While still at the NIH, I had sent a sample to Dr. Lennert Wetterberg, head of the psychiatry department of the Karolinska Institute in Sweden. The Karolinska has a rule that the chairman of a department can, at his prerogative, give a new drug to patients who have a fatal disease before the drug's been tested. Wetterberg gave Peptide T to four terminally ill men on a compassionate basis. Brain scans showed improvement of abnormalities attributed to AIDS, and all four had surprising rebounds in their various AIDS-related maladies.

But what I didn't know was one iota about business. Michael and I suddenly found ourselves—two people who had been shielded from having to deal with funding or budgets throughout their professional careers—sitting near the top of a multimillion-dollar bioscience venture. It was a reality check when our investors took out million-dollar insurance policies in our names and surrounded us with the various accoutrements of the private sector: competent, chic secretaries; car phones; business cards with titles like president and vice president; a board of directors on which we served but had no controlling interest. We had never played in this arena, and we proceeded to make a lot of mistakes.

One of the first had to do with our insistence on a futuristic fantasy lab, the one "nonnegotiable" item in our contract.

It was the eighties, and the whole venture was truly a go-go, biotech research dream come true. Topping it all off was a two-million-dollar state-of-the-art lab, which we named Peptide Design. It took months to get the details just right: pink walls and blue steel columns, expensive high-tech lighting, vaulted skylights, and rows of purple lab benches. Marking the building was a fabulous "Peptide Design" neon sign, made from the logo I'd designed and had executed by a local neon artist.

I had caught lab fever when, on a trip out West, I'd visited the lab of the Seattle scientist whose experiments using Peptide T had perfectly replicated our results, providing the positive evidence that led the Second Biggest Drug Company to take us on board. Inside that lab, everything was very sexy, soft beige and sophisticated, streamlined black—the opposite of how government labs are, drab, institutional, and gray-green—and the structure itself was perched on a hill with a spectacular view of the ferries coming and going across the Puget Sound. Many years before, I had seen the Salk Institute lab, located outside of San Diego, rising up from the beach at night like an illuminated cathedral of science, and I knew, even then, that someday I would have a gorgeous, inspiring lab in which to work.

To staff our fantasy lab, we assembled a dream team, twelve brilliant, mostly young scientists, including several smart, tough women I'd worked with at the NIH who probably were never going to get tenure, no matter how brilliant they were.

Once the lab was completed, we threw a big opening party, which our investors encouraged us to do, hoping to create a big splash right outside the Palace gates and to attract some talent to the private lands of biotechnology. I love to plan parties, and so I was thrilled to design this one, sketching the invitations, planning lavish decorations and food. I chose 8/8/88 as the party date, thinking it full of significant symbolism—this repeated digit is a Chinese indication of prosperity and also a graphic representation for infinity. But it caused a bit of a ruffle when one investor/board member and his lawyer had to fly back from their private island in Maine in the middle of August. Another board member was obviously discomfited by my rainbow-ribbon-cutting ceremony and festive, semimystical themes—especially when I asked him to cut the rainbow ribbon and announce: "It's now eight seconds, eight minutes after eight on eight, eight, eighty-eight" to begin the revelries.

As soon as we were settled at the Peptide Design lab, our investors, figuring Peptide T research was complete except for the trials (which were the responsibility of the drug company), began pressuring us to come up with another marketable product. But we did get a chance to pull together an important piece of Peptide T research during this early time in the lab, giving us a better understanding of how the HIV virus acted in AIDS. We were now seeing that once the virus entered the cell and was replicated by the cell's DNA, fragments of its viral protein envelope, gp_{120}, were spewed into the extracellular space and bound to a receptor or receptors on other body cells. These gp_{120} fragments, by occupying the receptor sites, blocked access to the cell by the body's own natural peptides, the substances intended to fit that particular receptor. The first such natural substance we identified was VIP (vasoactive intestinal peptide), but in recent years a number of other peptides that use the same receptors have been identified.

With our sensitive receptor bioassays, we could measure the gp_{120} in the blood of even those people in very early stages of the disease. This finding was a clue to us that it was the blocking of the body's receptors by these fragments, not the infection of the cell by the virus, that caused the symptoms of AIDS! Our neuropeptide receptor "blocking" theory was further strengthened by an understanding of just how the natural neuropeptide VIP functioned in the organism. First, VIP is active in the gut, where it regulates water flux, and in the brain, where it promotes the growth and health of neurons. When gp_{120} binds to receptors in the brain and preempts VIP activity, neurons die or, equally as damaging, lose their axons and dendrites, causing the dementia effects observed in an increasing percentage of AIDS patients. Second, VIP is also found in the thymus gland and bone marrow, where it controls the maturation of lymphocyte cells, which are responsible for a strong and healthy immune system. The gp_{120} receptor occupation causes apoptosis, or programmed cell death, which is another way of saying it shortens the normal life span of these critically important T4 lymphocytes, resulting in an impaired immune system and increased susceptibility to opportunistic disease, the cause of death for most people with AIDS.

This new understanding was in direct opposition, once again, to what most other AIDS researchers believed at that time, which was that it was the direct infection of the cell by the HIV virus that caused symptoms of AIDS. We saw it more simply, as the blockage of VIP, resulting in a failure of neurons to grow and immune cells to mature. (The wasting effects of AIDS—weight loss, failure to thrive—were to be understood much later, using our same theory, when we proved that the gp_{120} also fit and blocked the receptor for the growth-stimulating hormone GHRH.)

At the same time that we were doing this research, we were also trying to invent new drugs, which our investors were eager for us to do. So this was a period of daily brainstorming meetings, in which our handpicked staff of researchers would march into the huge octagonal conference room and lay out their data for all to see. At the time, the business of turning peptides into drugs—or finding nonpeptide analogs called *peptidomimetics*—had not yet emerged. Peptide T came ten years before its time, and while we believed we'd produced one of the first therapeutics to be custom-designed to the precise molecular requirements of the organism, most researchers were nowhere near as appreciative as we were of our new-paradigm child. Nonetheless, mimetics were our major interest, and the first one we began to develop was a drug that stopped brain damage due to severe head trauma or stroke. It was a hot topic then and is now, and quite a few other large drug companies had their people working to synthesize potential therapeutics. In patients with head trauma or stroke, the neurotransmitter glutamate gushes out of the neurons and, over time, kills the cells. If we could intervene and block specific glutamate receptors soon after the accident or stroke, we believed victims who now die or are permanently crippled from head trauma would be enabled to survive. Unlike the larger labs that were trying to create a drug de novo, we were looking for the natural endogenous peptide ligand.

I had quite a bit of familiarity with the mechanism involved in these often fatal conditions. Years earlier, in my lab at the NIH, Remi Quirion and I had set the stage for the current approach by mapping the PCP ("angel dust") receptor by autoradiography in rat brains, and in collaboration with Tom O'Donohue had even shown that its endogenous ligand was a peptide. PCP, a street drug that started life as a promising anesthetic, had its commercial development abruptly halted when it was observed that patients coming out of the anesthesia reacted like raving psychotics. My Bryn Mawr classmate Suzanne Zukin, now a full professor at Albert Einstein Medical School, had developed the first PCP-receptor binding assay with her then husband, Steve Zukin. In a rare example of old-girl networking in pharmacology (there're just too few in-the-know old girls around), Suzanne had passed on to me the results of the latest neurophysiological experiments, which suggested that the receptor for PCP was the same type of glutamate receptor that needed to be blocked to prevent further damage from glutamate leakage in head-trauma victims. Now I had the perfect project to launch in my own new lab! We would identify the endogenous peptide ligand for the PCP receptor, and then synthesize it for a "natural" therapeutic.

At the end of fifteen months of furious work, we cracked the structure for the endogenous ligand and took out a patent for the peptide we finally named Neuroprotectin for treating head trauma and stroke. (Our investors had wisely nixed our lab slang moniker, "Angel Dustin," a term we all had jovially used during the long nights of grinding up brains and running assays because we thought it was a clever punning name for a drug based on the brain's own PCP.) The long, tedious period of laboratory research made Michael and me realize how miraculous it was to have deduced Peptide T within just a few days—with no laboratory work, just a few intense hours at the computer.

It was right around this happy time that the Second Biggest Drug Company on the Planet made a sudden and unexpected executive decision to withdraw their support of further Peptide T development. We heard that a choice had been made, somewhere up at the top of the company pyramid, to pursue a new AZT-like drug that was showing great promise, and because of the high cost of shepherding a drug through the FDA regulatory procedures, the powers that be considered it economically unwise to develop both drugs. Peptide T was seen as a gamble, while AZT, the first drug shown to work against AIDS, although a highly toxic chemotherapeutic, was tried and true in the marketplace. Besides, AZT was the darling of the best-funded NIH institutes (NCI and NIAID) that were involved in AIDS research, and this new "son of AZT" was sure to be the same.

Within days of the decision being handed down, Peptide Design became Peptide Demise. Faced with the daunting prospect of having to pay for the further testing and development on their own, our investors wasted no time in dropping out. Our lab was shut down, and then the NIMH took away our license to develop Peptide T. Our staff, now numbering twenty-five people, was left without jobs (but we were able to help each and every one of them land new positions, many at the NIH).

The only bright spot in the general darkness was a phase I clinical trial we'd managed to get up and running at the Fenway Clinic in Boston, funded with money we might have used to postpone the lab's closure by a year. A group of thirty men and women received Peptide T in a trial to test for toxicity and improvements that lasted six months. The results were impressive; the disappearance of some AIDS symptoms without side effects could be well documented. When the trial was over, the Provincetown Positives, a group of HIV-positive men who believed in boosting their immune systems with nontoxic strategies like diet, food supplements, and exercise, fought to gain permission for friends to continue receiving the drug, and finally won. From that trial, mostly through the efforts of John Perry Ryan and the other Provincetown Positives, word of Peptide T's beneficial effects spread, and a number of national AIDS activist groups demanded to know what was going on with the successful nontoxic therapeutic Peptide T. But they received no answers. In a burgeoning AIDS drug underground, several small labs began to manufacture Peptide T and sell it through buyers' clubs in New York, Dallas, Atlanta, Washington, D.C., Los Angeles, and San Francisco, and while it was the number-one best-seller for a couple of years, the anecdotal evidence was of no use in furthering the drug's commercial development.

With the activists on our side, we fought to get back the license to develop Peptide T. But the NIMH stubbornly held on to it. During this phase when all decisions were going against me, I often grieved for the loss of the support of my powerful mentors. Finally, the government restored the license to Integra, our research institute, but in an odd and seemingly punishing move, they also gave joint licensing rights—unprecedented then and now for any drug—to a tiny Canadian company that had none of the qualifications we had. What the government had done, in effect, was to insure that we would not be able to get any mainstream funding for further development, as no drug company with the millions to invest would ever support research it could not own in its entirety, for obvious reasons.

Gradually, the magnitude of our loss began to dawn on us, along with the implications of that loss for the AIDS community. If Peptide T was going to be developed any further, which was now a matter of getting the FDA to approve phase II trials, then it would have to be Michael and I alone who would make it happen. And to do that we needed to raise more than ten million dollars. FDA approval is granted only when a pharmaceutical company or other sponsor has invested the millions of dollars that it typically takes to put a drug through clinical trials in this country. But with the divided license we now held, who was going to invest?

While we floundered, research being done in the labs of some of my professional colleagues was supporting our claims that Peptide T was a drug worth testing. Doug Brenneman's work at the NIH, for example, buttressed our claims when he showed that Peptide T could stop the death of neurons caused by gp_{120} blockage in an elegant brain-culture system he used. And data from a small NIMH-funded Yale double-blind study would have removed all remaining objections if it had been a larger study—and if science was all that mattered. For the first time we had solid evidence that the improvements in neuropsychological testing were caused by Peptide T. The subjects of the Yale study improved when they received Peptide T, and got sicker when they received the placebo.

The subjects in the Yale study were given Peptide T or a placebo randomly, and then, after a short time, were switched over, receiving the opposite. This type of study is a phase II study for clinical trials, whereas the Fenway Clinic study we'd done was a phase I trial, intended to test mainly for toxicity, which proved to be nonexistent. In fact, the lack of side effects was also accompanied by some marked improvements. However, in a phase I study, subjects know they are getting at least some of the drug, and so improvements in their condition could possibly be due to their expectations. (The placebo effect—the expectation of improvement actually changing outcome—is an extremely powerful demonstration of the involvement of the mind in healing. In his stunning book *Timeless Healing,* Dr. Herbert Benson explores this effect thoroughly in establishing his thesis that we are hardwired for faith.)

We wondered at the time whether Peptide T's lack of toxicity had become a problem for the Second Biggest Drug Company on the Planet. Certainly they'd never before had a nontoxic drug that was also effective. Most of their hot sellers were virtual atom bombs, anticancer drugs that destroyed the immune system itself in an attempt to save the patient. The idea of a nontoxic therapeutic could well have been a complete enigma to most of their top scientists, and Peptide T wasn't tailored to fit their preconception of what a winning drug for such a deadly disease should act like. In addition, no sizable phase II placebo trial to demonstrate improvement beyond a doubt had been carried out, and the studies in Boston and at USC were easy to dismiss as too small and of uncertain validity because of probable placebo effects. The fact that survivors of the Boston trial continuing to receive Peptide T had lived significantly longer than other people with the same T-cell level in 1989, when the Boston trial began, was harder to dismiss as a placebo effect.

But by then no one was listening. Peptide T was considered a joke, or worse yet a hoax, by the world of AIDS researchers—if they remembered it at all.

RETREAT

Expelled from the paradise of our Peptide Design lab, we retreated to an office in the basement of our home to lick our wounds and plan a strategy. Mike had thrown together a working space made up of a few pieces of furniture we'd managed to salvage from the lab and a jumble of assorted computer equipment we'd inherited. It was a sorry comedown from the 10,000-square-foot showcase where we had once, briefly, spent our days. Worst of all was the loss of all of our personnel. How could we function without Bernice Blade? We kept our wood-burning stove going all winter, trying to keep warm as we phoned and faxed every possible lead that came our way.

Besides leaving us labless to carry on our work, the drug company pullout had cast a shadow over our credibility and added to our difficulties in finding new sponsors. But we kept busy. For the next eighteen months I made the rounds to trot out my one-hour dog-and-pony act for potential investors, contacting over fifty large multinational drug companies and getting down on my knees (figuratively speaking) in at least a dozen boardrooms, only to hear the same depressing response: Yes, your science is very compelling, but why did the Second Biggest Drug Company on the Planet withdraw their support? And, what's this business about a split license?

But the drug company pullout and the split license, I had to admit, weren't the only reasons we were having such a hard time. One problem, Michael and I felt, was that Peptide T was just too good to be believed. After labs all over the country had spent millions to try and find the right piece of gp_{120} to block the virus, we reach out, see a rainbow, and get a structure off the computer that hits the bull's-eye. No wonder people thought it was hoax—or at least a blunder.

And then there was my reputation as a firebrand, a troublemaker, which still lingered all these years after the Lasker controversy, making people wonder what the "real" reasons were behind the drug company pullout that landed us on our fannies. Potential investors could practically be heard whispering to one another: "Pert? Wasn't she the one who was involved in that flap over the Lasker?"

It was a hard thing to face, but I had to come to terms with my own responsibility for the whole debacle. In addition to the very bad karma from the Lasker, I'd incurred further establishment wrath by being an unquenchable spitfire when the NIH had refused to support trials for Peptide T, and I had earned more than a few enemies by insisting that I had the answer, the only solution that could cure the AIDS virus. Hadn't I offended the private investors with my quasi-spiritual antics and extravagant demands? It was a bitter pill to swallow, but I was forced to take a long, hard look at my behavior, my very unpolitic lack of respect and consideration for forces that seemed to so fiercely oppose me.

As far as most of my former colleagues at the NIH knew, the ones who were running the major research being done at the time, Peptide T had never shown any promise. Now they could dismiss us all the more, criticize our work as absurd and worthless, and point to our investors' retreat as the supreme evidence. For a period of time after the crash, whenever we would apply to present our data at the big conferences, we were turned down or marginalized by only being permitted to display a poster on the last day of the conference. At the end of a talk, we used every opportunity to approach the microphone, because it was the only way we could get our data across, in spite of the sniggering moderators who tried to dodge or ignore our persistent requests. But we kept on, because at every meeting we'd usually manage to convince a few more researchers to try the drug, run it through their own systems, and see what Peptide T did to block gp_{120}. And each time they did, following our direction or getting new data from their own systems, they were able to replicate the results. Meanwhile, other labs were duplicating and extending our work and beginning to speak up and be heard, a fact that made us glad, even though we were kept from adding our two cents.

During this period of time, Michael and I were self-confessed business morons. We never really understood the requirements of the business world as we struggled valiantly but without skill to resurrect our research and development of Peptide T. Thelma, the tough investment banker we had hired to find us investors, used to lecture me on becoming a "CEO with guts," looking the part by wearing my hair in a bun, dressing in blue gabardine suits, not smiling at board meetings. And I made several attempts to assume the role in the last days as Peptide Design slowly drained away, but Thelma, who fiercely believed in us, was still unable to land us a new investor. Not that she didn't try. On what must have been one of our very last days in the lab, she called long distance and asked that we usher everyone into the richly furnished conference room.

"Your long nightmare is over!" Thelma's voice crackled from the West Coast over the little box on the conference room table. Our staff perked up and gathered round, their brows wrinkled with the stress of their secret job-hunting. "I have landed a major pharmaceutical company to develop Peptide T and fund your lab and other inventions for the next ten years. I am hoping to wrap up the deal and sign the contracts shortly!"

It was devastating when, just a few weeks later, a cruel and succinct turn-down arrived by fax at our dining table during a psychoneuroimmunology and cancer conference we were attending in Germany. As I read the news, tears rolled down my cheeks and I silently struggled to maintain my composure in front of our curious and concerned hosts. Sadly, it was a scene that would be repeated in different settings many times over the next couple of years.

The stress of the nearly constant disillusionment during this period was almost more than I could bear. The loss of the lab and the dying down of the incredible wave we'd been riding ever since leaving the Palace left us crashed upon a bleak and empty shore, stranded and on our own. It was the winter of 1990, and all we could do was hole up in our basement den while we made the calls and sent the faxes we hoped would enable us to carry forth what we believed in. Nothing seemed to be working.

Certainly I'd been through tough times before—the period after the Lasker, the frustrating last days at the Palace when we couldn't make any headway in publishing our results or getting funding for trials of Peptide T. I was no stranger to the abuse, the shutting out, the lack of support. But then I had been able to deal with it all by the sheer force of willing myself to, knocking over opponents and making my touchdowns in spite of incredible odds. Now what I experienced was the true nightmare, one in which all my old tactics for survival and success were failing me, and the drama was beginning to take its toll. In a period of just a few years, I gained over fifty pounds, using food as a sedative to alleviate the uncomfortable negative emotions that had arisen: the rejection, pain, and fear.

Michael was my saving light. As a young boy, he had been an altar boy, and because of his unflinching devotion, with time and service, he'd risen to the top as chief altar boy. Now he showed the same kind of steadfastness in support of me and the project we believed in, always treating me like a fellow scientist he respected and had confidence in, never acknowledging—perhaps not even to himself—just how close to the edge we really were. At times, he seemed to have an uncanny mental telepathy, anticipating what needed to happen next, making sure that the right people heard the right things, that faxes got sent, that calls were returned and appointments confirmed. He even voluntarily took over many house and child-care chores, which allowed me to catch a few moments to relax. I felt very grateful for his unceasing support.

HEALING

While this was a time of much tumult and suffering, it was also a period of immense personal growth for me, as I stretched the limits of my new-paradigm thinking to embrace new modalities of personal healing—physically, mentally, and emotionally.

Under the stress of disappointment and professional exile, I realized that I had been angry for years, harboring deep resentments that went all the way back to Sol and the Lasker, perhaps even further back. I had to face the fact that I'd never gotten past the Lasker, and I was still holding Sol responsible, not only for cutting me out of the prize but for refusing to mend bridges and give me the support I needed to gain acceptance for Peptide T. In my mind, Sol had become one of "them," a powerful player behind the scenes who, with his influence, did what he could to derail my efforts, bolster my critics, and generally wreak his revenge. In the Lasker days, when I began comparing what happened to me to the losses inflicted on Rosalind Franklin, I had only intuited that suppressing my emotions was dangerous and might lead to cancer, but now I had amassed enough hard scientific data to convince me that I needed to heal my emotions if I wanted to pull through this difficult time—alive and healthy.

Not that I hadn't tried over the years to make overtures to my former mentor, inviting him to parties at my house, attempting to get together in the hope of eventually healing our relationship. But my focus was always on getting Sol to forgive me, and while he was always polite in his responses, inevitably there came the moment of gentle rebuff. At rare times, when our work brought us into the same social circles, I tried to talk to him about my work with Peptide T and gp_{120}, but he professed not to understand any of it and changed the subject. When all my efforts failed, I would console myself by writing fanciful little notes, hand-decorated with colored hearts, in which I would express my apologies and ask for his forgiveness. But I never sent them.

Could I blame him for not wanting to be my friend? Hadn't my actions caused him to lose a shot at the Nobel? If so, how realistic was it to expect that he'd turn around and offer me a helping hand in my time of need?

Still, I was tormented by his shadow, and felt that without some kind of reconciliation between us, I would forever be in his thrall and under his power, a prisoner, a victim. For years after the Lasker incident, I went nightly to the local Y to swim laps, hoping that the exercise would turn my anger into energy, enabling me to let it go. But it wasn't working.

The demise of Peptide Design had brought much of this old feeling to the surface, and in my desperation I was more open than ever to doing whatever would help me to heal the old wounds. It even occurred to me that perhaps there was a link between this unresolved conflict with Sol and the political mistakes I had made while trying to develop Peptide T. Had Sol become such a bogeyman to me that I was still projecting my anger at him onto those who stood in my way, making it hard for them to accept my ideas and offending them deeply with my rash and sometimes confrontational, impulsive ways? Could my unhealed emotional traumas actually be altering "reality"?

There was no doubt I had a reputation for being a spitfire, someone who was so hellbent on her own path that other people often felt the best thing they could do was simply stay out of my way. For the first time, I seriously considered: Was *I* the problem? If I'd behaved differently, been a good girl, and played the game according to the rules, would Peptide T have made it to the market, saving the lives of people who were now dead because it wasn't available sooner?

It was in the midst of these depressing thoughts that I got a call from Eugene Garfield, my advocate during the Lasker days, requesting for the umpteenth time that I describe my research in a written statement for his publication, the *Citation Classic*. Garfield granted this honor to first authors of any scientific paper cited more than a thousand times in the literature. "Opiate receptor: demonstration in nervous tissue," the landmark paper Sol and I had published in 1973, had long since passed the qualifying mark, and a behind-the-scenes account of the discoveries that led to that paper was now long overdue. Actually, I had tried several times over the last few years to pull something together, but what came out was always either overly apologetic or angry and self-righteous. I knew what needed to happen in order for me to write the *Citation* piece was a deep and honest release of the anger and hurt I still carried and had been holding on to for the past twelve years.

Two events precipitated my being able to respond to Garfield's request with an integrity that finally put my inner feud to rest. One was the exploration of my Christian roots, and the other was my discovery of the healing power of dreams to bring about a near-magical resolution.

My attraction to the spirit and message of Christianity came to me through music. At one of my lowest points after the Peptide lab crash, I had been walking by a local church, feeling downhearted and nursing the usual grudges. The sound of voices carrying a beautiful melody drew me inside, where I found a choir rehearsing. When I spontaneously expressed my appreciation of their wonderful music, they invited me to join in, which I did. For several years after, I continued to sing alto in the choir.

My family thought I'd finally flipped. When I was growing up, religion was a more taboo subject than sex or money. I honestly thought that Jesus Christ was a curse word, because the only time I ever heard the name spoken was when my mother or father smashed a finger or couldn't pay a bill.

My parents had come from different religious backgrounds, and had solved the dilemma of their mixed marriage by avoiding the topic altogether. My mother was a Lithuanian/Ukrainian Jew whose own mother had been born in Russia—the old country, as we called it—while Dad came from a long line of Yankee Congregationalists and could trace his ancestry to John Beebe, who'd come to Connecticut in 1647. They eloped right after World War II, sending a telegram to Mom's parents announcing the marriage, a missive that hit the household like a second Pearl Harbor. In those days, a Jewish girl like my mom marrying a non-Jew was much less tolerated than it is today, regardless of whether the family was observant or not. Even so, the culture of Judaism held on, if the practice did not, and my mother always considered herself Jewish at heart. I remember asking my mother when I was about ten years old what religion we were, and her replying, "You're Jewish, and don't forget it." Having never seen the inside of a synagogue except once when attending a cousin's bar mitzvah, I had absolutely no idea what being Jewish meant. Later, as a young adult, I easily embraced the atheistic viewpoint that seemed to make the most scientific sense. It was the sixties and *Time* magazine had finally joined Nietzsche in proclaiming that

God was dead, reflecting the spiritual vacuum my generation experienced. But because the whole subject of God, soul, and spirit had been so suppressed in my upbringing, I was fascinated by it and found myself drawn to anything that touched on consciousness and dreams, which until then had meant the soaring sensibility of Romantic-era poetry and literature.

It was the strong Christian tradition on my father's side of the family that I now embraced. Many of his people had been ministers or, like my uncle Bill, who had played the organ at our wedding, church organists and musical directors with strong ties to the church. Soon I was attending services at the local church, throwing myself into the hymns and verses, trying to absorb the Christian ethic. I felt deeply moved by Jesus's message of compassion and forgiveness, knowing it spoke directly to what I needed to hear. Often, while I sang in the choir, tears would stream down my face for no apparent reason. I now understand that what I had found in the church, in the choir, in the music I was singing was a safe haven, a tremendous opportunity to heal my emotions. I could let go, finally, taking off the armor that I'd kept in place for years.

The Christian experience was the first step toward a resolution of my inner conflicts; interpreting my dreams was the second. I have believed in the importance of dreams ever since as a teenager I read Freud's *Interpretation of Dreams,* and began to pay attention to these messages from my subconscious. While I never dreamed about Sol, I did have one pivotal dream that was as clear as daylight and actually led directly to the step that allowed me to achieve forgiveness—at least in my feeling toward Sol, if not his toward me. In the dream, which was set in a *Wizard of Oz* movie, I, instead of Dorothy, threw a bucket of water on a witchy Sol, who shriveled up, shrieking, "I'm shrinking, I'm shrinking," just like in the movie, until he disappeared. What I realized on awaking was that *I* had given Sol his power over me, *I* had made him into a monster, the enemy, to the degree that his very existence tortured me beyond anything he may or may not have done to me.

I immediately wrote him a letter, one different from the little-girl letters I'd written but never sent. In this letter, I told Sol I forgave him and asked him to forgive me, making it very clear that I expected no response, no acknowledgment of my request. The slate was wiped clear, I told him, and that was how I truly felt. By doing this, I experienced a deep acceptance of the situation, a level of personal responsibility that led to a profound emotional healing. All these years I had been stuck in a figment of my own imagination, and, now, realizing this, I was becoming free. I learned that I could forgive regardless of whether the person I believed had harmed me acknowledged the forgiveness. It was something that happened inside me and didn't need to happen inside Sol. And although I have to admit I have temporary relapses in my absolution, and I'm not ascending to total sainthood as of yet, this act of forgiveness was a breakthrough that freed up tremendous energy for me to continue in my work and pursue my truth.

DEALING WITH STRESS

It was in 1991, at an all-time low in our Peptide T campaign, that I found myself in Boston speaking at a conference sponsored by Interface on the future of medicine. Interface was a forward-looking organization whose purpose it was to explore the ground where psychology and spirituality meet, and they had gathered several cutting-edge medical people to address this subject. After my talk, I joined a panel of the speakers, one of whom was Dr. Deepak Chopra, who at the time was just beginning to become well known for writing books that updated the ancient Indian Ayurvedic tradition for the Western mind. I had missed his talk, but was impressed with his responses to questions from the audience. He seemed to have an answer for everything. Perhaps that was what inspired me to ask a question of my own as the panel session was ending and the audience beginning to thin.

"Deepak, I don't know what's going on. I have a brilliant drug that can save people's lives. I've been working on it for years, and I can't get it out the gate. What am I doing wrong?"

He listened carefully and then, gazing calmly and deeply into my eyes, gave me a stunning answer: "You are trying too hard!" he said and then smiled.

I took this in for a moment and then responded.

"Trying too hard? But I've never heard of such a thing!" I exclaimed, incredulous that he would make such a suggestion. In the world I lived in, there was no such thing as "trying too hard." In fact, my whole life had been about trying hard *enough*, striving, ever striving to be the best at whatever I did, in spite of all the obstacles. The dual legacy from my parents—the Protestant work ethic and the fierce New York Jewish competitiveness—had driven me to graduate at the top of my high school class, to enter and succeed in the Ivy League college halls, and to plunge ahead, however ill-advisedly at times, in my journey to the very pinnacle of my profession. The idea of "trying too hard" was a concept as foreign to me as UFOs being real. I was genuinely mystified, and had absolutely no idea what he was getting at.

"Please, be my guest," I next heard him say, "and come to my health center in Lancaster, where I'd like to show you a few things." The idea that he had something to show me that might somehow lead to the acceptance of Peptide T was enough to get me packing. I accepted on the spot with a promise to make immediate arrangements, while what was left of the audience, which I'd completely forgotten about but which had been witness to the entire exchange, gave us an approving round of applause.

Within a few short weeks, I joined Deepak at the Maharishi Health Center in Lancaster, Massachusetts, where he was then medical director, and took up residence in the Barbra Streisand suite. I was fed elegantly exotic, tasty, vegetarian fare, and treated to a daily massage, complete with sesame oil being dripped slowly and luxuriously onto my forehead. The whole experience was utterly unlike anything I'd ever been exposed to.

A visiting Indian physician in full rishi garb visited me daily. "Fennel, she needs lots of fennel," he seemed to be saying beneath his breath after an examination that consisted of holding my wrist for a few seconds and reading my three Ayurvedic pulses in an extraordinarily mystical bedside manner that all the staff members seemed to have absorbed in their training.

But the most amazing benefits began when I was taught how to meditate, the heart of the healing methods offered at the Health Center. Deepak had a staff member teach me TM (transcendental meditation), a form of Indian meditation that had been packaged for the Western mind, and a method I knew about because the Beatles had taken it up in 1968. I remember thinking, If the Beatles did it, it's probably okay for me! (A longtime Beatles fan, I can still amaze my fourteen-year-old son Brandon by being able to sing every word of certain cuts off the *White Album.*) I easily learned the simple TM technique, which consists of a mantra, or Hindu sacred word, to be repeated over and over for twenty minutes, two times a day, and brought it home with me, and I've continued to practice it daily. In my quieter mental state, I could experience how events were unfolding quite naturally around me, without my having to make them happen.

I also started to become aware of synchronicity, to see connections between events and people happening simultaneously and then to act on this awareness instead of out of the more familiar linear cause-and-effect model. I had encountered the theory of synchronicity years ago in the work of Carl Jung, where it is defined as "the coincidence of events in space and time as meaning something more than mere chance." And while I didn't quite understand it then, it appealed to me intuitively. Now I understand that, as in the psychosomatic network, related events occur simultaneously in time and space, in spite of our perception of them as unconnected and independent. This allowed me to trust that life would unfold without me as the prime mover, the brain always leading the way!

When I first began meditating, I was besieged by visions of my father lying helpless and dying on his hospital bed, hooked up to IVs and devices, the paraphernalia of his Western medical "saviors." Other emotionally charged visions, some from childhood, seemed to percolate up into my conscious awareness as I continued meditating, as if these thoughts and feelings had been packed away in storage somewhere, waiting for me to stop everything, sit quietly and relax with a focused mind long enough to allow them to arise.

I marveled at this process and attempted to understand it in terms of physiology. I was especially interested in how meditation's effects on stress had an impact on immunity, and what this meant in terms of the brain-body connection I had seen in my laboratory research. At the time, I had read *The Relaxation Response,* Herbert Benson's first book written in the seventies, in which he attributed meditation's power to an alteration of the nervous system from sympathetic to parasympathetic pathways. But with my knowledge of the bodywide psychosomatic network, I was beginning to think of disease-related stress in terms of an information overload, a condition in which the mind-body network is so taxed by unprocessed sensory input in the form of suppressed trauma or undigested emotions that it has become bogged down and cannot flow freely, sometimes even working against itself, at cross-purposes. In the late fifties, when it was shown that tumors transplanted into rats placed in stressful situations grew more rapidly, we used to attribute stress-related disease to increased levels of steroids that acted to suppress the immune system. But our new understanding of neuropeptides and receptors has enabled us to see more of what is going on in conditions of stress. When stress prevents the molecules of emotion from flowing freely where needed, the largely autonomic processes that are regulated by peptide flow, such as breathing, blood flow, immunity, digestion, and elimination, collapse down to a few simple feedback loops and upset the normal healing response. Meditation, by allowing long-buried thoughts and feelings to surface, is a way of getting the peptides flowing again, returning the body, and the emotions, to health.

I came to think of my first experiences of meditation as experiments—experiments in the release of highly charged emotional memories that had been stored somewhere within the psychosomatic network. Since the only lab I had access to at the time was that of my own mind and body, I paid careful attention to these early experiments, and later found that my thoughts about what I was experiencing correlated with research that my colleagues were doing on how trauma and blockage of emotional and physical information can be stored indefinitely at the cellular level.

In addition to meditation, I was making practical, everyday changes in my lifestyle, getting regular massages, eating a healthier diet, doing more exercise. At this time, I also switched from relying strictly on allopathic, or Western, medicine. Having read an article about chiropractic, a onetime sizable and respectable branch of mainstream medicine that had been discredited by the drug and surgery branch, I couldn't help identifying with chiropractors as fellow underdogs, victims of what we call modern medicine. And then I met one!

The very day after reading the article, I was in a health-food store stocking up on fresh veggies when I thought I overheard a handsome young man say that he was a chiropractor. I got his attention and began a conversation, learning that he had a practice in town and often treated people who were dealing with stresses on the scale I was. But the odd thing was that he insisted he'd never mentioned that he was a chiropractor, and I must have been telepathic to think he was! When I went to his office for an appointment the next day and filled out the standard forms, I wrote in the section asking who had referred me: "God in the form of synchronicity." I became a regular at the office of Dr. Joseph Skinner, who introduced me to the power of chiropractic and who later became a close family friend.

Another of my early health "gurus" was Carolyn Stearns, a massage therapist recommended to me by a broad-minded M.D. who was treating me for a rib injury I'd sustained during a bike tour. Carolyn, who had been a professional dancer, writer, and poet, was now doing a form of "psychic" massage that she'd developed from her own intuitions and spiritual awareness. She put her hands on various parts of my body and "read" that even though I had spent my life in a left-brain profession as an analytical, rational scientist, I was an extremely spiritual, intuitive person. This part of me, she said, had been buried, shut down since childhood, and I instantly recognized the truth of what she was telling me. She was referring to that quiet inner voice that had been speaking to me for years, the one that had led to some of the biggest breakthroughs in my work. Now, with Carolyn's support and confirmation, I began to trust more and more in my inner voice, both personally and in my work.

I lost contact with Carolyn for a few years, but when I reconnected, she had moved on from doing psychic massage to teaching stretching, visualization, and therapeutic movement.

"What I do now is more powerful because I teach you to do it yourself," she told me on the phone when I called her. I began attending her classes regularly and greatly benefited from a series of deep stretches using an assortment of balls and props, a kind of "do-it-yourself" chiropractic that probably rearranged the peptidergic nerve bundles lying alongside the spine. Once we were stretched, she'd guide us through a soothing, rhythmic movement set to evocative music, allowing emotions to percolate up and be released into consciousness. As we lay on the floor deeply relaxed, Carolyn would read an inspired meditation or guided visualization to help us become more deeply aware of our emotions. One day she read us these words, which hit home in a very direct way: "If you look underneath your depression, you'll find anger. Look under your anger, and you'll find sadness. And under sadness is the root of it all, what's really masquerading all the while—fear."

I had experienced all of these emotions in my personal and professional journey, and now, as I struggled to deal with my stress and survive the rough times, I was beginning to understand Deepak's words. Finally, I was learning to stop trying so hard. It was through my experience with Carolyn, the meditation I learned to do, the many healers and alternative practitioners, open-minded M.D.s, massage therapists, and chiropractors I encountered that I moved closer and closer to what is my current, essentially spiritual outlook on life.

A NEW CROWD

During this period of labless existence, I increasingly accepted invitations to speak at conferences and meetings that convened for what I called the new-paradigm crowd—the practitioners and theorists, as well as the recipients, of alternative healing modalities. Though many of these modalities had a basis in Eastern philosophy and other non-Western traditions, which as a Western-trained scientist I would not ordinarily have known anything about, I had in fact had a limited exposure to Eastern ideas that dated back to the mid-eighties during my NIH lab days. Because of the growing public awareness of my research on endorphins and other neuropeptides, people from all kinds of unexpected backgrounds had sought me out at the time. A bearded yogi dressed in white and wearing a turban showed up at my office one day to ask me if endorphins were concentrated along the spine in a way that corresponded to the Hindu *chakras*. The chakras, he explained, were centers of "subtle energy" that governed basic physical and metaphysical functions from sexuality to higher consciousness. I had no idea what he was talking about, but, trying to be helpful, I pulled out a diagram that depicted how there were two chains of nerve bundles located on either side of the spinal cord, each rich with many of the information-carrying peptides. He placed his own chakra map over my drawing and together we saw how the two systems overlapped.

It was the first time I seriously considered that there might be a connection between my work and the Eastern viewpoint. Before he left, the yogi taught me some simple exercises for focusing attention at each of the chakra levels, which I experimented with and found highly enjoyable for the energizing effects they produced.

It was soon after that encounter that I experienced a whirlwind, California-style immersion in mind-body ideas when I spoke at an April 1984 symposium at Stanford University, sponsored by Eileen Rockefeller through the Institution for the Advancement of Health, on the theme "Can Positive Emotions Affect Disease?" I arrived at the event to find white-turbaned yogis mingling with buttoned-down medical researchers, one of my first glimpses into the new convergence of the Western, material viewpoint and the more Eastern, spiritual one (not to mention the convergence of the East and West coasts!). It was there that I began meeting people who had written and spoken on the interface between health and the mind, including Norman Cousins, whose book *Anatomy of an Illness* I had read not long before. As I listened to the various alternative theories and viewpoints on how emotions could modulate healing, I realized that back in my lab at Bethesda, I had been doing the science that explained many of the ideas that these Californians were exploring. Ideas that were familiar to them as ancient healing systems of the East were new to me, and they loved hearing me provide a scientific basis for what they had been experiencing and intuiting for some time.

When I arrived back home, the many bird eggs that had been precariously perched in nests on my front porch had all hatched, and the yard was filled with tiny, chirping offspring, eager and hungry for new life. It was a perfect reflection of how I felt: Ideas that had been gestating in my mind for years were now taking shape, asking to be born. I seriously began to put my science to work on discovering the possible molecular mechanisms of mind-body healing. But I didn't talk much about the ideas underlying my work with my fellow scientists (except for Michael and a few other like-minded colleagues), because they seemed too far out.

I "came out" on the East Coast in 1985, when an aide from the office of Senator Claiborne Pell knocked on my lab door to invite me to give the keynote address at a symposium cosponsored by his office and the Institute of Noetic Sciences. The theme, "Does Consciousness Survive Death?" made me hesitate, so afraid was I of sounding unscientific on my home turf, but the fat honorarium prompted a swift decision, and I accepted on the spot. Going public on the East Coast was a major breakthrough, forcing me to bring my theories out of the closet.

As a result of the noetic-sciences symposium, my popularity soared. It was my first science lecture to a mainly lay audience, and the first time I had let go of all my inhibitions about the philosophic and metaphysical implications of my work. The audio-taped presentation was edited and translated into written form by Harris Dientsfrey, appearing first in *Advances* as "The Wisdom of the Receptors: Neuropeptides, the Emotions and Bodymind" in 1986. A more technical presentation of this information had previously made its way into the *Journal of Immunology* as "Neuropeptides and Their Receptors: A Psychosomatic Network" in 1985. Because these papers were widely read by holistic and alternative practitioners, as well as by some more forward-thinking scientists and doctors, in the years that followed I began to receive more and more invitations to speak, many of them out West—in Los Angeles, San Francisco, Boulder, Seattle, even at Esalen at Big Sur.

By 1991, when I met Deepak, I had become a regular on the circuit, earning a reputation for myself as the "bodymind" scientist, meeting and benefiting from the thinking of such Western shamans as Stanley Krippner, Ernest Rossi, Stan Grof, Willis Harman, Fritjof Capra, Beverly Rubik, John Upledger, and Joan Borysenko. It was a stretch for my scientist mind to be open to their astounding theories and practices, but somehow I found myself able to straddle both worlds as I worked to integrate the best from each.

The ability to accept very diametrically opposite points of view is due, I believe, to the fact that I'm a woman. Because women have a thicker corpus callosum—the bundle of nerves that bridges the left and right brain hemispheres—they are able to switch back and forth from the rational, or left brain, to the intuitive, or right brain, with relative ease. With fewer nerves connecting the hemispheres, men tend to be more focused in one hemisphere or the other.

A high point in my speaking career occurred in 1991 when I spoke at the American Association of Holistic Medicine in Seattle. Arriving late for my talk, I was greeted by the smiling faces and open arms of many people I had heard of but not yet met, such as Jeanne Achterberg and Bernie Siegel. This instant and genuine acceptance of me and my work was in sharp contrast to the reception Michael and I got when we attended the many AIDS conferences, trying to win acceptance of Peptide T, and it made me feel totally at home with the new-paradigm crowd—as comfortable as I did with my more establishment friends and colleagues (as long as Peptide T wasn't the subject). The Seattle conference marked what I consider my personal merging of mainstream East Coast science and alternative California "healing" realms—and thanks to my generously endowed corpus callosum, I've been able to maintain an equal footing in both.

RESCUE

It was through a connection I made on the new-paradigm lecture circuit that a new investor for Peptide T was finally found.

I met Eckart Wintzen at a conference on "Medicine of the Future" in Garmisch, Germany, in late 1990. He had come to hear me audition as a potential speaker for his own conference, one he sponsored annually, to which captains of industry from all over Europe were invited. The current theme he was working on was simply "brains," and he'd been receiving pressure from people in his organization to add some women speakers to his conference roster. A fabulously wealthy Dutch business-man who had made his fortune in computer software and now altruisti-cally supported any number of advanced global projects, Eckart was a tall, slim man in his fifties, who wore his graying hair long and sported John Lennon glasses on his nose. After months of dealing with the cor-porate suits, I found him a breath of fresh air as well as an intriguing nov-elty—a rich, sophisticated European counterculture entrepreneur.

One of my two talks had focused on Peptide T, and afterward, over a lunch he'd invited me to, Eckart seemed especially interested in the progress of the drug and what its present commercial status was.

"Oh, it's fine," I lied. "We've got a great possibility coming up, a Japanese company that's about to come on line." I just couldn't bring myself to reveal how bad things actually were. But the truth was that we were at the end of a long string of potential deals that hadn't gone through, and our Japanese interest was looking dimmer every day. I men-tioned to Eckart that we had a deadline in a few months: We needed to come up with an investor that could prove it had the millions needed to bring the drug to market, or else the NIH Office of Technology would take the Peptide T license away from us again, perhaps permanently con-signing the drug to the limbo in which it had been languishing.

We finished lunch and strolled casually toward the restaurant exit.

"What's the date of your deadline?" he asked as he reached forward to open the restaurant door.

"April 4," I replied, and as we passed through the open door, he looked directly at me and said in an offhand way: "Well, give me a call if you need anything."

Eventually I did speak at his conference, but before that came to pass, our Japanese investor fizzled and the deadline was upon us. Michael and I literally had reached the end of the line, all hopes gone and all leads dead or dying. We were facing bankruptcy, and even the loss of the house we'd bought when we were riding high with Peptide Design. There had never been a gloomier hour. It was 9:30, the morning of the government dead-line, when the phone in our basement office rang.

"Hi, this is Eckart," a cheery voice said. My heart did a flip-flop. "How are you? Today's the day, isn't it, the deadline you told me you had for your drug? Did you find your company?"

"No," I said. "As a matter of fact, we haven't." There was a pregnant pause.

"Well then," he said, "tell me how much you need, and who my banker needs to notify that I have it."

With a simple fax that Eckart sent directly from his office in Holland, the government got what they needed, proof that Peptide T now had a major investor, one who could without a doubt supply the millions needed to support further research and development and successfully bring the drug to the marketplace.

We were back in business. True to Deepak's diagnosis of my problem, the solution had come only when I had stopped trying, for I had never gone after Eckart as an investor.

12

HEALING FEELING

THE MOUNTAINS of Southern California are spectacular in May, vibrantly green from the winter rainfall and dotted with bursts of colorful wildflowers and shrubs. My friend Nancy's trusty Volvo carries us up the steep and winding San Marcos Pass as we leave Santa Barbara and head for the Santa Ynez Valley. From the window, I look down the canyons to the coastal floor as we ascend, a dizzying sight of tiny houses and crisscrossing roads. The big, blue Pacific spreads out toward the horizon, where I can make out a spattering of distant, hazy islands floating on its surface. It's 1996 and we are on our way to an appointment with Dr. Robert Gottesman, an internist and alternative practitioner whose specialty is women's health.

How I love California! It's not just the stunning physical beauty but the style, the attitude, the healthfulness of the place. The pull that California exerts on me was recently put into focus by a note I received from Willis Harman, an electrical engineer/philosopher, known for his consciousness movement leadership at Stanford Research Institute and the Institute of Noetic Sciences. His note, in the form of a personal inscription written on the flyleaf of his book *Global Mind Change,* said that I was one of the few people he knew to have both an East Coast and a West Coast personality, depending on where my physical body happened to be located at the time! (Willis's book articulates what I believe to be the pivotal concept shaping the California movement, which is that consciousness creates reality, mind becomes matter, our thoughts precede our physical bodies, not vice versa. For many Asian thinkers this is a basic assumption, but for most Western Hemisphere types, it is an utterly foreign, revolutionary idea.)

As the car climbs steadily up the pass, I don't even try to contain my feelings of exuberance, thoroughly enjoying the moment and the rush of endorphins coursing through my blood. This is a trip that combines pleasure with business, although the two seem increasingly merged as of late. I am visiting first with my friend from childhood Nancy Marriott, and then plan to hop a plane to Anaheim, where I will present a talk at a conference playfully entitled "Medicine, Miracles, Music, and Mirth" at—how appropriately!—the Disneyland Hotel. Once again the token scientist, I'll be joining a roster of familiar alternative-health presenters, among whom will be my dear friend Carl Simonton, an oncologist who was one of the movement's earliest pioneers in his use of visualization, art therapy, and meditation to increase survival rates for cancer patients.

I've lost track of the number of times I've crossed the land to the far shore of California to speak at alternative-health conferences, meetings, and symposia, since the very first time, back in 1984, when I addressed the Institute for the Advancement of Health at Stanford University. Since that initiation into the holistic crowd, I have come to think of California as the vanguard in mind-body exploration, a place where Asian influences intermingle with the Western tradition, creating an easy acceptance of ideas that, for a long time, barely saw the light of day back on the East Coast.

Here in California there seems to be room for a consideration of the spiritual dimension of health, which can encompass such elements as prayer, energy flow, distant healing, and psychic phenomena, to name but a few. To me this seems absolutely the furthest edge to which my mind can go, but for many Californians, these are long-familiar ideas, dating back thirty years to the early days of Esalen in Big Sur, where Michael Murphy and his friends kicked things off.

Nancy has been my friend since elementary school, when her mother was my Brownie Scout leader, and we've remained close friends ever since. We were two girls who grew up together on Long Island, went to the same high school, came home to the same town during college years, and had the same dreams and apprehensions about the future. While I stayed on the East Coast, she headed west as soon as she'd finished graduate school at Columbia University, landing in San Francisco and becoming increasingly involved in the health and consciousness movement of the mid-seventies. Over the years, I made it a point to look her up whenever my lecture schedule brought me to San Francisco or, lately, to Southern California, where she now lives in Santa Barbara. Whenever we'd get together, we'd so often find that our lives had followed a parallel course—we gave birth to daughters a year apart, remarried around the same time, and kept converging on the same ideas at the same time—that we continued to feel an easy bond, a trusting resonance, regardless of the space and time that separated us. In fact, just knowing Nancy was living out West was comforting to me, especially during the ten-year period when I was crossing over from my establishment mentality to the more expansive "Californoid thinking."

As we drive and talk, we love to flash back on times when our paths intersected at pivotal points in our lives, often reflecting complementary stages in our personal growth. One of these times occurred in 1979, when Nancy was traveling with a few friends to New York and looked me up on her way through D.C. I was then in the midst of my post-Lasker trials and tribulations, and feeling very fragile, so Nancy offered to teach me what she called an "invocation for psychic protection," a ritual she'd learned from a spiritual teacher in California. It sounded hocus-pocus-y to me, to call on higher powers to protect me from evil, but on some intuitive level I sensed it might be helpful, and I certainly needed all the help I could get. At the Palace the next day, I recited the words she'd given me behind my closed office door, and immediately felt a sense of relief from the anxiety and tension I had been experiencing.

I continued to use the verses, especially at times when I felt particularly maligned and unsupported, gaining strength if only, I told myself, because of the psychological buffer they provided. But I now know that this odd ritual was a powerful form of prayer, and I can only surmise that it acted through some form of "extracorporeal peptide reaching," a form of emotional resonance that happens when receptors are vibrating together in seemingly separate systems. This was before the term *subtle energy* had been introduced to describe a still mysterious fifth force beyond the four conventional forces of physics—electromagnetic energy, gravity, and weak and strong nuclear forces—to scientifically explain anomalies such as the power of love. In this and in many other ways, Nancy was my pathfinder, introducing me to mystic and spiritual ideas that I then sought to understand within the context of my science.

NATURAL HORMONES

The rocky slopes and winding road soon give way to a panoramic vista of mountain caps dusted with snow and gently rolling, cow-studded fields. We have entered the Santa Ynez Valley and are rapidly approaching the tiny town of Ballard, where Robert Gottesman lives and practices medicine. Our visit has been prompted by the fact that as front-running members of the baby-boomer generation—we were both born in 1946—Nancy and I are now at an age to be making decisions about the latest hormone replacement therapies being offered up for menopause. With our female hormones on the wane, the question is whether to suddenly let nature take its course, when our recent ability to have sex without getting pregnant is not natural. A year earlier, Nancy had sent me a privately published book, *Natural Progesterone,* by Dr. John Lee, a Northern California physician/chemist whose pioneering work on natural hormone substances offers an alternative to the controversial, pharmaceutically manufactured hormone therapies such as Premarin, which are patented analogs of estrogen. As large numbers of our populous generation enter menopause, these drugs are selling in vast quantities. Premarin, for example, has jumped to the top of the charts in my hometown of Potomac, Maryland, replacing even Valium and Librium as the best-selling prescription drug. In fact, it is one of the top five most prescribed drugs in the United States.

Given that hormone replacement therapy is known to be associated with serious increases in breast cancer, and may pose other risks as well, this trend is disturbing. Why aren't doctors making available the natural, plant-derived forms of estrogen and progesterone, substances that are known to have fewer side effects than their laboratory-produced analogs? The answer reflects the economics of medicine: Since the natural substances are not patentable, there is no incentive for drug companies to study their benefits, and so the vast majority of M.D.s, who get their information about drugs from the drug companies, don't even know about them! Having heard that Gottesman had become a proponent of natural hormone treatment, following the lead of Dr. John Lee here in Southern California, Nancy has decided that this was the doctor we should be consulting.

WHEN WE arrive, Bob and his wife, Susan, greet us warmly and invite us into their small but stylish redwood-and-glass home. Buddhas and

Japanese water sculptures accent the modest space, providing a tranquil, natural setting, an East-meets-West flavor I rarely encounter back where I come from. Bob is a strikingly handsome man and, as I soon find out, another front-running baby-boomer like Nancy and me. His tall, slim build and sparkling blue eyes are offset by a shock of white hair, making him a study in contrast to his small-boned, dark-haired wife. Susan is a nurse and professional counselor who often works with her husband in their specialty, treating women with menopausal disorders. They are both familiar with my work, having seen me interviewed by Bill Moyers on the PBS special *The Healing Mind*, first aired in 1992, and make me feel as welcome as a member of their family.

Within minutes we find our common ground, which is an interest in mind-body medicine and its confluence with Eastern philosophy. But before we get too deeply into our conversation, Bob suggests we do the medical consultations first, after which we can reconvene in the living room for a talk and tea. We agree and Susan shows me to an office, where I am pleasantly surprised by the comfortable furniture and green plants and by the absence of antiseptic smells, instruments, and dressing gowns. Bob joins us and begins an extensive consultation consisting of a lengthy question-and-answer session about my current and past state of health. Although he is now in official doctor mode, he maintains the same warm, social demeanor he displayed when he first greeted us, listening thoughtfully to my answers and expressing a refreshing empathy and sensitivity. In a noticeable contrast to the procedures followed by more mainstream M.D.s, a large part of his professional assessment centers around my input about not only my physical status, but my emotional state as well. After the consultation, he recommends a variety of hormone and blood tests to check my existing levels of progesterone and estrogen—the two key hormones that are affected by the approach of menopause—and another consultation by phone once the results are in.

Bob gives me a jar of topical cream containing progesterone made from the wild Mexican yam, which I am to apply after the test results indicate what dosages would be appropriate. He explains that while both progesterone and estrogen start to fall off during menopause—in fact progesterone decreases for several years before menopause (perimenopause), because ovulation becomes very irregular during those years, and progesterone is released only after ovulation—many menopausal and perimenopausal women have an estrogen-dominant condition. This occurs because of a lack of sufficient progesterone to "oppose" the estrogen. Unopposed estrogen, Bob says, is the agent responsible for many of the symptoms of menopause: hot flashes, fibrocystic "lumpy" breasts, weight gain, and fluid retention. The topical progesterone he is giving me will be absorbed through my skin to protect me against these symptoms. Since progesterone is the "mother hormone," creating feelings of calm and nurturance (especially in pregnant and lactating women, who produce particularly high quantities of it), the cream may also alleviate the mood strains suffered by many women who are menopausal.

INFORMATION

Once our consultations are complete, Nancy and I find ourselves back in the cozy living room, where we are soon deep into a stimulating conversation about mind-body medicine, Eastern philosophy, information theory, and quantum physics. I am amazed to learn that Bob is the grandson of Karl Menninger of the Menninger Foundation, the forward-looking psychiatric clinic and teaching hospital in Kansas that supported work by biofeedback researchers Elmer and Alyce Green back in the 1970s. Gottesman is truly one of the new breed, an M.D. with one foot firmly planted in the Western tradition and the other stepping into the realm of alternative and Eastern traditions. And for him, it's all part of a family heritage.

Bob likes to philosophize.

"It seems to me that the way to heal the split between body and mind is to change metaphors," he suggests.

Metaphors? I am interested but wonder if we're going to be talking about poetry or science. It soon becomes apparent that he's combining the two in a unique and enlightening way that promises to offer clues to questions I've been thinking about for quite some time. A metaphor, after all, is just a way of looking at things, and could just as well be called a viewpoint or even a paradigm.

He continues: "The metaphor I propose to understand the mind-body question is one that uses information theory, a well-developed field with verifiable laws and theories that are very applicable to traditional sciences as well as to business and the humanities."

Now he has my attention. I myself have been theorizing for some time about a new mind-body biology, in which information is the bridge between mind and matter, psyche and soma. When Michael and I did our original research showing the link among immune, endocrine, and nervous systems, we chose words like *network*, *nodal point*, and *information molecules* in order to make the point that information-processing is what this system is all about, and we had later adopted Francis Schmitt's term *information substances* to refer to our neuropeptides and their receptors. So this is language that I understand.

"But first, I believe we have to make a distinction between the metaphor of matter and energy and that of information," Gottesman continues. "The older metaphor deals with matter, force, energy, and is expressed in Einstein's famous formula connecting those elements, $E=mc^2$. While these terms are useful for building locomotives and bridges, even atomic bombs, they are not so useful for understanding the human body. Physical processes aren't *things*, they are dynamic and take place in an open, fluid system, and therefore fit better with the metaphor of information than that of matter and force."

I'm beginning to understand what Gottesman is getting at. The older metaphor clearly belongs to what is still the reigning paradigm, a mechanical, deterministic view based on Newtonian physics. It was rigid rather than fluid, even macho in its reliance on force and control to accomplish goals, and could never apply to the nonhierarchical web of relationships that make up the biological systems in what we are calling the psychosomatic network—an altogether more feminine model.

Bob summarizes: "A generation ago, it was thought that the concept of matter and energy was the basis for understanding all phenomena. Today, the concept of information is replacing energy and matter as the common denominator for understanding all biological life and even environmental processes."

"Yes, and the neuropeptides and receptors," I say with new insight, "the biochemicals we call information molecules, they are using a coded language to communicate via a mind-body network. They are in the process of information exchange, having a two-way conversation—very different from what happens when there is a one-way push from behind, the way force works."

"Yes, and that brings me to another law of information theory," Bob continues, "which is that information transcends time and space, placing it beyond the confining limits of matter and energy."

We all look at him a little puzzled.

"To understand that," he explains, "let's backtrack and understand how Gregory Bateson defined information as 'the difference that makes a difference.' We all perceive the world by observing differences in our sensory fields, such as varieties of taste, texture, color, etc. For instance, a cow grazing in a meadow and a botanist strolling through the same meadow will both perceive the green grass as something that stands out from, say, the sky. But for a cow the grass will mean food, and for the botanist it will mean a possible sample to take home and study in the lab. The difference that makes a difference, then, is the difference *to the observer*. This is a very important concept in information theory, because including the observer in the equation admits a new level of intelligence to the system. In the old metaphor, we ignored the observer in an attempt to avoid any taint of subjective interference in determining reality. In the new metaphor, the observer plays an important role in defining the reality, because it is the observer's participation that makes the difference!"

I interrupt excitedly: "Oh, the consciousness of the observer—that's the link to quantum mechanics."

"Yes, exactly. Now, back to my point that information—the difference that makes a difference—does not change with time or space." To illustrate his meaning, Bob points to a cup of tea on the table. "The difference between me and the cup remains the same whether I'm here or I'm in Alaska. Information is not dependent on time or space, as is matter and energy, but exists regardless of these limits!"

I'm aware that Gottesman is moving onto some very radical ground with very profound implications. If information exists outside of the confines of time and space, matter and energy, then it must belong to a very different realm from the concrete, tangible realm we think of as "reality." And since information in the form of the biochemicals of emotion is running every system of the body, then our emotions must also come from some realm beyond the physical. Information theory seems to be converging with Eastern philosophy to suggest that the mind, the consciousness, consisting of information, exists first, prior to the physical realm, which is secondary, merely an out-picturing of consciousness. Although this is about as radical as my scientist's mind will let me get, I'm beginning to understand how such a view could coexist comfortably with the kind of science I've been doing.

"But let's move on to another important point of information theory, that of feedback," Gottesman continues. "According to Bateson, the greatest bite out of the apple of knowledge since Plato was the discovery of feedback. The idea of feedback comes from cybernetics, the scientific study of control processes in different systems. The word *cyber* derives from the Greek 'kybernetes,' meaning 'that which steers' or 'the helmsman'— the steerer of a ship. Now, the helmsman steers the ship by constantly adjusting the tiller in response to the information—or feedback—he is receiving from visual readings—via either sight or instrument. This is an example of a feedback loop."

"Yes," I interrupt. "I'm familiar with this concept from my brief experience of having crewed in sailboat races in the Chesapeake Bay. The common mistake of the inexperienced sailor is to anxiously trim the sails before receiving information about the boat's altered speed and direction. I had to learn to wait the seconds or even minutes until the sail catches the wind and the helmsman adjusts the tiller in response—then I could make use of the information, the feedback, to further trim the sails accurately.

"And the same principle functions in the psychosomatic network," I continue, "which is analogous to a boat sailing along as the result of a series of feedback loops. Cells are constantly signaling other cells through the release of neuropeptides, which bind with receptors. The signaled cells, like the helmsman or the sail trimmer, respond by making physiologic changes. These changes then feed back information to the peptide-secreting cells, telling them how much less or how much more of the peptide to produce. This is how both the body and the sailboat move forward, through a series of rapid feedback loops. A system is healthy—or 'whole,' a word that shares its origin with the word 'health'—when these feedback loops are rapid and unimpeded, whether they are occurring between peptides and receptors or between the helmsman and his tiller. I recently read in Fritjof Capra's new book, *The Web of Life*, how Walter B. Cannon, the famed physician-physiologist working in the 1920s, first formulated the concept of homeostasis as an inborn system of internal checks and balances to provide a relatively constant state within the body. Capra points out that Cannon's idea is perhaps the first vision of the organism as a closed circle of information flow."

"You're right," Gottesman says. "And I use the rapid feedback loop concept when treating patients. As you yourself have experienced during our consultation, I ask lots of questions to get my patients to pay attention to what's going on with them, to self-monitor. It takes time, which most doctors won't spend, but I do it because I want my patients to become aware of the difference that makes a difference to *them*. Those who are able to respond in this way, to do their own self-monitoring, get well faster, because they have more intelligence at work in their systems, more information to make changes that bring about improvement. So I think ultimately it saves time."

I'm thinking: This concept of the rapid feedback loop—it even explains the way I have done my science over the years. Most of the success my team and I have had resulted from a shortened feedback loop between performing an experiment and then using the results to make immediate changes or adjustments. In our AIDS research, Michael and I cracked the mystery of the peptide that fit the AIDS virus receptors by initiating a new experimental question each morning, getting the results in the afternoon, and then poring over the data every night for changes to be made the next day. This was part of my legacy from Sol Snyder—his philosophy of the rapid, one-day turnaround, what he called the "speedy flier."

"Oh, I understand," Nancy says, having thought over Bob's words and now ready to respond. "The faster or tighter the feedback loop, the more intelligence available to the system, whether it is used for health or for sailboat racing. Between doctor and patient, then, the more communication, the better, and the better the communication, the more health!"

We are silent for a moment. But suddenly Bob is off and running in a new direction:

"So, in my mind, all of this talk about metaphors and feedback loops points to a very basic question: Is the physical world of matter and energy 'real,' and do molecules really exist?"

I'm glad he's gotten back to what I call the ultimate chicken-and-egg question—whether consciousness precedes the physical or vice versa.

"I think it's best to look at molecules and the rest of physical phenomena as metaphors, devices we use in order to talk about something," Bob continues. "The equator doesn't really exist, but as a metaphor it's very useful, and in navigation, lives depend on it. I know these are uncomfortable realms for most mainstream scientists, but you seem quite liberal in your thinking, so I'm hoping you can absorb the more mystical significance of what I'm saying."

Could I ever.

Gottesman continues: "Consider that the body itself may be a metaphor, just a way of referring to an experience we all have in common. Maybe it's that we don't have consciousness, but consciousness has us."

Now we are on to a very Eastern drift, but I can follow with an open mind, thanks to the experiences I've had in the last decade or so.

Deepak's impact on me in such matters has been particularly profound, allowing me to make my own contribution to the give-and-take of this conversation.

"Let me tell you a story Deepak Chopra told an audience when he was introducing me for a talk at his institute," I offer. "He was on a visit to India when he met up with some wise men—the rishis, or sages, who are the spiritual leaders in that country. In the course of conversation, he attempted to explain my work to them, the idea that neuropeptides and receptors communicate as information molecules. But they could only shake their heads and give him very quizzical looks. Finally, the oldest and wisest appeared to suddenly get it. He sat straight up and, with an expression of great surprise, said, 'Oh, I understand. She thinks these molecules are real!' "

My tale gives us all a good laugh, and on that note we feel we can adjourn our discussion of the ultimate nature of the universe for the day. The warm sunlight that filled the living room at the beginning of our talk is now faded, and the sudden chill signals me it is time to be moving on.

Susan, who has been quietly absorbing our discussion up until now, turns to me and, while Bob and Nancy are walking toward the door, offers some wise, parting counsel: "You seem to be a person who has accomplished so much in the world of science. But I sense in you a desire to get more in touch with your soul, your spiritual self, your true femininity. Perhaps this may mean letting go of control, letting your husband take the lead in your joint research while you focus more on your health and your new way of being. There's a part of yourself that is longing to be born and requires only your nourishing attention."

She was right. Wasn't Susan offering me another version of the lesson Deepak had given me, to stop trying so hard, stop forcing the issue and pushing to develop my research, my AIDS drug? Instead, I should work to understand my life as a conversation, a web of interactions and relationships all connected with each other and all heading in the same direction, without my needing to push all the time. This was one of the lessons of meditation. Control belonged to the old metaphor, and was no longer useful in forwarding my goals, my quest. It was time to open up space to let Michael do more, and also to work with other scientists in joint research efforts, abandoning my stance as lone standard-bearer on the battlefield of AIDS research.

Our jars of hormone cream clutched firmly in our hands, Nancy and I said good-bye and started out on our journey back to Santa Barbara. As we rose from the valley and ascended to the pass through the mountains, we were both lost in our thoughts about the day's events.

WITH THIS new understanding of information science, I was beginning to see my theory that the neuropeptides and their receptors are the bio-chemicals of emotion in a new light. The emotions are the informational content that is exchanged via the psychosomatic network, with the many systems, organs, and cells participating in the process. Like information, then, the emotions travel between the two realms of mind and body, as the peptides and their receptors in the physical realm, and as the feel-ings we experience and call emotions in the nonmaterial realm.

Information! It is the missing piece that allows us to transcend the body-mind split of the Cartesian view, because by definition, informa-tion belongs to neither mind nor body, although it touches both. We must accept that it occupies a whole new realm, one we can perhaps call the "inforealm," which science has yet to explore. Information theory releases us from the trap of reductionism and its tenets of positivism, determinism, and objectivism. Although these basic assumptions of Western science have been ingrained in our consciousness since the six-teenth and seventeenth centuries, information theory constitutes such a new language—a rich language of relatedness, cooperation, interdepen-dence, and synergy rather than simple force and response—that it helps us break out of our old patterns of thought. Now we can begin to con-ceptualize a different model of the universe, and of our place in it.

BODYMIND

These were my ruminations as we pulled through the narrow mountain pass, our car hurtling us toward the pink sunset that spread out across the sky above Santa Barbara.

"So how was your consultation?" Nancy finally asked, breaking the silence.

"Like nothing I'd ever find back in D.C., I can tell you that," I responded. "I actually felt good after he was done, and not as if I were a piece of meat that had just passed inspection. In fact, I don't think I'll ever want to go to a regular doctor again—unless I fall off a building or my appendix erupts," I quipped, and we laughed.

Nancy laughed again. "I know what you mean. I felt the same way."

"Actually, it's been a long time since I've visited a regular doctor. I mostly go to chiropractors or get massage therapy in combination with nutritional advice for whatever ails me these days. Lately, I've been exploring Ayurvedic medicine, which is the traditional practice of India. Western-trained doctors like Gottesman are rare—an info-doctor!"

"Really, I liked what he was saying about the new metaphor," Nancy said. "It gives me a new way to think of myself. I'm no longer a machine made up of a body being pushed around by a brain, at the mercy of an electrical charge to keep my heart beating and my synapses crackling. Instead, I can now see myself as an intelligent system, one that involves a massive and rapid simultaneous exchange of information between mind and body. My cells are literally talking to each other, and my brain is in on the conversation!"

I had to agree. What Nancy was on to was not only a new self-concept, but a new sense of integrity, something I too was on the edge of embracing in my life, and which all my research pointed toward. The new self-image was one of an integrated body and mind, one with intelligence, an emotional intelligence, even a soul or spiritual component. And the undeniable implication was that each of us is a dynamic system with a constant potential for change in which self-healing is the norm rather than the miraculous. I nodded and waited for more.

"And now that I know my body has wisdom, this calls for a new kind of responsibility on my part. I can no longer act like a dumb machine and wait to be fixed by the mechanic—otherwise known as the doctor. Now I have the potential to consciously intervene in the system myself, to take an active role in my own healing. I'm both more powerful *and* more responsible in creating the health I experience than the dumb machine I used to think I was."

"Exactly!" I agreed, for I now saw the connection between Nancy's "conscious intervention" and what Deepak talks about in his books when he describes meditation as *intention*—the plan, the agenda, the focus. When I say this to Nancy, she gets it immediately. "Yes," Nancy said, herself a meditator for many years. "And that's how I understand meditation to work, as a process of putting forth my intention, having a purpose, and then simply paying attention to whatever that is by staying focused mentally. The focus could be a mantra, my breath, or some other point of concentration, such as healing my body or sending peace to the planet. I know I'm already intervening *unconsciously* in the system through my normal everyday thoughts, even sometimes to the detriment of the system—you know, overadjusting the sails, to use your phrase, or somehow gumming up the works, the natural balance—the homeostasis, as Cannon called it. And so I can choose to stop doing that, to intervene with intention."

I nodded. In the old reductionist model, chronic illness such as heart disease and cancer are seen as forces attacking the body, making us helpless victims, incapable of any response outside of high-tech medical treatments. But the concept of conscious intervention adds a new element to the equation, a scientifically valued intelligence that can play an active role in the healing process. Meditation is just another way of entering the body's internal conversations, consciously intervening in its biochemical interactions.

We arrive back in Santa Barbara shortly after sunset. I intend to go to bed early, still needing to catch up to West Coast time. But my body-mind is abuzz with the new ideas we've been tossing about, many of which I plan to explore and digest, perhaps sharing some of them with my conferees in Anaheim on the morrow.

HAPPINESS

Disneyland, the magical kingdom of eternal childhood, the promise of happiness everlasting!

When I received an invitation to present at a conference that would be held at the Disneyland Hotel, I was a bit surprised at first. Maybe it's my old-paradigm roots, but the incongruousness of having a medical meeting in a location associated with cartoons and fantasy—well, it struck me as odd. But as I read the title and schedule of speakers and events, I quickly caught on. "Medicine, Miracles, Music, and Mirth." This was to be a coming together of researchers, practitioners, musicians, even comedians to focus on how psychoneuroimmunology, nontraditional healing, and fun all interface.

I was actually thrilled by the prospect of being at Disneyland and staying right next door to the theme park of my childhood dreams at the Disneyland Hotel. Back in the fifties, when I was growing up in Levittown, New York, every normal, red-blooded American kid I knew dreamed of winning a trip to Disneyland. The closest I ever got was wearing my Mouseketeer ears while I sat for hours in front of the TV watching the Mickey Mouse Club. The idea of my family taking a trip to California and entering the fabled gates was unthinkable. No one we knew had ever gone to Disneyland. It might as well have been on Mars.

Now, forty years later, I am stepping out of the airport shuttle at the Disneyland Hotel. Above me, the monorail cruises along, packed with happy campers on their way to the fairy tale come true, and in the distance, the peaks of some mysterious castle loom, suggesting the land of make-believe is not so far away. It all seems a bit surreal to my scientist's mind, although the eight-year-old inside me is glad to be here.

And everyone seems so happy—the bellmen, the registration clerks. I know *I'm* supposed to be happy—I've finally made it to Disneyland! But, to be honest, I'm mostly feeling tired from jet lag, and the thought that my fourteen-year-old son Brandon is three thousand miles away and unable to share this with me makes me a little sad. And I miss Michael. I left D.C. in the midst of a crucial time for Peptide T, just as we had gotten word that an outside evaluator was coming to our lab at Georgetown to perform some new tests, which, if successful, might advance our work significantly. Leaving Michael to handle it on his own wasn't easy, but I console myself that it is part of my new attitude of trusting to the flow. Still, I can't help but wonder how things are going, and I have a hard time resisting the impulse to pick up the phone. Let it go, I tell myself. I'm here to have some fun!

I decide to get a bite to eat and check out the scene at one of the many outdoor cafes wrapped around the central waterway. Familiar Disney tunes waft through the air, piped in over an all-pervasive sound system, and I find myself singing along with many of my childhood favorites as I stroll: "When You Wish upon a Star," "Someday My Prince Will Come." Seating myself at a table, I order a meal and relax.

"Are we having fun yet?" The popular catchphrase runs through my thoughts as I observe the scene. How all-encompassing is the Disney myth of happiness! I grew up on it and so did my kids—and so will theirs, most likely. For us baby boomers, especially, the land of Disney looms large as a symbol of childhood happiness, the kind we were supposed to have, wanted to have, but often didn't have. Recently, I read a poll in which Americans were asked the question: Are you happy with your life? A surprisingly high percentage responded yes. Surprising because the statistics show that mood disorders such as depression and anxiety are on the increase, becoming more widespread all the time. Since clinical depression is a potentially fatal disease, depression-related suicides are also on the rise. I have to ask: If everyone's so happy, why is depression at near-epidemic proportions in our society? Are we all in denial, clinging to what we believe is the cultural norm, what is socially expected of us? Are we ashamed to admit we might be sad, unhappy, disappointed, and not altogether satisfied with life?

As discussed earlier, many people view depression as anger turned in on itself, unexpressed, buried below consciousness where it seems to be controlled but slowly implodes. As a culture, we keep our feelings hidden, afraid to express them honestly for fear others will be indifferent to our sorrows or alienated or hurt by our anger. Better to deny feelings, to suppress them, we tell ourselves, go through the motions of happiness and pretend to have fun—until the day the bottom falls out and the family physician hands us the diagnosis: depression.

It is this problem of *unhealed feeling*, the accumulation of bruised and broken emotions, that most people stagger under without ever saying a word, that the mainstream medical model is least effective in dealing with. When people do seek help, often what is offered through mainstream psychology and psychiatry is what I call "talk and dose" therapy: lots of talking and even more pills, which are supposed to make the unacceptable feelings go away. A treatment, yes, but one that really only Band-Aids the symptoms and consigns people to a drug dependency rather than directing them toward an opportunity for really healing feeling.

What is not given much attention by the mainstream is what it means to be antidepressed, i.e., happy. I believe that happiness is what we feel when our biochemicals of emotion, the neuropeptides and their receptors, are open and flowing freely throughout the psychosomatic network, integrating and coordinating our systems, organs, and cells in a smooth and rhythmic movement. Health and happiness are often mentioned in the same breath, and maybe this is why: Physiology and emotions are inseparable. I believe that happiness is our natural state, that bliss is hardwired. Only when our systems get blocked, shut down, and disarrayed do we experience the mood disorders that add up to unhappiness in the extreme.

I return to my hotel room after a brisk walk around the grounds and find a message from Michael, but it's too late on the East Coast to return his call. Sleep comes easily.

UNHEALED FEELING

The next morning, after my lecture, I'm making my exit toward the stairs leading up to the main lobby, when I hear a voice behind me.

"Dr. Pert? Do you have a moment?"

Barely. I've been invited to lunch by two women who approached me after the lecture. They were a most unusual duo, a physician and a psychic, friends who had come to the conference together. I was planning to sit in the sun for the next half hour before meeting them at an Italian bistro.

However, as I turn around I see the warm, smiling face of a very determined woman pursuing me up the stairs. "Sure I do," I say, finding her hard to resist, even though I have no idea what I'm getting into.

We make our way to a sunny spot alongside the centrally located watercourse and seat ourselves at a table. Happy kids and less happy parents pedal their brightly colored boat bikes along the water as we place an order for cold drinks. Marilyn, it turns out, is a licensed marriage, family, and child counselor, an MFCC, with a thriving practice in Northern California, where, she tells me, she's noticed a disturbing trend. She wastes no time getting to the point.

"It seems that ten years ago when I first began my practice, I rarely saw a client who was on antidepressant drugs," she tells me. "Maybe Valium or Librium occasionally, but those are relatively harmless muscle relaxants. Now I'm seeing Prozac, Zoloft, Paxil, Serzone, Tofranil. Most of my clients are on one or another."

I understand Marilyn's puzzlement about this sudden upturn in the use of prescription antidepressant drugs and I've talked to many psychotherapists who, like Marilyn, are confused and concerned about what appears to be the ubiquitous medical solution to the epidemic of depression.

"Just recently," she continues, "I asked a psychiatrist who's affiliated with my counseling group why so many of my clients are getting prescriptions for antidepressants. He explained that the drugs correct chemical imbalances in the brain that are the cause of depression, and, for some people, they are more effective than other kinds of therapy."

As Marilyn talks, my thoughts flash back to my dinnertime musings the night before. Antidepressant medications, along with many of the drugs given during childbirth, are examples of drugs that would be given much more sparingly if there were better knowledge of the full range of peptidergic activity in the bodymind.

"What are these drugs doing to our bodies and minds?" she asks. "Do you think so many people should be taking them?"

In my lecture that morning, I had explained how the psychosomatic network operates through a series of delicately balanced peptidergic feedback loops, and when the flow of chemical information is unimpeded, it results in homeostasis, or balance. The question of how legal and even illegal drugs enter that network and affect the natural homeostatic balance was only hinted at in my talk.

"Let's talk about what these drugs do first," I begin. "Basically they go to work at the level where brain cells are communicating with each other across the synapse. Chemicals are squirted out by one and bind to the receptor of another. If too much juice comes out, there is something called a 're-uptake' mechanism by which the cell reabsorbs the excess. The classical understanding of depression is that there is a shortage of the neurochemical serotonin secreted by the brain cells. To remedy this, an antidepressant drug is used to block the re-uptake mechanism, allowing the excess serotonin to flood the receptors, and thereby correct the imbalance."

"Sounds very precise, like they know just what to do," Marilyn interjects.

"Yes. But it's a false precision, because it doesn't measure what else is going on in other parts of the brain and body when these drugs are administered. Remember, we're dealing with an immensely complex psychosomatic network, one with trillions of shared components—the peptides and receptors—throughout many systems and organs. Your intestines, for instance, are loaded with serotonin receptors. What happens when these receptors get flooded with excess serotonin as a result of taking Prozac, for example? Well, it's known that people on antidepressants often have gastrointestinal disorders. And think what might be happening to cells in the immune system that also have these same receptors on their surfaces. We could be inadvertently affecting the ability of our natural killer cells to attack mutated cells that are on their way to becoming a cancerous tumor. But no one's doing the research to explore these kinds of effects."

"Certainly not the pharmaceutical companies," Marilyn says, quickly picking up on my drift. I nod.

"And the antipsychotic drugs—Haldol, Thorazine, Risperdal, Clozaril—work the same way and have many of the same side effects, only instead of blocking the re-uptake of serotonin, they block the receptors for dopamine, another neurotransmitter."

"Besides gastrointestinal disorders, what other kinds of effects can occur?" Marilyn asks, obviously concerned.

"There's a cascade of things going on, kind of like a waterfall that starts at the top but initiates changes all along the way to the bottom. For instance, when the dopamine receptors in the pituitary glands of women are blocked, prolactin is released, a hormone that stops ovulation during lactation, so women rarely get pregnant while breast-feeding. Women on these drugs stop menstruating and level off as long as they're on the drug, in a constant state of PMS, complete with water retention and weight gain."

"Sounds like they'd be good candidates for Prozac," Marilyn says wryly.

"And that's exactly what often happens, I'm sorry to say! These women are then given antidepressants on top of the antipsychotic drug. This is not uncommon, this treatment of what is called an "iatrogenic disorder," meaning a physician-caused disorder stemming from the treatment that's supposed to cure the patient."

Looking at Marilyn's now grim face, I'm getting the feeling that she will be on the brink of depression herself if I keep going down this road. "Why don't you join me and some new friends for lunch?" I ask her, trying to switch gears. "Maybe they'll be interested in hearing what the latest research is telling us about the good news—the potential for mind-body approaches to cure mood disorders."

NEW LIGHT ON DEPRESSION

We go in search of Kate and Dee, the doctor and the psychic, who by now are waiting in line for a table. Once seated, we all choose large salads from the menu, and enjoy our meal while getting further acquainted. I am intrigued by the idea of their friendship, and grill them politely. How do two people from such diametrically opposed paradigms meet, much less become close friends? The story unfolds that during her internship, Kate participated in a study that showed certain forms of hands-on healing and prayer were effective in speeding recovery rates for surgery patients, and Dee was one of the healers. Since then, they've become good friends, pursuing a common interest in the mind-body healing connection. Which gets us back to the topic Marilyn and I have been discussing, the concept of depression as a mind-body disease and the medical profession's tendency to overprescribe and overmedicate it, while ignoring possible side effects.

"My sister has been on Prozac for years," Kate blurts out. "I don't think we know enough about the stuff, and I've told her so. But she's convinced that the drug is the answer to her problems. And she refuses to deal with any of the underlying issues that I believe to be at the heart of her problems."

"And right you are to be concerned," I begin, delighting in the chance to take my listeners a little deeper into the science. "Just recently, researchers at the National Institutes of Health have found a link between depression and traumas experienced in early childhood. Studies have shown that abused, neglected, or otherwise unnurtured infants and children are more likely to be depressed as adults, and now we have a way to understand the link between the experience and the biology. It all relates to something called the hypothalamic-pituitary-adrenal axis."

Our waiter arrives just as I am getting launched. A platter of chocolate mousse, cream puffs, and cheesecake is thrust under our noses by a beaming young man. We admire the goods politely, but no one takes him up. Ah, California, land of the healthy and conscious! He leaves and I get back on track.

"Very simply, the hypothalamus is part of the emotional brain, the limbic system, and its neurons have axons that extend into the pituitary gland, which sits below it. There axons secrete a neuropeptide called CRF—cortical releasing factor—which controls the release of another informational substance. Thus, when CRF hits the pituitary gland, it stimulates the secretion of ACTH, an informational substance that then travels through the bloodstream to the adrenal glands, where it binds to specific receptors on adrenal cells." I am getting some puzzled looks. "Is everybody with me so far?"

"Adrenal glands—don't they have something to do with adrenaline? The 'fight or flight' response?" Dee asks me.

"You've got it. Adrenaline is what causes the fight-or-flight alarm response, which is the body's natural, unconscious reaction to threats, either real or imagined. Our ancestors put it to good use when the saber-toothed tiger threatened to leap from a precipice and have them for lunch. It's often characterized by an energy rush, dilated pupils, a racing heartbeat—all conditions that enable us to deal effectively with perceived danger. But another thing that the adrenal glands do, which is what happens when ACTH hits them, is that they begin to make steroids. These are not the steroids associated with sex and the reproductive system, however.

"The steroid they make is corticosterone, a substance that is necessary for healing and damage control when an injury has occurred. You've all probably applied cortisone cream for a skin rash or had a cortisone shot to treat poison ivy or oak.

"Now, here comes the connection to clinical depression. Ever since studies done thirty years ago, we've known that stress increases with increased steroid production. Depressed people typically have high levels of these stress steroids. In fact, depressed people are in a chronic state of ACTH activation because of a disrupted feedback loop that fails to signal when there are sufficient levels of steroid in the blood. So the CRF-ACTH axis just keeps pumping out more and more steroids. Autopsies almost always show a tenfold higher level of CRF in the cerebrospinal fluid of those who killed themselves compared to those who died from other causes."

"Sounds like this CRF is the depression peptide," Kate says. "Assuming that a given peptide has a specific tone, that is," she continues, referring to some of the more speculative excursions of my lecture.

"It sure looks that way. We could say that CRF is the peptide of negative expectations, since it may have been stimulated by negative experiences in childhood. There are animal studies, for example, showing that monkey babies deprived of maternal nurturing, neglected or abused, in effect, have high levels of CRF and so have high steroid levels. Remember, it's a feedback loop that is out of control. Depressed people are stuck in a disruptive feedback loop that resists any kind of drug therapy that aims at suppression of the steroids. Eventually, there is so much CRF in the system that fluctuations of other peptides throughout the organism are curtailed, leaving ever fewer possibilities in the range of behavior. In baby monkeys, this takes the form of failure to groom or repetitive behaviors that don't seem to have any purpose. In humans, the result can be extremely limited patterns of behavior and response, which eventually drive people into an emotional black hole."

"My sister is convinced that if only her husband hadn't left her, everything would be okay—she can't seem to get past that," Kate interjects.

"Yes, and the reason we can get stuck like this is because these feelings get retained in the memory—not just in the brain, but all the way down to the cellular level. This is how it works: As CRF levels increase in highly stressed infants and children, the receptors for CRF become desensitized, shrinking in size and decreasing in number. These changes happen when receptors are flooded with a drug, whether it's a drug your body produces naturally or a drug you buy at a pharmacy. The memory of the trauma is stored by these and other changes at the level of the neuropeptide receptor, some occurring deep in the interior of the cell at the very roots of the receptor. This is taking place bodywide. Although such changes can be reversed and need not be permanent, this takes time."

"So what's the latest remedy from the researchers?" Marilyn asks. "More drugs to decrease CRF production or block CRF receptors?"

"Unfortunately, drug therapy is the main direction for research at this time. But the good news is that these findings let us see the potential for nondrug interventions, new kinds of treatments for mood disorders. Remember the stressed-out monkey babies? In another study to determine maternal influence, a group of monkey babies was raised by a fake monkey mother, a wire-and-cloth structure with milk bottles instead of breasts. The babies were fed but not touched, cuddled, or held. They soon had all the signs of trauma and depression, as would be expected in light of all we've just talked about. But they were cured—the stress symptoms reversed—when researchers brought in what they called a 'monkey hug therapist,' an older monkey who constantly hugged and cuddled the stressed-out baby monkeys. So what was going on? The hugging broke the feedback loop, sending the message 'No more steroids needed,' damage over and done with! The chronically elevated CRF levels came down."

"So when we see those bumper stickers 'Hugs Not Drugs,' we should take them more seriously!" Dee points out. We all laugh at the realization that the science I've been laboring at such length to explain is obvious enough to have become popular bumper-sticker wisdom.

So obvious, I'm thinking, but not obvious enough to change the agendas of pharmaceutical companies or the mainstream medical model. As a researcher on the drug frontier for over twenty years, I have to depart from the opinion of most of my colleagues in the mainstream and say that *less is best*. The implications of my research are that all exogenous drugs are potentially harmful to the system, not only as disrupters of the natural balance of the feedback loops involving many systems and organs, but because of the changes that happen at the level of the receptor.

Each of us has his or her own natural pharmacopoeia—the very finest drugstore available at the cheapest cost—to produce all the drugs we ever need to run our bodymind in precisely the way it was designed to run over centuries of evolution. Research needs to focus on under-standing the workings of these natural resources—our own endogenous drugs—so that we can create the conditions that will enable them to do what they do best, with minimal interference from exogenous sub-stances. But when they can't do their job, such research will also enable us to create mimetic drugs that imitate the natural substances and cause minimal interference with the bodymind's balance because they have been developed with an awareness of the whole psychosomatic network.

"Of course, I'm not suggesting that hugs alone can cure all our major ills," I say. "The prescription drugs do serve a purpose, and I recognize that they save people's lives. If I have a bad infection, I'm going to use an antibiotic. If I have a serious clinical depression, I will take an anti-depressant. But from my research with the endorphins, I know the power of touch to stimulate and regulate our natural chemicals, the ones that are tailored to act at precisely the right times in exactly the appro-priate dosages to maximize our feelings of health and well-being."

I had experienced this personally when, during the early days of our relationship, I jokingly called Michael my "monkey hug therapist," because we were in one gigantic, continual hug most of the time, and we felt happy and high most of the time. We did some of our most exciting work "under the influence" of those hugs. Later we relied on them for solace. In fact, sometimes I wonder how we would ever have withstood the stress that we encountered in the early days of our Peptide T strug-gles without those hugs.

"I get it," Kate says thoughtfully. "You're saying that by bringing touch into the healing process, we may be able to offer another kind of help to people with mood disorders. It's the other half of the equation: Just as we can harness the power of our minds for physical healing, so can we do physical things to help heal our feelings."

"So that's why people feel better when they get a good massage or other kinds of hands-on healing," Dee joins in.

But Marilyn seems disturbed by this turn of the conversation. "Well, the sorry fact is that most of us in the mental-health professions, espe-cially the psychiatrists, the M.D.s, who prescribe the drugs, would get our licenses revoked if we touched our clients!"

"You're right," I respond. "Mainstream medicine is notoriously touch-phobic for the most part, dating back to the original days of the Cartesian split and perpetuated by ignorance of how sensory informa-tion is processed in the psychosomatic network. There's a whole history, beginning with Freud, who laid the cornerstone of modern psychiatry as a no-touch affair back in the Victorian era when people were so uncomfortable with their bodies that any kind of touch was—God forbid—considered part of the sexual domain. Other contenders, such as Wilhelm Reich and Alexander Lowen, made attempts at introducing a more body-linked approach, believing the body to be a gateway to the mind and working with different forms of emotional release. But they were ferociously marginalized, even persecuted in Reich's case.

"In recent years, however, there have been loads of animal and human studies that show the benefits of touch, not only for depression but for illnesses that have physical symptoms as well. And I'm glad to report that this knowledge is entering the medical mainstream, if rather slowly."

Our bills arrive just then, and on an upbeat note, we decide to move on. Outside, the crowds have thinned, with most of the hotel guests having left to spend the day at the park, and there is a peaceful ambience to the place. Together we stroll back toward the convention hall to catch the remaining presentations before the afternoon workshops begin. But halfway there, I tell the others to go ahead without me. I'm listening to my bodymind, and the message I am getting is to kick back, find a warm concrete slab out of the way of the traffic, and stretch out in the sun.

BODY PSYCHOTHERAPY

Our lunchtime talk has touched on a frequent theme in my thoughts of late, the question of *healing feeling*, something so desperately needed in our society, as reflected by both the rising numbers of people on antidepressant medications and the escalating use of illegal drugs. In my mind, both kinds of user—the one who gets the drugs from a doctor and the one who buys them from a dealer—are doing the same thing: altering their chemistry with an exogenous substance that has widespread effects, many of which are not fully understood, in order to change feelings they don't want to have.

My research has shown me that when emotions are expressed—which is to say that the biochemicals that are the substrate of emotion are flowing freely—all systems are united and made whole. When emotions are repressed, denied, not allowed to be whatever they may be, our network pathways get blocked, stopping the flow of the vital feel-good, unifying chemicals that run both our biology and our behavior. This, I believe, is the state of unhealed feeling we want so desperately to escape from. Drugs, legal and illegal, are further interrupting the many feedback loops that allow the psychosomatic network to function in a natural, balanced way, and therefore setting up conditions for somatic as well as mental disorders.

But the idea of the network is still too new to have affected the way mainstream medicine and psychology deal with our health and our illnesses. Most psychologists treat the mind as disembodied, a phenomenon with little or no connection to the physical body. Conversely, physicians treat the body with no regard to the mind or the emotions. But the body and mind are not separate, and we cannot treat one without the other. My research has shown me that the body can and must be healed through the mind, and the mind can and must be healed through the body.

The so-called alternative therapies that focus on somatic-emotional release understand this, and it is through them that we can complement what is offered by the mainstream. In the case of treating mood disorders and other mental unwellness, the mainstream misses a lot by excluding touch, by ignoring the fact that the body really is the gateway to the mind, and by refusing to acknowledge the importance of emotional release as a mind-body event with the potential to supplement or even sometimes replace talk cures and prescription pills.

I first came in contact with somatic-emotional release approaches, also called "body psychotherapy," at Esalen in California, when I gave a talk there in the early eighties. The Greeks and Romans had their baths, their spas, their temples of healing at places like Epidaurus, and we have Esalen, where beautiful, natural springs come from deep within the earth to fill the pools perched high on a bluff over the Pacific at Big Sur. When I spent some time in the baths of Esalen, I met many massage therapists, chiropractors, and theorists who saw my research as a confirmation of what they were seeing in their practices. I was very impressed by their power to simultaneously access the emotions through various kinds of body work while enlisting the power of the mind through talk, thus creating what seemed like a loop of healing.

This exposure opened me to other kinds of alternative healing modalities that aim to release emotions through different processes, but always by involving some form of touch. One of the most dramatic experiences I had occurred in 1985 as a result of a chance encounter with an old friend and Bryn Mawr alumna, Caroline Sperling, a psychologist who founded her own cancer foundation. It was shortly after my divorce from Agu. When she asked me how I was doing, I told her I was fine, that everything was amicable and civilized. But she stopped me in midsentence.

"You're lying," she said bluntly. "How can you not have pain?" I was taken off guard. "Don't you know?—that's how people get cancer. By burying their emotions, denying and repressing them."

Intuitively, I knew she was right, and I listened. Caroline, who had been living with cancer for three years, told me about a practice she'd developed that combined techniques from Janov's primal scream with Lowen's bioenergetics to bring about emotional release through movement, hugging, and screaming. Soon afterward I attended one of her daylong sessions and found that it enabled me to unleash the torrent of anger and hurt I'd been bottling up inside of me ever since the divorce. I returned home eager to tell Michael all about it, but was so exhausted I went straight to bed and slept for almost twenty-four hours.

It was at a Common Boundary meeting in 1988 that I met Bonnie Bainbridge-Cohen, who introduced me to her bodymind centering technique, an approach that grounds mental, emotional, and spiritual elements in the physical body. (The common boundary that gives its name to the organization is that shared by psychology and spirituality.) I was impressed with Bainbridge-Cohen's very accurate understanding of how trauma and stress are forms of information overload. She used the mechanism of nerve reversal to explain how impulses are rejected by the brain and bounced back to other areas of the central nervous system, where they are stored in both the autonomic and somatic tissues. Bonnie's approach uses movement and body work and is based on these psychological and physiological principles.

More recently, I have discovered a new breed of chiropractors who differ from the conventional ones in that they bring an awareness of energetic, emotional levels into their healing. One of these is Donald Epstein, who founded the school of Network Spinal Analysis Chiropractic and wrote a book entitled *The 12 Stages of Healing: A Network Approach to Wholeness*. I have had some profound experiences while being treated with this method, involving the release of traumatic stored memories from the autonomic ganglia on either side of the spinal cord. Often visual images related to the trauma have surfaced in my consciousness as part of the emotional release, which I can then talk about with the practitioner.

Another healer I respect tremendously is John Upledger, creator of craniosacral therapy, a modality that aims to balance the cerebrospinal fluid through gentle manipulations of the cranium. He talks about "somato-emotional cysts," pockets of blocked emotion held in the body, causing a breakdown in the energy flow and general health. Epstein and Upledger both refer to "feeling the energy" as they do their work, while other practitioners report actually seeing energy move in the body as the emotions are released.

What is this "energy" that is referred to by so many alternative healers, who associate it with the release of emotion and the restoration of health? According to Western medical terms, energy is produced strictly by various cellular metabolic processes, and the idea that energy could be connected to emotional release is totally foreign to the scientific mind. But many ancient and alternative healing methods refer to a mysterious force we cannot measure with Western instruments, that which animates the entire organism and is known as "subtle" energy by metaphysicians, *prana* by Hindus, *chi* by Chinese. Freud called it libido, Reich called it orgone energy, Henri Bergson called it *élan vitale*. It's my belief that this mysterious energy is actually the free flow of information carried by the biochemicals of emotion, the neuropeptides and their receptors.

When stored or blocked emotions are released through touch or other physical methods, there is a clearing of our internal pathways, which we experience as energy. Free of the Western dualism that insists on disanimated flesh, healers from various Eastern and alternative modalities can literally see the mind in the body, where it does indeed exist, and are adept at techniques that can get it unstuck if necessary. In fact, almost every other culture but ours recognizes the role played by some kind of emotional energy release, or catharsis, in healing.

Approaches that manipulate this kind of energy are almost unanimously rejected by most of Western medicine, with the possible exception of acupuncture, a discipline still looked on with suspicion. Yet the effectiveness of acupuncture has been clearly documented in numerous studies, including ones I myself have been involved in. Back in 1980, I did some work with my husband Agu and Larry Ng, a Chinese psychiatrist and neurologist in the Western tradition, which was published in *Brain Research*, showing that acupuncture stops pain by stimulating the release of endorphins into the cerebrospinal fluid. We were able to demonstrate that it was indeed the flow of endorphins that caused the pain relief, because when we used an endorphin antagonist (naloxone) to block the opiate receptors, the pain-relief effects of acupuncture were reversed. As interesting as this work is, however, it only begins to address the manifold implications of the psychosomatic network and its potential for healing. The body psychotherapists, people who know how to help us tap into this network, are showing us many other uses for the "info-energy" that coordinates all our systems. We need to listen to them, learn from them.

PLAY

I feel restored by my rest—even energized—and lift myself off the sun-warmed concrete slab and head back to the lecture hall to catch the last of the presentations. I arrive to find that it is the closing session and the last speaker is winding down. Soon, a playful group takes over the stage, singing, dancing, and playing musical instruments, and there I spot my buddy Carl Simonton, who signals me to join the crowd. I hesitate a minute, not quite sure whether such antics fit with my scientist self-image, but I easily overcome my resistance and find myself hopping up on stage, where I sway rhythmically with my fellow conferees. It's fun, it's Disneyland, and I'm ready to get some quality stress-reduction after my busy day. But play is more than simple stress-reduction. It serves an important function in both animal and human life. We see this in young animals who regularly engage in mock battles as an important part of their development. Like them, we can use play in many ways—to act out our aggressions, fears, and griefs, to help us gain mastery over these sometimes overwhelming emotions. When we are playing, we are stretching our emotional expressive ranges, loosening up our biochemical flow of information, getting unstuck, and healing our feelings.

It's play that does it for me, lets me fully express myself and prevents me from taking myself too seriously. And that's what I do for the rest of the evening, greatly assisted in my effort to have fun by Carl's two young children, who are having the time of their lives with Mickey and friends.

For one brief evening, I too have been a child again. The thoughts of our business deals with Peptide T are far behind me, and I put on hold all my anxieties about the outside evaluators coming to our lab, the teenaged son who's now too old to play with his mom, the necessity to play the role of important scientist. I laughed, which Norman Cousins calls internal jogging, an exercise to keep us in emotional shape, I played, I let the emotions—and the peptides—flow.

There's much more to health than play and progesterone, of course. And achieving optimal health involves more than just minimizing our drug dependencies, or maximizing our capacity for self-expression. The final chapter of the book and the Appendix explore many more of the ways you can put the practical implications of my research to work in your lives.

13

TRUTH

WELLNESS

I'm circling high above the General Mitchell Airport in Milwaukee, my plane preparing for the descent. Looking out my window, I see the great plains of the Midwest stretched out beneath me toward the horizon and, off to the left, a huge and glorious body of water, which, I'm surprised to learn, is Lake Michigan. This is the heartland, the geographical center of the country, and a psychological center as well. Here folks don't care much for extremes, showing little interest in either the cerebral intensity of the East Coast or the laid-back, touchy-feely attitude so prevalent on the West Coast.

My plane lands, and after collecting my bags, I find my way to the prearranged meeting point for my ride to northern Wisconsin. I'm glad to be here in the Midwest, to breathe in the fresh, sweet air and leave the fast lane in D.C. behind me, the current triumphs and frustrations with Peptide T firmly in Michael's hands for the time being. It's the summer of '96, and I have come to Wisconsin to present at the Twenty-first Annual National Wellness Conference.

The Wellness Conference is the product of the "wellness movement," a grass-roots effort that took off back in the seventies when a group of exercise physiologists, nutritionists, psychotherapists, an ex-priest, and an ex-nun found common ground in their commitment to fitness and healthy lifestyles as the optimum route to disease prevention. The movement was the Midwest's answer to the alternative medicine/

consciousness movement originating at places like Esalen on the West Coast, only minus the New Age patina. Although the initial impetus was local, advocates of health and wellness from all across the nation are now part of the effort, coming together in a huge annual conference sponsored by the University of Wisconsin every summer at its Stevens Point campus.

My first contact with the wellness folks was in Ithaca, New York, a year earlier, when I was invited to speak at one of their smaller events. I was impressed by how down-to-earth and relaxed the organizers were, dressing in shorts and sneakers instead of power suits and high heels, and providing plenty of opportunities in between the proceedings to party. I learned that their focus is on wellness as opposed to sickness, a concept that emphasizes the positive by defining health as something more than the absence of illness. This is a trend I've been encountering in many quarters, most recently when I ran across a group of American chiropractors and nutritional biochemists who call their approach "functional medicine" to emphasize optimal functioning of all organ systems, rather than simply the absence of disease. For the wellness folks, prevention—as in "an ounce of prevention is worth a pound of cure"—is a primary focus. This leads them to promote *self-care*, which requires patients to take increasing responsibility for their health by making lifestyle choices that minimize or eliminate the use of drugs, alcohol, and cigarettes, and promote behaviors that are life-enhancing.

As an outcome of this emphasis on self-responsibility, the wellness movement has spawned a team of "wellness consultants," trained specialists who go into major public and private corporations to design and implement on-site health and fitness programs. The point is to help employees reduce stress and improve health by making lifestyle changes, thus saving employers' insurance plans from the skyrocketing costs of cardiac operations and other dire medical interventions. Here in the Midwest, the wellness people are quietly but effectively doing their part to change the face of health care and medicine in America.

ENVIRONMENTAL MEDICINE

I'm met at the airport by Dr. Norman Schwartz, a doctor from Milwaukee who has offered to drive me to Stevens Point, some 200 miles north. Norm, whose specialty is environmental medicine and who also has a degree in physics, is one of the new breed of alternative or complementary M.D.s, the midwestern counterpart of Bob Gottesman, who also treats his patients with an understanding of the whole bodymind. We met earlier this year in Arizona when I presented at a gathering of physicians concerned with chemical sensitivity and diseases caused by environmental factors. Afterward, we talked a few times on the phone, exchanging expertise in our respective fields, and I was impressed with his knowledge of nutrition and vitamin therapy. Many of his suggestions sounded plausible, and so I decided to try some of them, making changes in my diet and lifestyle, and self-monitoring for the results. Our long drive gives us plenty of time to catch up, and I'm eager to hear the latest about nutrition, the environment, and the bodymind.

The last time we talked, Norm was concerned about the impact of environmental pollutants and toxins on people's health, and, if anything, his concern has deepened. I'm shocked when he reels off statistics indicating that cellular levels of heavy metals and dioxins from herbicides and pesticides are 300 to 400 times greater than they were when first measured, and every year, hundreds more chemicals are added to the 80 to 100,000 chemicals that already exist in our environment.

I knew that environmental pollutants could enter into the cell membrane and change the shape of the receptor, making it looser and sloppier, and often wondered how this might affect the transfer of information so necessary to run the delicately balanced systems. It had to have some effect on what is essentially a self-organizing system, one that is processing tremendous amounts of information at incredibly rapid speeds.

Norm has his own theory about how some very basic life processes are interrupted and altered by these pollutants. He explains how electrons flowing—the classical energy currency of all biological life-forms—through cell membrane gradients is what normally allows the mitochondria, the energy-generating component of cells, to transfer energy at about a 98 percent efficiency rate. But the pollutants suspended in the cell membrane are altering and interrupting that electron flow, causing "energy starvation" and resulting in conditions like chronic fatigue, allergies, and chemical sensitivity.

Particularly alarming is Norm's belief that accumulated environmental pollutants within our bodies are mimicking and disrupting the action of our sex hormones—estrogen, progesterone, and testosterone—which run the male and female reproductive systems. And he's far from alone in this belief—despite the medical establishment's apparent lack of interest in the subject, as exemplified by how little attention it has paid to the link between toxicity and breast cancer. A recent report on receptor binding in *Science,* for example, has shown that environmental toxins have estrogenlike effects and bind to estrogen receptors, where they can stimulate breast cancer tumor growth. Similarly, various toxins can act like testosterone in the male body and stimulate prostate cancer, which is embryologically similar to breast cancer. Although this has been suspected for a long time, only recently have we gotten the hard proof that the accumulation of these toxins in our bodies chronically stimulates our estrogen and testosterone receptors, putting them into a state of overdrive and leading to cancer.

And as if this weren't enough, Norm continues, he believes it's very probable that increased levels of environmental toxins are causing our immune response to lose much of its resilience. To be effective, the immune system needs to be in a state of constant readiness to fight off the many viruses and other invading pathogens we encounter daily. When it's overloaded and diverted by high toxicity, it gets "tired," failing to stay on its feet, so to speak, which is possibly why we're seeing so much suboptimal health such as vague complaints of fatigue, not to mention more serious immune-deficiency diseases.

As Norm talks, I'm thinking of the biblical saying "As ye sow, so shall ye reap." It is becoming apparent that people's overall state of health today is a direct reflection of the ecological mess we've inflicted on our planet, a mess that has been created in blind ignorance and disregard for what turns out to be the essential relatedness of all life. How can we expect to be healthy when our water is foul, our air dirty, our food poisoned? And it seems to me that much of the current disaster can be traced to old-paradigm thinking, which views each of us as an isolated entity, separated from others and from our environment, living apart from the whole and not connected to it. It is this erroneous belief that has permitted the poisoning of our environment through the development and careless manufacturing of toxic chemicals for farming and industry. Fortunately, the new breed of M.D.s, such as Norm Schwartz and Bob Gottesman, and doctor-friends such as Jim Gordon and Nancy Lonsdorf on the East Coast, are taking seriously the fact that we are all part of the earth's ecosystem, and they are willing to look at what needs to be done to protect and cleanse our bodyminds in this late-twentieth-century reality.

Norm is happy to pass on some recommendations for ridding the system of toxins and maintaining a relatively pollutant-free bodymind.

To begin with, he tells me, high doses of Vitamin C (1,000 mg. or more) should be a part of everyone's basic nutritional defense kit. He also has some simple rules for a healthy, pollution-free diet, which include the following: Only eat food that has been around for at least six thousand years—no processed foods! Don't eat something if you can't pronounce the ingredients. Try to buy fruits and vegetables grown organically—or even start your own garden! And stay away from poultry, meat, and dairy that has been pumped full of antibiotics, a common practice in animal farming today. Instead, choose products from animals allowed to graze freely ("free-range") and less susceptible to the diseases antibiotics are used to prevent. As part of his own practice, Norm offers a clinical test to measure toxicity in the liver, the organ that acts as the initial filter to prevent harmful substances from getting to the bloodstream. If the liver tests as having high toxicity, he prescribes a variety of dietary, vitamin, and herbal therapies that will help cleanse it and restore its full functioning.

I listen, but I'm well aware that the establishment mocks the concept of detoxification and makes no room in its offerings for nutritional and cleansing approaches. Where I come from, the biomedical research community, these modalities are seen as "fringe," or irrelevant, for there has been very little good research done on the subject. However, the results of several small but well-designed trials showing the ability of nutritional supplements to support the liver for well-being and health have impressed me enough to experiment on myself. I've used liver vitamin products like Ultra-Clear, which work by altering the chemical structure of toxins in the liver so that they can be excreted through the urine or the large intestine. The results have convinced me that the possibility of turning back the clock and increasing energy levels by removing years of accumulated toxins is not "pie in the sky." I only wish that there were more studies, other than manufacturer-funded ones, in this very important area. (Of course, the biggest liver-cleansing boost of all is to go off drugs, including alcohol.)

As we pull into Stevens Point, I am thinking about what special contribution I can make, based on my research and theories on emotion and the psychosomatic network, to help people achieve the goals of the wellness movement. I feel challenged: What is it that I in particular have to tell people about wellness, about self-care and healthy lifestyles? A bodymind lifestyle for the twenty-first century—what would that look like?

• • •

JOTTING DOWN some notes later that evening about what I will tell people in my talk, I imagine myself standing before my audience and confessing: "I'm from the sickness movement, the mainstream biomedical establishment. We're the people who do all the high-tech interventions that are costing billions in insurance dollars while patients are allowed to smoke and drink their way to the operating room and the intensive-care unit." I cringe at the thought. While I don't ally myself with the disease-focus of the mainstream, people still tend to see me as a representative of it.

In mainstream medicine, the importance of lifestyle in preventing disease is still largely ignored in spite of inroads made by such heroic and outspoken doctors as Bernie Siegel, Dean Ornish, Christiane Northrup, Larry Dossey, and Andrew Weil, to mention just a few. A recent article in *Parade* magazine made clear the party line. In "What Medicine Will Conquer Next," fourteen leading biomedical researchers, many of them Nobel Prize winners, were asked to predict the advances they see occurring in the next fifty years. What follows is a paean to the glories of high-tech, high-cost modern medicine, with particularly glowing praise heaped on genetic research as the answer to all our ills. Noticeably absent in the article is any reference to, or acknowledgment of, the necessity of people taking responsibility for their own health—until the very end, when Dr. Bernadine Healy, former NIH Director and now a professor in the medical department at Ohio State, gets the last word: "Changing your lifestyle can make a difference. Genetics is a big factor in determining a person's susceptibility to disease, but a healthy lifestyle may be just as important." One lone female voice pointing to what we all know to be true: The choices we make about how we live are at least as valuable as the high-tech interventions of the biomedical establishment—especially because they may minimize the need for such interventions.

Part of the reason that more lifestyle advice is not forthcoming from the laboratories is that we scientists, in general, do not see our role as that of advicegiver, having been taught by our training that pure science is not necessarily practical science. We tend to get uncomfortable when pressed to talk about how to put the conclusions we draw from our research into practice. I can't deny that I, too, would prefer to hide behind the laboratory door, tinkering away at my bench and letting everyone make of my work what they will. Pure science! Isn't that what I got into it for? But something has happened to me along the way. The research I have done has profoundly transformed me, letting me see what a healthy life looks like from a perspective that was not available before I did my work. It is from this perspective, coming out of a radical, transformative paradigm shift, that I wish to share some advice on healthy living with my wellness audience. I believe it's a unique angle I offer, based on a new understanding that can help us all to live happier and healthier lives, because it acknowledges that the body and mind are not separate, but really one system coordinated by the molecules of emotion.

LIFESTYLES OF THE HEALTHY, WHOLE, AND CONSCIOUS: AN EIGHT-PART PROGRAM

For most of us, the very words *healthy lifestyle* conjure up images of low-fat meals, daily exercise regimens, and the elimination of alcohol, tobacco, and recreational drugs. While these are all good, health-enhancing strategies—and I'll have something to say about them from the point of view of the peptidergic network—what's missing for most of us is any focus on ongoing, daily, emotional self-care. We tend to deal with the physical aspects of keeping ourselves healthy and ignore the emotional dimension—our thoughts and feelings, even our spirits, our souls. Yet, in light of the new knowledge about emotions and the psychosomatic network, it's obvious that they, too, are a part of our responsibility to manage our own health.

The tendency to ignore our emotions is *oldthink,* a remnant of the still-reigning paradigm that keeps us focused on the material level of health, the physicality of it. But the emotions are a key element in self-care because they allow us to enter into the bodymind's conversation. By getting in touch with our emotions, both by listening to them and by directing them through the psychosomatic network, we gain access to the healing wisdom that is everyone's natural biological right.

And how do we do this? First by acknowledging and claiming all our feelings, not just the so-called positive ones. Anger, grief, fear—these emotional experiences are not negative in themselves; in fact, they are vital for our survival. We need anger to define boundaries, grief to deal with our losses, and fear to protect ourselves from danger. It's only when these feelings are denied, so that they cannot be easily and rapidly processed through the system and released, that the situation becomes toxic, as discussed earlier. And the more we deny them, the greater the ultimate toxicity, which often takes the form of an explosive release of pent-up emotion. That's when emotion can be damaging to both oneself and others, because its expression becomes overwhelming, sometimes even violent.

So my advice is to express all of your feelings, regardless of whether you think they are acceptable, and then let them go. Buddhists understand this when they talk about nongrasping, or nonattachment to experience. By letting all emotions have their natural release, the "bad" ones are transformed to "good" ones, and, in Buddhist terms, we are then liberated from suffering. When your emotions are moving and your chemicals flowing, you will experience feelings of freedom, hopefulness, joy, because you are in a healthy, "whole" state.

The goal is to keep information flowing, feedback systems working, and natural balance maintained, all of which we can help to achieve by a conscious decision to enter into the bodymind's conversation. I'd like to explore a number of different ways of using awareness and intention to tap into the psychosomatic network, in order to prevent disease and maximize health.

One: Becoming Conscious

Most lifestyle choices involve things we do or don't do. But I'd like to consider a choice that has more to do with *being* than *doing*—after all, we are human *beings*, not human *doings*—and this is the decision to become more conscious. Full consciousness must involve awareness of not just mental but emotional and even basic physical experiences as well. The more conscious we are, the more we can "listen in" on the conversation going on at autonomic or subconscious levels of the bodymind, where basic functions such as breathing, digestion, immunity, pain control, and blood flow are carried out. Only then can we enter into that conversation, using our awareness to enhance the effectiveness of the autonomic system, where health and disease are being determined minute by minute.

Just how powerful consciousness can be to intervening at the level of our molecules and making significant changes in our physiology was brought home to me by an encounter I had in 1986 at Lake Arrowhead, California. I was attending a conference of researchers in the emerging field of PNI, organized by Norman Cousins, and was fortunate to spend some time with Evelyn Silvers, the widow of Phil Silvers (of *Sergeant Bilko* fame). Evelyn had been a practicing psychotherapist for years and had recently become so caught up in PNI that she went back to UCLA to get her Ph.D. in a related field. She knew of my work and had already sought me out on the East Coast a year before, arriving in a limo at my NIH office to take me and Michael out for lunch. I was fascinated to hear about her therapy, which combined relaxation, autohypnosis, and visualization in a guided technique to help a person direct her or his own healing. At Arrowhead, I got to experience it. After a brief counseling session, in which I confided about the stress I was experiencing in my efforts to develop Peptide T, she said she thought I might benefit from an increase in my endorphin levels. She offered to put me into a light trance, and we began a process of guided visualization.

"Which is the most potent of the endorphins and where is it most highly concentrated?" Evelyn asked, once I had relaxed into a pleasant altered state. I told her it was the beta endorphin, which is found most abundantly in the pituitary gland.

"Good," she encouraged. "Now I want you to close your eyes and focus on your pituitary gland. Do you know where it is?"

It took a few moments, but I, of all people, knew exactly what to look for and had very little trouble bringing the pituitary into sharp visual focus. I nodded.

"Great, now can you see the beta endorphin molecules in there?" she coached me further. The beta endorphin was clear on my inner screen, all thirty-one of its amino acids strung together in a bead chain and stored tightly in tiny, balloonlike pouches at the ends of the cell axons, ready for launch.

She continued: "I want you to listen as I count backward from ten, and when I get to one, you are going to release the endorphins out of your pituitary into your bloodstream."

I did exactly as she directed and felt an instantaneous rush, a feeling that accompanied what I knew was the outpouring of endorphins from my pituitary as they began swimming and binding receptors all over my body and brain to work their magical effects.

It was clear that the knowledge I had of physiology—just where the endorphins were located and how they were secreted—had enabled me to consciously intervene and intentionally change my molecules. I wondered if this same kind of knowledge could help others who might benefit from the release of certain biochemicals in their systems. Later I got a chance to try that idea out on a group of female heroin addicts at the Baltimore County Jail. Some of my colleagues were running an experimental program there, which offered auricular acupuncture—three needles in the ear a day—to alleviate the craving for heroin and diminish the body-wide pain that makes withdrawal so difficult. The researchers knew of my work showing how acupuncture stimulates the release of endorphins to create analgesia (pain relief), so they invited me to come and see their program at the jail.

My visit took place on Mother's Day, not a happy time for incarcerated women who were missing their children as well as suffering from withdrawal. In simple terms, I explained to them how they all had a natural form of heroin, the endorphins, in their brains and bodies, and that as a result of continually shooting the artificial substance, the flow of the endorphins had been diminished. They found this an amazing concept. I explained that the craving they felt would cease when the natural flow of the chemical was restored, and that exercise and orgasm were two means of enhancing that natural flow. Out of this discussion came a new way for these women to think about their addiction. Although no long-term studies have been done to explore how visualization to release endorphins might benefit drug addicts, I saw in the faces of the women that day that the very idea was empowering to them, giving them hope for an eventual healing as well as a new respect for the powers of their own bodies.

Two: Accessing the Psychosomatic Network

Because of my consciousness—awareness of the anatomy and bio-chemistry—I had been able to access my psychosomatic network and enter the bodymind's conversation to redirect it. The nodal point I used was the frontal cortex, a part of the brain that is rich in peptides and receptors. Also known as the forebrain, the frontal cortex is unique to humans and sits behind the forehead. It is the location for all the higher cognitive functions, such as planning for the future, making decisions, and formulating intentions to change—which is what I did, in the case of releasing the beta endorphin. In short, the frontal cortex is what makes us truly human. Chimpanzees have 99 percent of the same genetic material as we do, but lack a developed frontal cortex. In humans, this part of the brain does not fully develop until some time in the early twenties—a fact that helps us understand, and hopefully be more patient with, our teenagers!

Interestingly enough, the frontal cortex is just as dependent on the free flow of the peptides of emotion through the psychosomatic network as any other part of the bodymind. In order to function at a level that allows it to perform the kind of conscious intervention into the bodymind's conversation that I am talking about, the frontal cortex needs adequate nourishment. The brain's only food is glucose, which is carried to the brain in the blood. It's glucose that supports the ability of the neurons to store and secrete all the messenger chemicals—neurotransmitters and neuropeptides—and glucose that fuels the brain's glial cells, which perform many essential functions. Acting as "cleaner-up" cells, the glials are peptide factories that move around macrophagelike, sometimes destroying and sometimes nurturing nerve endings in an ongoing sculpting of connections, literally making up our minds. Only when there is enough blood flow to bring plentiful supplies of glucose to the brain will the neurons and glial cells be able to carry on their functions and ensure full consciousness.

Blood flow is closely regulated by emotional peptides, which signal receptors on blood vessel walls to constrict or dilate, and so influence the amount and velocity of blood flowing through them from moment to moment. For example, people turn "white as a sheet" when they hear shocking news, or "beet red" when they become enraged. This is all part of the exquisite responsiveness of our internal system. However, if our emotions are blocked due to denial, repression, or trauma, then blood flow can become chronically constricted, depriving the frontal cortex, as well as other organs, of vital nourishment. This can leave you foggy and less alert, limited in your awareness and thus your ability to intervene into the conversation of your bodymind, to make decisions that change physiology or behavior. As a result, you may become stuck—unable to respond freshly to the world around you, repeating old patterns of behavior and feeling that are responses to an outdated knowledge base.

By learning to bring your awareness to past experiences and conditioning—memories stored in the very receptors of your cells—you can release yourself from these blocks, this "stuckness." But if the blockages are of very long standing, you may need help in achieving such awareness, help that may come in many different forms. I would include among them psychological counseling (hopefully, with some kind of touch!), hypnotherapy, touch therapies, personal-growth seminars, meditation, and prayer. Any or all of these can teach you to respond to what is actually occurring in the present, which is in large part what consciousness is all about.

Three: Tapping into Your Dreams

One of the best ways I know to integrate awareness of emotions into lifestyle is to develop the daily habit of recalling and transcribing your nighttime dreams. Dreams are direct messages from your bodymind, giving you valuable information about what's going on physiologically as well as emotionally. Becoming aware of your dreams is a way of eavesdropping on the conversation that is going on between *psyche* and *soma*, body and mind, of accessing levels of consciousness that are normally beyond awareness. This allows you to get valuable information and then, if necessary, to intervene, making appropriate changes in your behavior and your physiology.

What is happening when you dream? Different parts of your bodymind are exchanging information, the content of which reaches your awareness as a story, complete with plot and characters drawn in the language of your everyday consciousness. On a physiological level, the psychosomatic network is retuning itself each night for the next day. Shifts are occurring in feedback loops as peptides spill out into the system (in greater or lesser numbers) and bind to receptors to cause activities necessary for homeostasis, or return to normalcy. Information about these readjustments enters your consciousness in the form of a dream, and since these are the biochemicals of emotion, that dream has not only content but feeling as well.

We have seen how strong emotions that are not processed thoroughly are stored at the cellular level. At night, some of this stored information is released and allowed to bubble up into consciousness as a dream. Capturing that dream and reexperiencing the emotions can be very healing, as you either integrate the information for growth or decide to take actions toward forgiveness and letting go.

Classically, Freudian psychology uses dream analysis to help people understand motives, desires, behavior: "Aha, you dreamt you killed your mother? Must mean you're harboring resentment that you haven't admitted—the source of your neurosis!" But from the bodymind viewpoint, your dreams can relate not just to your mind but to your body as well. Dreams can be your own early-warning system, letting you know if a medical condition is developing and helping to bring your conscious attention to the problem area. The body may be discussing this condition with the mind, and you can get in on the conversation by consciously recalling the dream. Although it's hard to translate such a conversation for anybody else—to say that the dream of an army invading means that a cancer is growing or some arbitrary equivalency like that—I *can* tell you that once you make the decision to pay attention to your dreams, they will start to speak to you, and you will understand them with ever-greater fluency over time, with practice.

I have been writing my dreams down in a notebook for years. I started during the period when I didn't have a laboratory, while Peptide T was awaiting further development and Peptide Design had been disbanded, and I think my unconscious motivation was that I could use myself as a laboratory (as I also did with meditation). Some of my most valuable insights leading to growth are the results of dreams I recalled during that time. The pivotal dream I had of Sol, in 1986, in which I threw water on him and he shrank up—my nemesis, the monster I myself had created—gave me the courage to write him a letter of forgiveness that allowed me to let go of resentments that had been eating away at me for years.

I keep my dreambook the way I keep a laboratory notebook, writing the narrative content on the right-hand page, where I usually record the steps of a lab procedure, and the emotional content on the left-hand page, where I usually jot down calculations and comments. The bodymind is a laboratory for each of us in which we are all participating scientists, seeking to better understand and affect behavior and physiology. In this way, we are all seekers of the truth! Just as the bench scientist does when evaluating a series of experiments, you may want to review a particular dream at a later time to see what kinds of emotions were being processed. The review process may reveal patterns that add to your awareness.

I often hear people say, "I can't remember my dreams," as if they were helpless to initiate this process. But that is the crucial first step, the simple decision to remember your dreams, which the frontal cortex of your brain enables you to do. From that decision all else will follow. Once you have decided, prepare yourself by placing a pen and notebook by your bedside. Deepak Chopra talks about *intention* and *attention* in *Seven Spiritual Laws*, and this is a perfect illustration of what he means. The intention is the decision to capture the dream in writing, the attention is the focus, the readiness to carry out the action created by intention, in this case, writing down the dream! By consciously applying your attention and intention in this way, you will be able to cultivate the habit of recalling your dreams and gain greater access to your bodymind's information system.

When you wake in the morning, stretch, yawn, and reach for your dreambook. Write whatever comes to mind, no matter how fragmentary, and try not to filter or edit any of the content. If associations arise—*Aha!* That yellow car is the same one my dad had when I was ten!—write them in parentheses. Even more important than the content, however, are the feelings and emotions you experienced in the dream. Always ask: How did I feel? And include these observations in the writing. Sometimes the emotions are contrary to the action, such as dreaming about a tragedy and feeling happy. The feeling is the clue. Even if the feelings are disturbing or uncomfortable, force yourself to write them down. This is good practice for becoming more aware of both your waking emotions and those you experience in a dream, and to become less judgmental of your own inner processes.

Write down dreams that you recall only partially, too. When I first started my dreambook, I would often have just a glimmer to write about, but I found that if I wrote even the tiniest fragment, it started a process of deeper recall, oftentimes bringing back the dream in its entirety. Be sure to write down even the most insignificant-appearing dreams, because by discounting a dream that seems dull or boring, you may be preventing yourself from getting an important message. Often the apparent banality is only a mask for material that you are resisting because of its disturbing or discomforting content. Once you write down the boring part, other parts will surface into memory.

Like our emotions and thoughts, our dreams follow the laws of information, existing on a level that is beyond time and space. Many tribal peoples recognize this and credit their dreams as coming from the spirit world, treating them with utmost respect. While Jung's collective unconsciousness is as close as we come to such a belief in our own culture, we can put information theory and ancient wisdom together and start giving more validity to dreams as a low-cost, no-drug form of psychotherapy. If we are looking for some practical, low-tech self-care on our journey to wellness, dream work can make a tremendous contribution.

In my more mystical moments, I like to consider that dreams are just one more way God whispers in our ear, delivering messages to us via the psychosomatic network.

Four: Getting in Touch with Your Body

But lest we get too philosophical, I want to return to the level of the physical body and how through it we can access the mind and emotions for wellness. Dream work and other forms of conscious intervention are important, but we need to acknowledge other places of access as well—the skin, the spinal cord, the organs, which are all nodal points of entry into the psychosomatic network. As such, they are all used by the touch therapies that have grown out of alternative medicine: Epstein's network spinal analysis, Bainbridge-Cohen's mind-body centering, the synergism of Illana Rubenfield, Lowen's bioenergetics, the new identity process, and massage. I highly recommend these and other forms of body work that use movement or touch to heal emotions. (See the Appendix for more information on such therapies.)

But you can also do much simpler forms of body work to equally good effect. Feeling low and sluggish? Take a walk. Feeling anxious and jittery? Run! Feeling worthless? Get a massage, a spinal adjustment, or a lot of good hugging, and see what happens. Your mind, your feelings are in your body, and it's there, in your somatic experience, that feeling is healed.

Five: Reducing Stress

No discussion of lifestyles and wellness would be complete without mentioning stress-reduction. In my experience, the most effective method for reducing stress is meditation, because it allows us, even without conscious awareness, to release emotions that are stuck in modes that subvert a healthy mind-body flow of biochemicals. I believe all forms of meditation are useful, but the one I have personally used is transcendental meditation, or TM. TM requires sitting in a comfortable position with the eyes closed twice a day for twenty minutes while silently repeating the same word, called a mantra. The teaching and practice of TM have been standardized and many scientific studies show strong evidence of physical benefits such as lowering high blood pressure, reversing autoimmune diseases, and stimulating a wide range of antiaging effects.

Another form of meditation that is gaining popularity is "mindfulness," as introduced by psychologist/researcher Jon Kabat-Zinn of the Medical Center Stress Reduction Clinic at the University of Massachusetts. This simple method is based on an Eastern technique known in Buddhist tradition as *vipassana*, in which you simply bring your attention to your breath, sitting or lying down, eyes open or closed. By breathing consciously in this way, you enter the mind-body conversation without judgments or opinions, releasing peptide messenger molecules from the hindbrain to regulate breathing while unifying all systems. Studies have shown that mindfulness meditation can dramatically reduce pain and improve mood for people who live with chronic pain by allowing them to exist in the present moment rather than in the constant fear that their pain is "killing" them. With this different awareness, they can carry on daily activities despite discomfort. Kabat-Zinn's approach is presented in his books *Full Catastrophe Living* and *Wherever You Go, There You Are*.

An easy way to sample the benefits of meditation is by listening to any of the many "relaxation music" audio tapes available, some of which use guided imagery to help you project your consciousness into your psychosomatic network through programmed affirmations, or positive statements, about health, prosperity, relationships, etc. Some use Eastern terms or instruments, while others have unassuming titles like "A Trip to the Beach" or "A Walk in the Forest." I have found that when I listen to the music and the words, the pattern of my breathing shifts and becomes deeper, slower, bringing about a profound sense of relaxation. (I have several tapes that are so deeply relaxing, I've yet to listen to them all the way through without falling asleep.) Interestingly, these changes are not just short-term. My experience indicates that this kind of relaxation can bring about a basic reconditioning of my breathing patterns, so that even when I am not listening to a tape, my breathing tends to be more relaxed, freer.

For some, meditation provides a direct link with the spiritual world, but practitioners need not have this as their goal to benefit. Perhaps the key mechanism of meditation is that of simply being in the present for a period of time. Shifting the mind from *shoulda, coulda, woulda* types of thinking promotes self-regulation and healing on all levels. In the race of modern life, we all tend to adjust our sails far too frequently, running this way and that, always in a hurry, not pausing long enough to see the effect of our trimming on the course of our lives. Meditation provides an opportunity to stop and wait for some feedback before charging ahead on an uninformed course, a chance to let the body catch up with the powerful transforming effects of our natural information flow.

A simpler, less formal practice than meditation, but equally effective at stress-reduction, is the habit of self-honesty. By self-honesty, I mean being true to yourself, keeping your word to others as well as to yourself, living in a state of personal integrity. There is a profound physiological reason why honesty is stress-reducing. We have seen how the emotions bring the whole body into a single purpose, integrating systems and coordinating mental processes and biology to create behavior. Walking is an example: You have a thought or intention, which is then coordinated with the physiology to produce a behavior, walking. If I have a purpose, such as finding a cure for cancer, then every system in my body gets behind that intention and does what needs to be done, be it increasing my appetite for protein, mobilizing my gastrointestinal system to digest a protein better, sending blood flow into digestive organs to produce necessary enzymes for maximum absorption, or whatever. There is a physiological integrity and directness about this process that is the result of my clarity about my own intentions. When I am at cross-purposes, however, going through the motions but not really committed to my goal, saying one thing and doing another, then my emotions are confused, I suffer a lack of integrity, and my physiologic integrity is likewise altered. The result can be a weakened, disturbed psychosomatic network, leading to stress and eventually to illness. Always tell the truth, I have said repeatedly to my children for years, not just that it's the moral thing to do, but because it will keep you on a healthy path and disease-free! My friend Maggie McClure, a practicing spinal-analysis chiropractor, puts it this way: "I never tell a lie, because it takes too much energy to keep everything straight—energy I'd rather use doing other things!" Honesty, it seems, is supported by our biochemicals, and it only slows us down to choose otherwise.

One last word about stress: *Play!* Having fun is the cheapest, easiest, and most effective way I know to instantly reduce stress and rejuvenate mind, body, and spirit. The source of most people's ongoing, daily stress, I believe, is the perception of isolation and alienation, being cut off from the company of others. Engaging in play is the antidote because it gets our emotions flowing, and our emotions are what connect us, give us a sense of unity, a feeling that we are part of something greater than our small and separate egos.

With this new understanding of the importance of consciousness, emotions, and blood flow, we can look at exercise and diet in a new way.

Six: Exercising

For the vast majority of us still mired in old-paradigm thinking, exercise is drudgery, something we do in fits and starts by motivating our "physical machine" in a carrot-and-stick fashion for such benefits as a slimmer figure or harder muscles. But with our new concept of ourselves as emotion-laden information systems, exercise can be easier and more fun. Whenever I exercise, I always try to engage my emotions by putting on a headset and listening to my favorite rock tunes. Walking is a breeze this way, as the music helps to loosen up stuck emotions and puts me more in touch with my body so I can "hear" what it is telling me to do. This way I don't push myself beyond my capacity to a possible injury or give up prematurely because of missing important feedback about how good it feels to work muscle and bone. This is valuable information that encourages me to keep going—the difference that makes the difference!

I've learned a couple of tricks to help me move my body in a way that helps to enhance the communication flow throughout the bodymind. For example, I let the opposite hand swing forward with each step as I walk, without music this time. Somehow this sets left brain–right brain information flowing, breaking up old patterns of worry and rumination. I've found that it's impossible to stay stuck in unproductive old thought patterns when I move my body this way.

Remember—the value of exercise has less to do with building muscles or burning calories than it has to do with getting your heart to pump faster and more efficiently and thereby increase blood flow to nourish and cleanse your brain and all your organs. Of course, if you exercise hard enough to break a sweat, you'll also get the benefit of a mood improvement prompted by the release of endorphins (and other as-yet-to-be-discovered peptides).

Yoga is a particularly health-enhancing form of exercise. Any kind of conscious breathing accompanied by relaxation and body awareness is yoga. But the best way of learning yoga, if you are unfamiliar with it, is through classes—often at the local YMCA these days!—or through numerous tapes and books available at local bookstores.

My own favorite style of yoga is to use conscious breathing as I engage in well-defined rhythmic movements and postures, such as those I use in walking or swimming. To experience the power of this when you walk or swim, try breathing in on two counts and breathing out on four, or whatever ratio of in-to-out breath feels comfortable. Just make sure you double the beats on exhaling and stay in this rhythm for ten minutes, if you can. The effect is one of increased energy and good feeling, both elements that reflect the good work I'm doing and make it enjoyable to keep going.

Seven: Eating Wisely

Diet is another lifestyle area that can be reframed in light of our understanding of the emotions.

Eating, because of its survival value, has been wisely designed by evolution to be a highly emotional event. (All processes that impact on survival—sex, eating, breathing, etc.—are highly regulated by neuropeptides, and thus are emotionally directed. The simple emotions of pain and pleasure, signaling us to move either toward something or away from it, have been key determinants in whether an animal or human survives and evolves.) Our large and small intestines are densely lined with neuropeptides and receptors, all busily exchanging information laden with emotional content, which is perhaps what we experience when we say we have a "gut feeling." There are at least twenty different emotion-laden peptides released by the pancreas to regulate the assimilation and storage of nutrients, all carrying information about satiety and hunger. Too often, however, we ignore that information, eating when we're not really hungry, using food to bury unpleasant emotions. Nervous eating, depression eating—these are the resultant behaviors.

By tuning into your emotions as information about your digestive process, you can develop your ability to know what your body needs in the way of nourishment and when. Remember, it's the peptides that mediate satiety and hunger, and we can't hear what our peptides are telling us when we are disconnected from, or in denial about, our emotions. Ask yourself: Do I feel hungry?—and wait for a feeling of hunger to occur before eating. A great idea I learned from the Hindu Ayurvedic tradition is to slowly sip hot water, which will satisfy any false hunger and help excrete partially digested food. If you are really hungry, however, it will prepare the body to digest a meal completely.

We are all made frantic by the continually changing advice from the nutrition gurus. While I don't advocate disregarding basic nutritional principles, I am calling for more reliance on the wisdom of the body when making eating decisions. A craving for something sweet may be a signal that your brain needs fuel, so bite into a piece of fruit; a desire for a hamburger may be telling you that your body needs more protein, so add more animal and/or soy products to your diet. The benefits that come from eating according to your feelings—not your impulses—are greater than any particular food rules you may be following to build muscle or lose weight. If impulses dominate you, therapies that get at the source of the emotions involved, like the body psychotherapies (see Chapter 11), can be effective in putting you back in touch with honest, reliable emotions.

The environment in which you eat has a lot to do with your emotional experience at mealtimes. I try always to eat in a peaceful, stress-free environment and enjoy good company. Eating while emotionally upset or rushed can be very detrimental to the peptide-regulated process of digestion. The thoughts and feelings you bring with you to the table are just as important as the sensible balanced meal you sit down to. Are you worried, tense, thinking about the grocery bill or the calories or the toxicity levels? Are you scarfing down each bite as if your meal might disappear at any moment? Or perhaps you "go unconscious" when you eat, leaning over a newspaper or sitting in front of the boob tube while you mindlessly shovel food into your mouth. This is a kind of disintegration, a mind-body split that will lead to weight gain and disease conditions caused (according to Ayurvedic tradition, which I believe to be scientifically valid) by incomplete digestion.

Come into full consciousness as you eat and feel thankful for your food, savoring its tastes and textures. Blessing your food doesn't have to be a religious ritual. It can be as simple as saying: "Mmmm, this is good for me, I'm grateful, I'm nourished." I do this even when I "slip" and have a chocolate-chip cookie or two, because I know that the emotion that accompanies eating is as vital an ingredient as the vitamins and minerals in the food.

And a word about sugar. I consider sugar to be a drug, a highly purified plant product that can become addictive. The body naturally produces sugar in the form of glucose, which is the only fuel the brain needs to function. We have seen how the chemicals of emotion regulate blood flow to bring nourishment to the brain, but our emotions and their corresponding chemicals also control the availability of glucose. In states of high excitement, such as panic or hysteria, the liver metabolizes glycogen, the form in which glucose is stored, and releases it into the bloodstream where it is transported to the brain, making us alert to handle an emergency if need be.

Relying on an artificial form of glucose—sugar—to give us a quick pick-me-up is analogous to, if not as dangerous as, shooting heroin. The artificial substance is utilized by the body in the same way as the natural form, but, like a drug, it floods and desensitizes receptors, thereby interfering in the feedback loops that regulate the availability of instant energy, such as glycogen release from the liver. Over eons of time, our bodymind has evolved a system for supplying the brain with the fuel it needs, and we would be wise to respect it.

Eight: Avoiding Substance Abuse

For the same reasons that it's best to avoid overconsuming sugar, I want to warn against the dangers of alcohol, tobacco, marijuana, cocaine, and other drugs. All of these substances have natural analogs circulating in our blood, each of which binds to its very own receptor bodywide. Alcohol, for example, binds to the GABA receptor complex, which also accommodates Valium and Librium, common prescription drugs for quelling anxiety, providing an antianxiety effect, but only in the short run. When we ingest these exogenous ligands, they compete with the natural chemicals that were meant to bind with the GABA receptors, oftentimes flooding them and thereby causing them to decrease in sensitivity and/or number. The receptors then signal a decrease in peptide secretions, as I explained in my talk to the women heroin addicts in prison. All drugs can alter the natural flow of your own feel-good peptides, and so, biochemically, there is no difference between legal and illegal ones: They are all potentially harmful, they can all be abused, and they can all contribute to suboptimal health in one form or another, including chronic depression.

When multiple drugs are taken, such as when a person is smoking marijuana regularly and taking antidepressants—a common situation that is often missed by the prescribing doctor—their side effects interact, and natural feedback loops of the system can collapse, leaving only a small number active.

The good news is that the physiological effects resulting from substance abuse are reversible: The heroin addict can be cured, the chronic pot smoker can kick the habit, those who think life wouldn't be worth living without antidepressants may find that they have healed sufficiently to do without them. But it can be a very slow and sometimes painful process before the receptors return to their original sensitivity and number and the corresponding peptides get back into bodywide production and flow. In recovery, what is often overlooked is that many systems— gastrointestinal, immune, and endocrine to name a few—have been affected, not just the brain. Drugs put a tremendous strain on the liver, the organ responsible for providing enzymes to metabolize the drugs and dispose of their toxic waste products. While the liver is thus overburdened and distracted, toxicity from other sources builds up, predisposing the bodymind to disease. Recovery programs, both formal ones and those we institute for ourselves, need to take into account this multisystem reality by emphasizing nutritional support and exercise. Eating fresh, unprocessed foods, preferably organic vegetables, and engaging in mild exercise like walking to increase blood flow through the liver can speed the process up.

What causes people to consume legal and illegal drugs—one of the central problems in our society, I believe—is emotions that are unhealed, cut off, not processed and integrated or released. Trauma and stress continually lodged at the level of the receptor block nerve pathways and interrupt the smooth flow of information chemicals, a physiological condition we experience as stuck or unhealed emotions: chronic sadness, fear, frustration, anger. Reaching for that drink or cigarette or joint is usually precipitated by some disturbing and unacceptable feeling that we don't know how to deal with, and so we get rid of it in ways we know "work." The frustrated cigarette smoker, the depressed alcohol drinker, the hyper marijuana smoker—what if we stopped and checked in with our feelings to ask ourselves what emotions are present before using an artificial substance to alter our mood? If we can bring this level of awareness to our habitual use of substances, then we have a chance, a possibility, of making another choice. By continually ignoring feelings, we have none. Perhaps we'll find that it's a matter of a communication that needs to happen, a feeling that needs to be expressed, a need satisfied, a problem solved—all potential actions to get our own endogenous juices flowing for a natural, peptidergic "feel good" state. Or it could simply be that movement, in the form of exercise or a walk, could shift our mood.

CONNECTION

I'm jotting these thoughts into a rough outline for my talk the next day when I'm interrupted by the ringing of the phone. It's Naomi Judd welcoming me to the Wellness Conference and inviting me to join her and a few other presenters on a panel she will moderate the next morning.

The immensely popular country western singer, who with her daughter Wynonna toured as the Judds, has been traveling in my circles lately. Now retired from the stage, she found a new interest, psychoneuroimmunology and alternative therapies, which was spurred by a diagnosis of life-threatening hepatitis. In the last few years, she has turned from the country circuit to what might seem to be its diametrical opposite, the alternative healing circuit, which I call complementary medicine. Like me, Naomi straddles two very different worlds, standing on a bridge between the two, and I feel a sense of camaraderie with her because of this. The fact that we were mothers before the age of twenty and had to struggle to raise children while achieving success in our respective fields is a further bond between us. We met once before, when she had invited me to dinner at an earlier wellness conference. She had exhausted all traditional approaches to curing her condition and, having heard of my work, hoped I might help her to understand her illness from a mind-body point of view. Sometime later, she sent me an autographed copy of her book, *Love Can Build a Bridge*, in which she mentions me and my work. As I read her book, I was touched by her life as a mother and a performer, but what really impressed me was that here was a person who was genuinely living her spiritual beliefs, using them to heal herself and heal her family.

"So I'm all ready with a talk about wellness and lifestyles," I tell Naomi, after a brief, friendly, personal update.

"Well, of course we all want to hear that," she says enthusiastically. "But I've got something else up my sleeve, something I think you'll be just as interested in!" And then, not skipping a beat: "I think what people really want to hear about is how we bring spirituality into the mind-body question. Now, don't you think that's so?"

"Well, sure." I hesitate. Naomi is pushing the edge. Scientists don't talk about spirituality publicly, and even though I'm considered something of a revolutionary, I still consider myself a mainstream scientist and feel uncomfortable talking spirit and metaphysics, even to the holistic crowd. But the opportunity, at the same time, intrigues me. I have certainly thought a lot about how spirit plays into the mind-body equation, and have even been able to see how the science I've done could support this idea.

"Good, then you'll tell all those people how God has been found alive and well in your laboratory, and that He's really gonna heal us through those neuropeptides?" she jokes.

If only it were that simple, I think. "I'll do my best," I finally tell her. "But you're going to have to help out, Naomi. I'm a scientist, not a guru . . . It makes me very nervous . . ."

"Oh, shucks, Candace, they'll love anything you say," she shoots back.

We hang up and I return to my outline. But it's late, and I'm more in the mood to relax and call it a night. I wonder what I'll say tomorrow on the panel. It occurs to me that I've been interested in the role of spirit—consciousness, the soul, the psyche—my entire life, ever since I was a young child growing up in a mixed-marriage family and wondering what religion I was.

The wellness crowd may be getting a surprise tomorrow, I muse as I drift off to sleep. I am ready for a new level of uninhibited expression. The panel moderated by Naomi Judd might be the perfect place to formalize my understanding of how spirituality fits into the bodymind picture, and to deliver a radical new message from the laboratories of mainstream science.

SPIRITUAL HEALING

The Wellness Conference events for the day are to be held in the campus fieldhouse, a structure more suitable for holding athletic events than for showcasing presenters of the latest update on health and wellness. But the climate is casual, and the sawdust on the floor is a reminder that folks here don't get out too far ahead of themselves. I report to a large room just off the main stadium for the 10 A.M. panel and take my place on a raised platform with the other panel members. The room is packed with what looks to me to be about five hundred people.

Moderating a panel for the first time, Naomi is in her glory, a sincere and gracious performer, perfectly at home with her audience and exuding country charm. She opens the session and introduces the panelists, Dr. Brian Luke Seaward, David Lee, Elaine Sullivan, and myself, the first three as mind-body advocates who have been advancing wellness in their multiple roles as psychotherapists, lecturers, consultants, counselors, and authors, and me as the token scientist. The introductions are long and laudatory, crammed full of details obviously taken off our CVs and prepared with an eye to public relations, making us all appear highly significant and our work remarkable. During the introduction, Naomi, in earnest, comments that she fully expects me to win the Nobel Prize in biochemistry one day. But when Naomi comes to herself, she jokes, "Oh, just skip it!" The audience loves her humility, and it's no secret that Naomi is the draw for this event, the one everyone has come to hear.

"Me, I'm just the goofiest woman in country music," is about all she'll say by way of self-introduction. "I'm enjoying being here with my girlfriends, Candace and Elaine, and with you boys, Brian and Dave," she says, her southern twang giving away her Kentucky roots. Striking the broadest, most inclusive note, she begins: "I think one of the most powerful things we do here at the Wellness Conference is offer an opportunity for community. Like the feeling I get from my congregation or my friends in country music. It's the unity and support I feel here, being with y'all." We all feel warmed, somehow included—although a few audience members seem to be wondering if perhaps they're in for a revival rather than a wellness panel.

"I've got questions here." Naomi shifts abruptly, adjusting her reading glasses in readiness to tend to the business at hand. She refers to a list of questions she has prepared for the panel, intending to get the conversational juices flowing before opening up the discussion to a general Q&A from the audience. "But first, did anybody see the issue of *Time* magazine that came out a few weeks ago?" Naomi flashes the magazine for all to see, the cover picture depicting an angelic-looking modern-day creature, with the words "Faith in Healing—Can Spirituality Promote Health?" printed below it. I'd read the article and was reminded of a similar article in *Newsweek* in 1988, in which my colleagues and I were interviewed about the possible spiritual implications of the latest PNI research. While now I was pleased to see some national recognition of this theme, back then I was horrified at the sight of the four-letter word "soul" in such close company with my own comments about science and medicine.

"It amazed me that in this entire article, there was not one mention of the greatest faith healer of all time, Jesus Christ," Naomi is saying. "So the first question I'd like to put to our panel is: Why is the spiritual aspect of healing sometimes overlooked in mind-body health?"

Elaine gets the first chance to respond and goes right to the heart of Naomi's question.

"We rarely talk about spirituality in mind-body health because it's too difficult to describe and more difficult to research. In addition, people equate spirituality with religion, which often divides us and brings up strong disagreement. In contrast, I believe spirituality is about a deeper search for meaning in our lives, and brings us together in a commonality of purpose—however each of us finds that. I believe that spirituality is coming full force into our culture because we all know that unless we tap into this deeper strength and community, we won't make it on this planet."

David Lee is next: "Within the past ten years, there's been a shift in psychotherapy, and therapists have begun to realize that psychology is not big enough—there's so much more to human experience. It's been slow, but we are gradually integrating spirituality into the practice of psychology."

He pauses, and I take a turn: "My feeling is that there is no scientific reason to leave spirituality out of medicine. It's a habit that our culture has gotten into ever since the seventeenth-century philosopher René Descartes declared body and soul to be distinct, separate entities, entirely unrelated to each other. But the truth that I have learned through my own late-twentieth-century science is that the soul, mind, and emotions do play an important role in health. What we need is a larger biomedical science to reintegrate what was taken out three hundred years ago."

Naomi is beaming, the audience is applauding, I'm feeling pleased with myself for using my status as a member of the mainstream medical community to advance the bodymind movement just a little further, when suddenly I hear Naomi divulging a bit of personal history—my own. "Candace is a basic scientist who sings in a church choir," she says. "Do you want to tell them all how that came about?"

A little uncomfortable to be suddenly so intimate with my audience, I nonetheless try my best to answer. "Yes, I belonged to the Methodist Church choir for years, having joined when I came to a place in my personal journey where I was unable to forgive someone who I felt had betrayed me deeply. The music of the Christian church drew me in, and I started attending services and singing in the choir. I heard sermons about the teachings of Christ, which I'd never been exposed to before, and especially clear was the message of forgiveness. I was beginning to let go of my grievances, and I sought out other approaches to heal myself as well—meditation, dream work, massage therapy to bring about a healing of my past experience. Forgiveness is a key Christian concept, but it is also a key bodymind concept."

Was I really saying this? A little part of my mind is surprised, but it feels right to be integrating my personal life with my science—even in public!

Brian takes my lead: "What I think we're talking about here is love. All healers—the shamans, the wisdom keepers—tell you they're tapping into a higher source of energy they call love, and that they are sharing this love with whomever they're healing. Jesus' message was about love and compassion, both of which follow forgiveness. And I agree with Candace and Elaine that these elements need to be included in our

Western medical mindframe. I see a growing grass-roots movement aimed at bringing spirituality back into healing. This is reflected in the huge increase in the number of people turning to alternative therapies, many of which speak to spiritual as well as physical and emotional issues. People are saying *enough!*—we're tired of the body-is-a-machine approach, and we want to get back to the original equation."

"Brian," Naomi interrupts. "I read in your new book that eighty percent of illness and disease is caused by stress, and I've heard it said before that stress is a condition that results from spiritual isolation. What do you have to say about that? Are we really separated from God and our soulfulness by modern technology and materialism?"

"Yes, Naomi, I believe that stress is a disconnection from our divine source—or, more accurately, the *perception* that we are separated, because the truth is we're really always connected. At times, though, we forget and don't feel connected to our source. One reason for this is we are so cut off from nature in our society. I think the closest some people get to the outdoors is the Discovery Channel!"

"We are certainly spiritual beings in a physical body and not the other way around," Naomi comments, her country western accent an anomaly in such New Age–speak. "Now, another question: How can we best communicate for healing at the mind-body level?"

David picks this one up: "The approach that I've been trained in, traditional talk therapy, doesn't seem to impact the mind-body level. We often hear our patients say, 'I know I shouldn't feel this way, but I *do!*' Knowing something doesn't always impact how we feel, and we may have to get past purely verbal communication to access our emotions. Some of the approaches I have found effective at getting to deeper, more fundamental levels are storytelling, hypnotherapy, neurolinguistic programming, and any of the expressive therapies that employ visualization, music, and art. In the past, these kinds of therapies were thought of as alternative approaches, but now we're seeing them included more as complementary approaches. Very shortly, I believe we'll be calling them integral approaches, indicating that they are completely integrated into the mainstream."

Naomi turns to Elaine: "Elaine, how can healers and helpers use this new mind-body research?"

"Well, a technique I use with my clients is journal-writing, because it helps them to give a literal structure to their internal world. Writing gives us an awareness of what our patterns are so we can change them if necessary. Studies have shown that when trauma victims write about their experiences, physiological changes occur, such as increased blood flow and a boost for the immune system that can last for up to six months. Also, I recommend meditation. When I heard Dr. Larry Dossey say that a half hour of meditation was as powerful a stress-reliever as a half hour of jogging, I was greatly relieved, because I'm not a jogger!"

I can hold back no longer: "I certainly agree with both Elaine and David, but an element I think we are skipping in our discussion of practical applications for mind-body health is body work: the touch therapies of massage, chiropractic, and any other modality that includes the body as a means of healing the mind and emotions. It's true, we do store some memory in the brain, but by far, the deeper, older messages are stored in the body and must be accessed through the body. Your body is your subconscious mind, and you can't heal it by talk alone."

There is a thoughtful pause, until Naomi cracks, "Well, I saw a few mouths drop open on that one!"

"But it's true," Brian reflects, and then poetically: "The body becomes the battlefield for the war games of the mind. All the unresolved thoughts and emotions, the negativity we hold on to, shows up in the body and makes us sick. Forgiveness is about opening up the heart and learning to love, which is why I think we're all here on this planet. So simple, yet such a difficult lesson to learn."

"Absolutely," Naomi says, "and I found that out through a therapy I utilize called network spinal analysis. It was started in New York about fifteen years ago, by a man named Dr. Donald Epstein. It works by using very gentle manipulations to remove interference from the nervous system. As you all know, I have a chronic illness. Wynonna had a ruptured disc, Ashley has sinusitis, and we all use network for these things."

Naomi turns toward the panel and reads another question. "Explain subtle energy and subtle anatomy, including the human energy field or aura, chakras and meridians. How does all of this fit into the mind-body healing formula?"

Brian is ready with a response: "I believe subtle energy is a kind of universal life force that flows through us from the divine. According to Eastern wisdom, everyone has an aura, a field of this energy surrounding the physical body, and flowing through it as well, traveling along lines called meridians by the Chinese or radiating out from the seven body centers called chakras by the practitioners of yoga. Westerners even have a version of it in Christianity, as the halo surrounding the crown of the head, often depicted on saints and angels in medieval art. But you don't have to be a saint, an angel, or a yogi to acknowledge this subtle energy. It's in everyone and can be a force for healing."

Yes, I'm thinking, but the church certainly co-opted that one over the centuries, making it the singular province of holy men and divine beings.

"There's a wonderful book called *Vibrational Medicine* by Richard Gerber, which describes different modalities of mind-body energy healing, ranging from therapeutic touch and massage to mental imagery and subtle energy," he concludes.

"Candace?" I'm called back by the sound of Naomi's voice. "You've been awfully quiet during this discussion. What do the scientists say about all this subtle energy stuff?"

"As you know, Naomi," I begin, "I've spent a good part of my professional life trying to integrate these Eastern ideas with science. One thing I can tell you, you won't find anything about chakras in a biology textbook! For me the key concept is that the emotions exist in the body as informational chemicals, the neuropeptides and receptors, and they also exist in another realm, the one we experience as feeling, inspiration, love—beyond the physical. The emotions move back and forth, flowing freely between both places, and, in that sense, they connect the physical and nonphysical. Perhaps this is the same thing that Eastern healers call the subtle energy, or *prana*—the circulation of emotional and spiritual information throughout the bodymind. We know that the way health occurs in the physical body has to do with the flow of the biochemicals of emotion. My work has taught me that there is a physical reality to the emotions."

"And how true that is," Naomi says, addressing the crowd in her matter-of-fact, down-home manner. "I've known this all along, and so have most of you. Before I was a country western star, I was a nurse, and I could always tell which patients were going to get better and which weren't by seeing who would laugh at my corny jokes! But I want to show you all an example of bodymind unity. And to do that, I need a volunteer from the audience—someone I can pick on," she says mischievously, and then lights on a young man from the front row. He looks confused as she beckons him to come up on the panel platform.

"Good! You're squirming already!" Naomi says playfully. "Now I'm going to ask you a question: How many guys does it take to change the toilet paper roll?" The young man, at a loss for words, smiles awkwardly and looks at his feet. "I don't know," Naomi shoots back after a pause. "It's never happened!" The audience lets out a roar, but Naomi interrupts to make her point: "Now look how he is blushing, I want you all to see. His face is red as a beet!" She puts her arms around his shoulders and gives him a good-natured hug, thanking him as she sends him back to his seat.

"Candace, I have just put your entire life's work into a single joke. A thought started in that young man's mind and instantaneously it was translated into a physical reality. He turned bright red with embarrassment! Now, there's your neuropeptides at work, for all to see!"

Naomi has certainly done it, translated my research in such a simple, graphic way that everyone could get it. I'm grateful for her unabashedly frank demonstration, her ability to convey in simple, everyday terms what I can only speak about medically, philosophically.

She continues: "Now, we've known all along about this. But we needed the research for the validation. And that's what I'm so grateful to you for, Candace."

Now it's my turn to blush, as all eyes are on me. But it's true that this has been the role I've been able to play in recent years—saying that many alternative medical approaches have validity equal to that of establishment medicine—and I'm grateful to have been given this opportunity. Thanking Naomi, I begin to address the crowd.

"It's a fact that healing approaches that incorporate emotional and spiritual elements have been around for years, in one form or another. But the mainstream has treated them like poor stepchildren and relegated them to the fringes of alternative medicine. The argument that they are untested, and therefore can't be taken seriously, is not valid. So much of mainstream medicine itself is totally unproven—yet we do it anyway. I'm afraid that we are holding the alternative therapies, those that use mind-body and spiritual techniques, to a higher standard than we apply to mainstream medicine. And, furthermore, just because we may not understand the mechanism of a particular technique, that shouldn't necessarily stop people from using it. For thousands of years, we've known that when you're sick, you need bed rest and warmth, and yet this advice has never been studied or published in a medical journal. Every once in a while a piece of folk wisdom does get tested. We now know, for example, that chicken soup really does have therapeutic effects for the common cold! I guess what I'm saying is—learn to trust yourself."

Naomi begins to wrap it up. "I'd like the panel to discuss this one last question before I turn the mike over to the floor and take questions from the audience," she announces. "How can we reconcile religion and such healing processes as yoga, meditation, biofeedback, aromatherapy? I myself am a practicing charismatic Pentecostalist, the Middle American, flag-waving girl next door, and I have people ask me all the time, 'Naomi, how can you sit on the advisory board for Deepak Chopra's institute and then go on Christian talk-radio shows? How can you go to church and then go home to meditate or do yoga?' Elaine, you're an ex-nun, so I'd like to give that one to you."

"Yes, Naomi, you're so right," Elaine begins. "There is a lot of confusion out there, and I think it stems from fear and misunderstanding. It all goes back to the mind-body split, which, as Candace has pointed out, is an arbitrary division that current research has shown once and for all to be invalid. We have learned to distrust our bodies and our feelings, to place our trust in outer authorities instead of our own inner power. I find that meditation in no way contradicts my faith, because there are many routes to the spirit."

The discussion that follows is lively and to the point. Naomi was right, people are eager to explore the role of spirituality in healing, even here in the Midwest, where I would have thought it was too controversial, too *woo-woo*, too Californoid.

The panel discussion has been an incredible experience, giving me the chance to synthesize all of my ideas about spirit, the emotions, and science, and leaving me with a profound understanding of my own transformation. Most amazing is that all of this was catalyzed by my new girlfriend, Naomi Judd, who is not a scientist, not a mystic, but the down-home, country western sweetheart of whitebread America! Getting to know Naomi has made me feel a sense of deep connectedness and unity that is unquestionably spiritual. Her simple message of healing through spirit embraces all my science, making it easy for everyone to understand.

HOLY SPIRIT

The next day I catch a predawn prop plane out of Milwaukee back to Washington, D.C. As the small craft inches through the pink and purple sky, I watch from my tiny window as the growing light slowly obscures Venus, the morning star. Suddenly, the round disk of the sun looms over the horizon and the colors fade, transforming the sky into a flood of soft blue.

I can't stop thinking about how brilliantly—and simply—Naomi illustrated the principle of the mind becoming matter, preceding matter, organizing matter, by singling out the young man in our audience and planting a thought in his mind that made him blush. Thoughts and emotions came first, and the peptides followed, causing the blood vessels in his face to open. As Deepak's sages in India understood, the non-stuff, the "no-thing," is the source; and the stuff, the material phenomenon, manifests from there.

This is such a fundamental shift for the Western mind, but one that science can help us understand. Originally, we scientists thought that the flow of neuropeptides and receptors was being directed from centers in the brain—the frontal cortex, the hypothalamus, the amygdala. This fit our reductionist model, supporting the view that thoughts and feelings are products of neuronal activity, and that the brain was the prime mover, the seat of consciousness. Then, as a result of my own and other people's work in the laboratory, we found that the flow of chemicals arose from many sites in the different systems simultaneously—the immune, the nervous, the endocrine, and the gastrointestinal—and that these sites formed nodal points on a vast superhighway of internal information exchange taking place on a molecular level. We then had to consider a system with intelligence diffused throughout, rather than a one-way operation adhering strictly to the laws of cause and effect, as was previously thought when we believed that the brain ruled over all.

So, if the flow of our molecules is not directed by the brain, and the brain is just another nodal point in the network, then we must ask— Where does the intelligence, the information that runs our bodymind, come from? We know that information has an infinite capability to expand and increase, and that it is beyond time and place, matter and energy. Therefore, it cannot belong to the material world we apprehend with our senses, but must belong to its own realm, one that we can experience as emotion, the mind, the spirit—an *inforealm!* This is the term I prefer, because it has a scientific ring to it, but others mean the same thing when they say field of intelligence, innate intelligence, the wisdom of the body. Still others call it God.

Although it's a simple concept, it's hard for the Western mind to understand. But I recall one person who was able to grasp it instantly, a cameraman working on the set of Bill Moyers's PBS special *Healing and the Mind.* As I was groping to explain how the innate intelligence, generated by subtle energies from flowing biochemicals, all converged in the inforealm, what came out was puzzling to Bill, but not to the cameraman. When the taping was over and everyone was packing up, the gentle, soft-spoken man approached me and said, almost whispering it in my ear, "You were talking about the Holy Spirit, weren't you?"

Feeling a bit embarrassed, I had to admit that, yes, maybe I was.

Reductionists will always argue that the molecules come first, are the primal force, and that thoughts and emotions follow as a kind of epiphenomena of the molecules. And they've got good evidence: Doesn't the flow of peptides *change* the physiologic responses, which then create the feelings we experience? Doesn't the chemical release of endorphins *cause* the feeling of pain relief or the euphoria of runner's high?

I don't deny this, but what I'm saying is that we must recognize that there is a two-way system of communication at work. Yes, the release of endorphins can cause pain relief and euphoria. But, conversely, we can bring about the release of endorphins through our state of mind, as I experienced so vividly when working with Evelyn Silvers. I like to think of mental phenomena as messengers bringing information and intelligence from the nonphysical world to the body, where they manifest via their physical substrate, the neuropeptides and their receptors.

HOME

I'm happy to be back in my office at Georgetown University Medical School, where both Michael and I now hold positions as research professors, and where we are able to continue our research on Peptide T and its effects on gp_{120} and the AIDS virus. It's a spiritual place with a legacy, Georgetown, founded by the Jesuits in the eighteenth century, and I am comfortable here because it is a setting that reflects my current state of bodymind: mainstream, decidedly, but with an added dimension of the spiritual.

I have been talking with Father Sweeney about infusing holistic approaches into the medical school hospital, bringing the mind/body/spirit reality into the medical setting and setting the Georgetown hospital apart from others with an apt slogan—"The whole person comes first!" This could be a boost to the ailing financial status of the hospital, which has suffered from the influx of HMOs and from other difficulties that are part of the current health care scene. I suggest to Father Sweeney that hospi-

312 ≈ MOLECULES OF EMOTION

tals have to be competitive these days, to offer something unique to gain the competitive edge. *Holistic, whole, healthy,* and *holy*—all words stemming from the same root, the Saxon *hal,* and all linked in meaning. The Jesuits are a holy order, so it makes sense for them to be part of the holistic movement.

The spiritual viewpoint, Father Sweeney tells me, confirming my feeling about the religious aspect of holism, is one that sees the unity of all things, that allows us to experience our oneness with all others and with God. I can understand this on a scientific level: Yes, we have a biochemical psychosomatic network run by intelligence, an intelligence that has no bounds and that is not owned by any individual but shared among all of us in a bigger network, the macrocosm to our microcosm, the "big psychosomatic network in the sky." And in this greater network of all humanity, all life, we are each of us an individual nodal point, each an access point into a larger intelligence. It is this shared connection that gives us our most profound sense of spirituality, making us feel connected, whole.

As above, so below. To think otherwise is to suffer, to experience the stresses of separation from our source, from our true union. And what is it that flows between us all, linking and communicating, coordinating and integrating our many points? The emotions! The emotions are the connectors, flowing between individuals, moving among us as empathy, compassion, sorrow, and joy. I believe that the receptors on our cells even vibrate in response to extracorporeal peptide reaching, a phenomenon that is analogous to the strings of a resting violin responding when another violin's strings are played. We call this *emotional resonance,* and it is a scientific fact that we can feel what others feel. The oneness of all life is based on this simple reality: Our molecules of emotions are all vibrating together.

This is where I have come to in my personal life, to an acceptance of my part in the bigger play, and an awareness of how in science, we are all working together to bring about the truth. Now it turns out that Peptide T is not just part of the latest efforts to cure AIDS, but a possible broad-spectrum antiviral drug that shows remarkable results in many other chronic conditions as well. This new application is based on the discovery that the virus acts at an additional receptor other than the T4, a "coreceptor" called the chemokine receptor, and it is a discovery made by the very researchers I believed were trying to stop me from developing Peptide T as an AIDS drug. Now they welcome me as I return to the fold, and I feel the forces of synergy and cooperation at work in the universe.

I'm relaxing with these thoughts in the familiar comfort of my office, still small but not as tiny as that early cubicle I inhabited at the Palace, the one Biff Bunney entered to console me in a much darker hour. It's an attractive space, decorated with my favorite rainbow art, one wall covered with photos and mementos given to me by people I have met over the years. A large bulletin board frames my "famous-person montage," where I've arranged the photographs I've collected of myself with various public figures. One shows me with the Pope, who granted me an audience when I presented at a conference in Rome in 1985. The Pope! How strange—my work here at Georgetown is to bring forth the new paradigm, integrating the new with the old, and the Pope watches from my office wall, this time in alliance with *me*, not Descartes! Suddenly I remember that I have a new photo, one of Naomi and me taken at Stevens Point, and I pull it from my briefcase to place it in my gallery alongside the Holy Father.

As I admire the new addition to my collection, my eye is caught by the colorful rug hanging on the adjacent wall. My mystical rug, I call it, because the design has mystical overtones: a dawning sun surrounded by yellow sunflowers and large black birds. But its real significance is that it first appeared to me in a dream the night before I left for a gathering sponsored by Brigham Young University—The Psychobiology of Health and Wellness, a Conference on Healing and the Mind—which was held in 1995 in Provo, Utah. I dreamt I was on my way to some very important place, to make a presentation perhaps, when suddenly I realized I was completely naked. I was feeling very frightened and dejected, as if I'd been cast out into the wilderness with no protection, when, magically, as if from thin air, a rug appeared and wrapped itself around me. I immediately felt better, my confidence and sense of purpose restored as I continued on my journey in my new cloak. When I awoke, I could make no sense of the dream but recorded it anyway.

The next day I was in Provo, where I gave my usual talk for a group that had been hosting mind-body health meetings for the past twelve years, presenting an assortment of speakers from backgrounds in spirituality and psychology. Joining me on the roster was Bob Ader of early PNI fame, the two of us representing the organizers' first attempt at bringing some hard science into their program. The audience, largely made up of Mormons, was quite serious, and even though they barely laughed at my many jokes, I liked these very healthy and hardy-looking people whose ancestors had forged the early Church of the Latter Day Saints. While I knew very little about their religion, I could tell they were survivors, coming from a stock of tough pioneers whose lives were guided by spirit, and for this I admired them.

After the sessions ended, all the speakers shared an exquisite walk in the surrounding mountains, led by a group of elder churchmen, and I had a chance to interact more closely with some of them. My talk had caused quite a stir, it seemed, not only for the science I had presented, but also because I had talked about the role of the emotions, mind, and spirit in health and how I had come to an understanding of these elements in my own life. This is what had fascinated them, they told me, the idea of a person who had been transformed by her work, who had come to a spiritual place from scientific truth-seeking. I thanked them for their flattering words, and later, in my room that night, thought about how true I felt this to be.

The next morning, I was driven to the airport by one of the younger Mormons. After I had checked my bags and we were saying good-bye, he handed me a large package and explained shyly that the conference sponsors had wanted me to have it as a token of their appreciation. I unwrapped it right there, and literally got chills all over my body as I recognized it as the rug from my dream of a few nights before. This was my new mantle, the symbol of my spiritual transformation through scientific truth-seeking, which the Mormons had somehow recognized.

When I'd returned from Provo, I brought the rug into my new office at Georgetown, intending to spruce up the empty room with some personal touches. It remained on the floor until one day I decided to hang it on the wall where I could see it better. Now it serves as a daily visual reminder of my purpose here at Georgetown, symbol of the role I aspire to play as truth-seeking scientist and catalyst in the mindbodyspirit revolution in modern medical science. It is an honor and a gift I will never forget.

SCIENCE: THE SEARCH FOR TRUTH

For me, science has been a quest to understand nature—both human and Mother. As I have known it in its purest, most exalted form, science is the search for truth. It was this belief that drew me to science, and, through all my naivete and despite all my many false turns, it's what has kept me on the journey.

The heart of science is feminine. In its essence, science has very little to do with competition, control, separation—all qualities that have come to be associated with science in its male-dominated, twentieth-century form. The science I have come to know and love is unifying, spontaneous, intuitive, caring—a process more akin to surrender than to domination.

I have come to believe that science, at its very core, is a spiritual endeavor. Some of my best insights have come to me through what I can only call a mystical process. It's like having God whisper in your ear, which is exactly what happened on Maui when I stood up with a slide of the HIV receptor in the brain and suggested a new therapy for AIDS, only to hear an inner voice say to me: *"You should do this!"*

It's this inner voice that we scientists must come to trust. We must stop worshiping a dispassionate "truth" and expecting the experts to lead us to it. There's a higher intelligence, one that comes to us via our very molecules and results from our participation in a system far greater than the small, circumscribed one we call "ego," the world we receive from our five senses alone. New understanding from quantum physics and information theory points us away from the cool, detached, solitary genius, the one who has the answers that others don't have, as if the truth could be owned, and toward a more collegial, participatory model of knowledge acquisition. The rational, masculine, materialistic world we live in places too much value on competition and aggression. Science at its most exalted is a truth-seeking endeavor, which encompasses the values of cooperation and communication, based on trust—trust in ourselves and in one another.

EPILOGUE

PEPTIDE T—THE STORY CONTINUES

MICHAEL AND I call it "Peptide T weather"—ice storms, heat waves, hurricanes passing close or hitting hard, shutting down electricity and closing roads. Whenever our research takes a pivotal turn, it seems, meteorological aberrations abound. Rainbows, too, have appeared as mystical heralds announcing crucial moments in our Peptide T adventures—Maui, Puerto Rico, Provincetown.

So, it's no surprise to either of us, in September of 1996, to be caught in the worst local flood of the century on our way to Baltimore for Dr. Robert Gallo's annual AIDS meeting. The Potomac River is overflowing its banks as a result of the combined rains of hurricanes Fran and Hortense, making freeway travel all but impossible. As might be expected, we are late for this fifth day of a week-long meeting in which promising new AIDS research has been the focus, research that we hope will be the beginning of a full vindication of our claims about Peptide T.

We arrive soggy but spirited at the hotel in downtown Baltimore. A directory in the lobby tells us where to go for the talk we have come to hear: "New Concepts in Immuno-Pathogenesis of AIDS." We race off to join our colleagues in hopes of learning more about a development that was first reported in the *New York Times* a few months earlier, causing quite a stir at the midsummer International AIDS conference in Vancouver.

The big news is that there has been a shift in attention by the mainstream AIDS researchers away from the HIV virus itself and toward the cells that the virus attacks—in particular, a certain protein studding the surface of most immune cells. It is this protein, a peptide receptor, that has been discovered as a new mechanism by which the virus enters the immune cells.

Scientists have been fighting for a decade about how the virus actually kills T4 cells: Does the virus enter the cell and kill it by literally imploding within it, or does another, more indirect mechanism cause the disease, that of apoptosis, a programmed cell death? And then there's the theory of Candace and Michael with their child of the new paradigm—no longer a radical theory—which says it's not the virus itself that causes the damage, but the blocking of the receptor by the gp_{120} viral fragments, and the resulting denial of access to the cell by natural peptides vital for cell and whole organism functioning and health. One thing we all agree on now—the T4 (CD4) receptor on the immune cell is necessary, but not sufficient, for the HIV virus to enter the cell.

Now, in this latest development, it had been found that a new receptor, the chemokine receptor, was part of this process. Five major labs had made headlines by simultaneously reporting that the HIV virus was using not only the previously understood mechanism of the T4 (CD4) receptor to enter cells, but one of the chemokine receptors as well, now considered a "co-receptor." The labs had elegantly demonstrated that both receptors were required for the virus to enter the cell, working together as a kind of double-docking system.

Gallo and his team were acknowledged extravagantly for doing the seminal work that led to the finding. As part of their ongoing research, they had been following up on an observation made by researchers that some patients infected with the virus remained healthy for years, seemingly impervious to the infection in spite of continued exposure. Dr. Jay Levy of the University of California at San Francisco had shown that the immune cells of these long-term survivors secreted a substance that seemed to block the virus from entering the cell, but because the molecular structure of the active ingredient had never been cracked, it had been difficult to make much progress.

With the help of his team, particularly Tony deVeco, Gallo was able to isolate the activity from the immune cells of long-term survivors, then crack the structure of the factor, only to find it had been discovered before as a peptide ligand for chemokine receptors. The focus on chemokines had already heated up when they were observed to mediate inflammation, a key process in many diseases from Alzheimer's to psoriasis. Now, with its newly realized relevance to AIDS, this field was about to become positively scorching!

Epidemiologists had shown that a small percentage of people who lack the chemokine receptor (due to a rare genetic mutation) never got AIDS, no matter what high-risk behavior they indulged in. Bingo! This was the solid clinical proof in favor of the theory that the chemokine receptor is not just another heartbreaking laboratory artifact, but instead, a viable vulnerable spot in the virus's life cycle. In its absence, the virus is prevented from entering the cell and causing the symptoms of AIDS. Clearly the race is on to find new receptor blockers as the next generation of anti-AIDS drugs.

When I'd heard the news back in June, I went straight to the library and quickly compared the peptide sequences of the various relevant chemokines with that of Peptide T. I was thrilled to find out that there was a possible match. Could Peptide T, invented six years before the term chemokine was even coined, be an antagonist to the chemokine receptor? In a synchronicity almost beyond belief, we learned that the key activity test for putative chemokines was the very chemotaxis assay that Michael and I had used to link mind and immunity fourteen years ago!

Over the course of the Gallo meeting in Baltimore, my colleagues have been approaching me, asking in half teasing, half dead-serious tones: "Do you think Peptide T binds to the chemokine receptor?" I field their inquiries cautiously and keep my lip zipped, refusing to say a word. A bit wiser from my experience, I'm content to wait until a ton of data is available to make an iron-clad case for what I believe will be a positive answer. After all, Michael has the pivotal assays up and running at our Georgetown lab, the crucial chemotaxis assays, which will soon give us the data to know for sure.

THE FACE OF AIDS has changed. Thank God for the protease inhibitors! At last there is an effective long-term treatment for AIDS. The new protease inhibitors, when started in a cocktail with the older anti-virals (which slows the onset of drug resistance) is enabling people to live longer. Invented by scientists at several companies, the drugs were rapidly and efficiently tested and approved with lots of AIDS activist–inspired cooperation.

But sadly, unexpected trouble is appearing a year or two after treatment in the three-quarters that can tolerate the "triple drug cocktail." Despite the many effective treatments doctors devised for preventing, diagnosing, and treating opportunistic infections, patients still never regain all the weight they lost during their last infection. About a quarter of the long-term AIDS survivors are steadily losing weight—muscle mass—and the data show they will die with "wasting" when their weight falls to 65 percent of normal. During all this suffering, their virus levels often remain so low as to be undetectable.

Is gp_{120} causing AIDS symptoms like wasting? Gp_{120} is so potent that just a few cells infected with viruses resistant to the new powerful anti-viral cocktails could secrete enough gp_{120} to poison several kinds of peptide receptors. It is so potent that an assay to routinely measure the tiny but damaging levels of gp_{120} found in HIV-infected patients hasn't yet been found. Scientists believe infected cells—which secrete gp_{120}—are lurking in "sanctuaries." Sanctuaries include places like the brain into which the current drugs penetrate poorly. Neuro-AIDS is also on the rise, another black cloud, along with wasting, on the horizon of the relatively brighter picture of AIDS we see today.

URGENCY quickens the pace of our Georgetown team . . . still further. Other scientists in the physiology department, experts in wasting and neuropeptide receptors, are now working on Peptide T as a gp_{120} antagonist. Now, we must compile our findings to present as impeccable scientific papers. Our presentations at meetings have been helpful. They have allowed us to hone the meaning of our data and hear suggestions of experiments necessary to fill the gaps. But talks at meetings are merely published as "abstracts"—only the publication of full reports in scientific journals will be considered to provide the solid scientific rationale needed to expand human testing of Peptide T.

Growth hormone, which promotes lean muscle mass and testable strength performance, is the only therapy even provisionally approved for AIDS wasting. Gp_{120} injected into the brains of rats produces weight loss! Peptide T restores the secretion of growth hormone to rat pituitary gland cells, which has been reduced by gp_{120}. These experiments had been started two years ago when our chairman at Georgetown, noticing a peptide sequence homology between Peptide T and a peptide which releases growth hormone, had invited us to join the department.

We've handed off the chemokine peptides to Doug Brenneman's NIH team for collaborative experiments—they protect against gp_{120} neuronal cell death just like VIP and Peptide T! Chemokine receptors are on brain as well as immune cells where gp_{120} binding must certainly contribute to Neuro-AIDS and other inflammatory brain diseases.

Now we've got the chemotaxis data that show Peptide T is a chemokine receptor antagonist. To complete the story, we need to get the Peptide T receptor binding assay to work so we can see if it can be displaced by chemokines, as predicted. With yet another eerie synchronicity, we've learned that chemokine receptors, VIP receptors, and growth-hormone-releasing hormone receptors are all from the same biochemical family—the one to which the opiate receptor belongs! So far I can't get a signal, but I should be able to figure out how to get that binding assay to work. We need to get those papers out.

While we focus on our daily experiments in the lab, we are awaiting further results of Peptide T clinical trials for Neuro-AIDS with great hope and prayer. Hard as it is to do a simple one-day lab experiment perfectly, the ethical issues central to experiments with human beings and the resources such experiments demand intensify the difficulties enormously. But despite their difficulty, I believe only scientifically sound human trials can teach us the potential of Peptide T—or any other drug or mind-body treatment for that matter. In the end all the rosy anecdotes and uncontrolled data are almost worthless. Last week we heard that some renowned AIDS clinicians will recommend that Peptide T be added to some NIH trials. We need more trials. We need more good science. We need to get those papers out. We push on.

We can hardly believe that it's been over ten years on this AIDS project. It seems only to get richer and more interesting as it continues to unfold, now coming forward into receptor-based treatments, where we started. Now nearly everybody wants to know what part of gp_{120} binds to the chemokine receptor, what peptide sequence will block it.

Hold on to your horses, this Peptide T baby is about to bolt. It's very much a time of renewed focus and determination. All sorts of very smart, well-funded labs are taking up these questions. The scientific understanding of AIDS is truly leading to a new understanding—with new strategies for curing—of many other diseases. Mike and I, at times, smile knowingly at each other, sharing feelings that are somewhat like those we have for our young but suddenly mature daughter, who graduated from college last week. Childhood ends, and so begins a new phase, with all its risks and hopes.

APPENDIX A

PREVENTION-ORIENTED TIPS FOR
HEALTHFUL, BLISSFUL LIVING

We must take responsibility for the way we feel. The notion that others can make us feel good or bad is untrue. Consciously or—more frequently—unconsciously, we are choosing how we feel at every single moment. The external world is in so many ways a mirror of our beliefs and expectations. Why we feel the way we feel is the result of the symphony and harmony of our own molecules of emotion that affect every aspect of our physiology, producing blissful good health or miserable disease.

We have the hubris to think that just because we invented electric lightbulbs, we can keep any hours we want. But neuropeptide informational substances link our biological clocks to the motions of the planets, which is why your quality of sleep—and wakefulness—is likely to improve the more closely your retiring and your rising are linked to darkness and daylight. If you get to sleep between 10 and 11 P.M., most of you will be able to wake up naturally and rested with the sunrise, if not before.

Meditation practiced early morning and early evening, routinely, even religiously, is, I believe, the single quickest, easiest, shortest, and cheapest route to feeling good, which means being in sync with your natural feelings.

The early morning is a great time to enjoy, to consciously envision a wonderful day. It's a great time for the conscious mind to reenter the body with bodyplay (exercise sounds dreary), which may be gentle stretching or yoga one day, a brisk walk with dancing or a run to break a sweat the next. See how you feel before you decide. It seems natural—what the body was designed to do— to move a bit on arising, before eating or climbing into a car. Our foremothers and forefathers would almost certainly have started their days with movement.

For those of you interested in weight loss, another reason to get moving early on is that we are designed to be able to turn on the fat-burning neuropeptide circuitry in our bodyminds with just twenty minutes of mild aerobic exercise at the beginning of the day. Research by exercise physiologists has shown that after twenty minutes of elevated heartbeat and the deeper, more frequent breathing that naturally comes with it, our bodyminds enter a smooth, fat-burning mode that lasts for hours. The alert and calm feeling that settles in after an initial feeling of exhilaration usually goes hand in hand with a reduced appetite.

Spend some time in nature every day, longer on days off. Being outside is being in nature regardless of whether you're in a forest, at the beach, or downtown in a large city. Look at the sky! Even cities have skies over them. Bad weather is no excuse—invest in warm clothes, good shoes, and a waterproof outer layer.

When to eat is as important as what you eat. Don't starve yourself all day and eat late. In fact, your biggest food intake should be your midday meal, as it is in every nonindustrial culture, and as it used to be in our own. Eating at midday allows the food plenty of time to be completely and wholesomely digested before you retire for the evening. It also means that the molecules of nutrition will be carried to sites in the bodymind where they will reinforce conscious, vigorous, waking activity rather than being deposited as fat, which happens more readily when we eat too late. If you've never observed such a schedule, you may be amazed at the jolt of mental and physical energy you will feel—which is the way you are supposed to feel.

Avoid exogenous ligands that perturb the psychosomatic network so much that they warp its smooth information flow, producing "stuck" information circuits that prevent you from experiencing your full repertoire of potential experiences, and instead cultivate feedback loops that restore and maintain your natural bliss. Translation: To feel as good as possible all of the time, avoid doing drugs, legal or illegal. Question any chronic prescription: If you have to have it, make sure you are taking the lowest possible dose that does the job. Under the supervision of your doctor or other medical consultant, consider taking a "drug holiday" every once in a while to see if you really still need that sleeping pill, antidepressant, antiulcer, or high blood pressure medication. Experience how amazingly responsive and resilient, lively and blissful the natural undrugged state is. Being drug-free allows your system to focus on healing your own bodymind rather than compensating for drug-induced alterations and expending bodymind effort on detoxifying and excreting drugs.

Think of sugar as a drug with chronic effects right up there with more acknowledged "drugs of abuse." Sucrose, the white powder isolated from acres of green plants (sugarcane or sugar beets) turns into glucose, a key metabolic regulator of your bodymind, which acts on glucose receptors to control the release of insulin and numerous other neuropeptides from the pancreas, drastically altering how we feel—sluggish or peppy, low or high—and how we metabolize our food. Satisfy sweet cravings with fruit, which has a different kind of sugar, fructose, which less readily causes the release of insulin. Refined white sugar changes the profile of peptides released from the pancreas (in addition to insulin), which results in a sluggish, fat-storing mode. In general, work on exploring the impact of what you eat on the way you feel.

Drink eight glasses of unchlorinated water every day. So often we eat when we're really thirsty rather than hungry. Our internal signals have gotten confused because we evolved eating whole, natural foods (fruits and vegetables), which have a much higher water content than our current diets of chips and dips and numerous other packaged, processed foods and junkstuffs.

Aim for emotional wholeness. When you're upset or feeling sick, try to get to the bottom of your feelings. Figure out what's really eating you. Always tell the truth to yourself. Find appropriate, satisfying ways to express your emotions. And if such a prescription seems too challenging, seek professional help to feel better. I believe the alternative or complementary therapies are a form of professional help much less likely to do harm and more likely to do good than conventional approaches. They work by shifting our natural balance of internal chemicals around, so we can feel as good as possible. They are often particularly helpful for alleviation of the many chronic maladies that currently have no good medical solutions (see Appendix B).

Consciously and lovingly acknowledge each family member before sleep. That is, say goodnight. Don't program your bodymind with images of death, destruction, and the bizarre before retiring. Translation: Never wind down with the nightly news. Instead, try a book, a relaxing hobby, a hot bath, or even light housework.

Last, but definitely not least, health is much more than the absence of illness. Live in an unselfish way that promotes a feeling of belonging, loving kindness, and forgiveness. Living like this promotes a state of spiritual bliss that truly helps to prevent illness. Wellness is trusting in the ability and desire of your bodymind to heal and improve itself given half a chance. Take responsibility for your own health—and illness. Delete phrases like, "My doctor won't let me . . ." or, "My doctor says I have [name of condition], and there is really nothing I can do" from your speech and thought patterns. Avoid unscientific beliefs about your need for medications and operations.

APPENDIX B

BODYMIND MEDICINE:
RESOURCES AND PRACTITIONERS

Although I am a scientist and not a clinician, I am of the strong opinion that complementary therapies can not only help with chronic degenerative diseases like cancer, arthritis, heart disease, and auto-immune disorders, but also with other health problems, such as asthma, pre-menstrual syndrome, headaches, sinusitis and gastrointestinal disorders, for which conventional medicine can offer only incomplete or no help. Complementary therapies can also enrich our lives so that we can live them more fully and joyfully. Following is a list of resources related to bodymind medicine to help you find further information about therapies or locate a practitioner in your area.

For this section, I am most fortunate, and very grateful, to have had the collaboration of Jacqueline C. Wootton, M.Ed., my friend and neighbor as well as colleague, and a long-standing expert in alternative medicine information resources. The therapies, treatments and approaches represented by the following organizations are those I think are most relevant for impacting on the molecules of emotion, some of which I've touched on in the book. You can reach Jackie's database of resources organizations on her World Wide Web site (http://www.clark.net/pub/AltMedInfo/). Thanks also to Karen Sullivan for additional research in the UK.

Jackie has organized the list according to the categories relevant to the theme of the book. She would point out that this categorization is not definitive or complete; several therapies could be grouped differently. Specific biological/pharmacological treatments have not been included, nor have homeopathy and herbal medicine, except where they are part of a complete system of traditional medicine. Some sections, such as cancer care and meditation, have been expanded to include some treatment centers or training facilities.

We offer you this information in the spirit of promoting personal responsibility for health. Inclusion in this list does not constitute our endorsement. You will find that some of these organizations give conflicting recommendations, so you would be wise to research a range of different viewpoints and recommendations in order to reach your own health and well-being decisions.

Please note, also, that the following information was as accurate as we could make it at the time we went to press. Unfortunately addresses and other aspects of this listing are subject to change so please be understanding when using this material for reference.

BODYMIND

Bodymind medicine is based on the recognition of the relationship between mind and body, the body's innate healing potential, and the partnership of patient and healer in restoring the body to health.

British Holistic Medical Association
Trust House
Royal Shrewsbury
Hospital South
Shrewsbury
Shropshire
SY3 8XF
Tel: 1743 261155

Holistic and Creative Therapy Association
2A Burston Drive
St Albans
Herts AL2 2HR
01727 674567

Holistic Health Foundation
2 De La Hay Avenue
Plymouth
Devon PL3 4HH
01752 671485

British Association for Holistic Health
179 Gloucester Place
London NW1 6DX

British Complementary Medicine Association
39 Presbury Road
Cheltenham
Glos GL52 2PT
01242 226 770

Council for Complementary and Alternative Medicine
206-8 Latimer Road
London W10 6RE

ABT – Association for Analytic and Bodymind Therapy and Training
8 Princes Avenue
Muswell Hill
London N10 3LR

The Zero Balancing Association UK
36 Richmond Road
Cambridge
CB4 3PU

Association of Physical and Natural Therapists
68A The Avenue
Worcester Park
Surrey KT4 7HJ

The Awakened Mind Limited
9 Chatsworth Road
London NW2 4BJ

Association of Therapeutic Healers
Neals Yard Therapy Rooms
2 Neals Yark
London WC2

Energetics Association
72 Dumbarton Road
Lancaster LA1 3BX

Relaxation for Living
29 Burwood Park Road
Walton on Thames
Surrey KT12 5LH

Bodymind medicine encompasses a great many modalities, some of which are listed below.

BIOFEEDBACK
Biofeedback is used to train patients to control brain-wave activity so they can modify their own autonomic body processes. This technique may be used to retrain cardio-vascular and respiratory functions.

British Association for Holistic Health
179A Gloucester Place
London NW1 6DX

International Stress and Tension Control Association
18 Aldury Ride
Cheshunt
Herts DN8 8XF
Helpline: 01992 633100

Suppliers of Biofeedback equipment include:

Alpha One Limited
The Old Courthouse
Bottisham
Cambridge CB5 9BB

Biodata Limited
10 Stocks Street
Manchester M8 8QG

BODY PSYCHOTHERAPY
Body Psychotherapy, which is conducted in private or group sessions, uses simultaneous touch and talking to empower the patient, or it may simply involve a quiet, cathartic experience of holding one another. Later, those who participate, discuss the experience.

British Association for Counselling
1 Regent's Place
Rugby
Warwickshire CV21 2PJ

Alternative Health Information Bureau
12 Upper Station Road
Radlett
Herts WD7 8BX

Health Practitioners Association
187A World End Lane
Chelssfield
Kent BR6 6AU

British Register of Complementary Practitioners
PO Box 194
London SE16 1QZ

GUIDED IMAGERY AND VISUALIZATION
This technique enlists the imagination to aid diagnosis and promote physiological functioning.

British Register of Complementary Practitioners
PO Box 194
London SE16 1QZ

Brahma Kumaris World Spiritual University
98 Tennyson Road
London NW6 7SB

Vanita Miller
3 St Elmo Court
London Road
Hitchin
Herts SG4 9ET

Guild of Complementary Practitioners
Alpha House
High Street
Crowthorne
Berks RG11 7AD

HYPNOTHERAPY

A hypnosis-induced trance state can be used as part of a treatment for many different kinds of conditions. Under hypnosis, a patient may be relieved of anxiety, pain and stress; posthypnotic suggestions to the unconscious mind can be made to promote continued healing.

British Hypnotherapy Association
67 Upper Berkeley Street
London W1H 7DH

National Register of Hypnotherapists and Psychotherapists
12 Cross Street
Nelson
Lancs
BB9 7EN

National Council of Psychotherapists and Hypnotherapy Register
24 Rickmansworth Road
Watford
Herts WD1 7HD

Central Register of Advanced Hypnotherapists
28 Finsbury Park Road
London
N4 2JX
Send an sae for a register

British Hypnosis Research
St Matthew's House
1 Brick Row
Darley Abbey
Derbyshire
DE22 1DQ
Send an sae for a register

Corporation of Advanced Hypnotherapy
PO Box 70
Southport
PR8 3JB

European Society of Medical Hypnosis
3 Troy Road
Morely
Leeds
LS27 8JJ

UK College of Hypnotherapy and Counselling
10 Alexander Street
London W2 5NT

British Society of Medical and Dental Hypnosis
42 Links Road
PO Box 6
Ashtead
Surrey
KT21 2HT

MUSIC, ART, DANCE AND HUMOUR THERAPY
Use of the arts to heal, maintain, and improve a person's physical and mental health.

The Association of Professional Music Therapists
Chestnut Cottage
38 Pierce Lane
Fulbourne
Cambridgeshire
CB1 5DL

British Association of Art Therapists
11A Richmond Road
Brighton
Sussex
BN2 3RL

British Association of Dramatherapists
5 Sunnydale Villas
Durlston Road
Swanage
Dorset
BH19 2HY

Association for Dance Movement Therapy
c/o Arts Therapies Department
Springfield Hospital
61 Glenburnie Road
London
SW17 7DJ

British Psychodrama Association
8 Rahere Road
Cowley
Oxford
OX4 3QG
01865 715055

Institute for Arts in Therapy and Education
2-18 Britannia Row
Islington
London N1 8QG
0171 704 2534

Laban Centre for Movement and Dance
Laurie Grove
New Cross
London
SE14 6NH

British Society of Music Therapy
25 Rosslyn Avenue
East Barnet
EN4 8DH

QI GONG
Self-healing traditional Chinese practice using movement, meditation, and controlled breathing to balance the body's vital energy force, qi or chi, to promote health, fitness and longevity.

British Council for Chinese Martial Arts
28 Linden Farm Drive
Countesthorpe
Leicester LE8 5SX

Linda Chase Broda
The Village Hall
163 Palatine Road
Manchester M20 2GH

Zhi-Zing and Zhen-Di Wang
Chinese Heritage
15 Dawson Place
London W2 4TH

London School of Acupuncture and Traditional Chinese Medicine
60 Bunhill Row
London EC1Y 8QD

British Acupuncture Council (BAC)
Park House
206-208 Latimer Road
London W10 6RE

College of Integrated Chinese Medicine
19 Castle Street
Reading
Berks RG1 7SB

Tse Qigong Centre
Qi Magazine
PO Box 116
Manchester M20 3YN

CANCER AND COMPLEMENTARY CARE

The following organizations offer a range of services: research findings and information; advice; treatment and residency programs. This list is not exhaustive, but should help patients and professionals with their own investigations into complementary programs as an adjunct to cancer care.

British Cancer Help Centre
Grove House
Cornwallis Grove
Clifton
Bristol BS8 4PG

Breast Cancer Care
Kiln House
210 New Kings Road
London SW6 4NZ

BACUP
3 Bath Street
London EC2A 3JR

Cancer Care Society
21 Zetland Road
Redland
Bristol BS6 7AH

Cancer Link
17 Britannia Street
London WC1X 9JN
Information service: 0171 833 2451

Cancer Relief Macmillan Fund
Anchor House
15-19 Britten Street
London SW3 3TZ

New Approaches to Cancer Care
Egham
Surrey TW20 9RB

ALTERNATIVE AND COMPLEMENTARY HEALTH CARE SYSTEMS

ACUPUNCTURE AND TRADITIONAL ORIENTAL MEDICINE
Ancient Chinese medical system of balancing the flow of vital energy through the body's meridians; used to alleviate pain, enhance natural immunity, and treat many health problems. Acupuncture is characterized by the insertion of very fine needles, a usually painless and often pleasurable procedure, followed by a deep rest or even a nap.

Register of Chinese Herbal Medicine
21 Warbreck Road
London W10 8NS

Association of Practitioners of Tibetan Medicine
45 East Trinity Road
Edinburgh
EH5 3DL

Register of Chinese Massage Therapy
PO Box 8739
London N28

American Association for Oriental Medicine

4101 Lake Boone Trail, Suite 201

Raleigh, NC 27607

(919) 787-5181

Web site: http://www.aaom.org/aahome.htm

Professional association for nonphysician acupuncturists; publications; membership list

National Acupuncture and Oriental Medicine Alliance
(The National Alliance)

14637 Starr Road SE

Olalla, WA 98359

(206) 851-6896

Fax: (206) 851-6883

E-mail: 76143.2061@compuserve.com

National membership association representing the diversity of practitioners of acupuncture; information; publications; membership list

Traditional Acupuncture Institute, Inc.

The American City Building, Suite 100

Columbia, MD 21044

(301) 596-6006 (DC)

(410) 997-4888 (Baltimore)

Fax: (410) 997-4793

AYURVEDA

Ancient Indian medical system meaning "science of life." Based on the harmonization of body, mind, and spirit.

Ayurvedic Institute

11311 Menaul NE, Suite A

Albuquerque, NM 87112

(505) 291-9698

Fax: (505) 294-7572

Information; publications

College of Maharishi Ayurveda

PO Box 282

Fairfield, IA 52556

(515) 472-8477

Fax: (515) 472-7379

Health-education programs; physician referrals

NATUROPATHY

Eclectic health care system based on current biochemical studies that utilizes a wide range of healing practices to enhance the body's own restorative capacities. Training requires a rigorous four years and emphasizes a gentle approach that "does no harm." Naturopathy relies on natural therapies and supplements and ancient wisdom.

American Association of Naturopathic Physicians
2366 Eastlake Avenue East, Suite 322
Seattle, WA 98102
(206) 323-7610
Web site: http://infinity.dorsai.org/Naturopathic.Physician
Information on licensing; referrals

Homeopathic Academy of Naturopathic Physicians
PO Box 12488
Portland, OR 97042
(503) 795-0579
Training; publications; referrals

TRADITIONAL HEALING SYSTEMS

Other traditional systems, such as shamanism, use a variety of spiritual practices for healing, achieving well-being, and helping others.

Foundation for Shamanic Studies
PO Box 1939
Mill Valley, CA 94942
(415) 380-8282
Training; information

Institute for Traditional Medicine (ITM)
2017 SE Hawthorne Boulevard
Portland, OR 97214
Please do not phone ITM; write, enclosing self-addressed stamped envelope.
Web site: http://www.europa.com/~itm/index.html
Information on Chinese, Tibetan, Ayurvedic, Native American, and Thai traditional medicine

Professional Association of Traditional Healers (PATH)
190 E. 9th Avenue, Suite 290
Denver, CO 80206
(800) 735-PATH
Fax: (303) 830-2346
E-mail: path@holistic.com
Web site: http://www.holistic.com/path
Research and information resources

DIET, NUTRITION, AND PREVENTIVE MEDICINE

Nutritional supplementation is used to alleviate a variety of health problems, maintain physical and psychological health, promote longevity, and prevent chronic complaints.

American College for Advancement in Medicine (ACAM)
PO Box 3427
Laguna Hills, CA 92654
(714) 583-7666

The Council for Nutrition Education and Therapy (CNEAT)
1 The Close
Halton, Aylesbury
Buckinghamshire HP22 5NJ

Health Education Authority
Hamilton House
Mabledon Place
London WC1H 9TX

Society for the Promotion of Nutritional Therapy
PO Box 47
Heathfield
East Sussex TN21 8ZX
Helpline: 01435 867007
(Send an SAE plus £1 for a copy of the register)

Dietary Therapy Society
33 Priory Gardens
London N6 5QU
0181 341 7260

Institute of Optimum Nutrition
5 Jerdan Place
Fulham
London SW6 1BE

Women's Nutritional Advisory Service
PO Box 268
Lewes
East Sussex BN7 2QN

MACROBIOTICS
A way of eating and living that recognizes the natural order in all things physical, emotional, mental, ecological and spiritual.

Jon Sandifer
Freshlands
194 Old Street
London EC1V 9FR

Worldwide:

The Kushi Institute
PO Box 7
Becket
MA 01233, USA
The Kushi Institute will provide you with a list of recognized practitioners around the world. Send an SAE.

WELL ELDERLY
Promotion of healthy ageing.

Counsel and Care for the Elderly
131 Middlesex Street
London E1 7JF

Age Concern England
Astral House
1268 London Road
London SW16 4ER

ENVIRONMENTAL MEDICINE

Practitioners and patients work together to uncover relationships between their environment and health problems such as allergies. Sometimes called clinical ecology.

British Ecological Society
Burlington House
Piccadilly
London W1V 0LQ

Society for Environmental Therapy
521 Foxhill Road
Ipswich IP3 8LW

National Society for Research into Allergies
PO Box 45
Hinckley
Leics LE10 1JY

British Society for Allergy and Environmental Medicine
The Burghwood Clinic
34 Brighton Road
Banstead
Surrey SM17 1BS

HOLISTIC HEALTH CARE (GENERAL)

Philosophy of medical care that sees the individual as a whole, seeks to integrate body, mind and spirit, and encourages personal responsibility for a total healthy lifestyle.

British Holistic Medical Association
Trust House
Royal Shrewsbury
Hospital South
Shrewsbury
Shropshire
SY3 8XF
Helpline: 01743 261155

Holistic and Creative Therapy Association
2A Burston Drive
St Albans
Herts AL2 2HR
01727 674567

Holistic Health Foundation
2 De La Hay Avenue
Plymouth
Devon PL3 4HH
Helpline: 01752 671485

British Association for Holistic Health
179 Gloucester Place
London NW1 6DX

British Complementary Medicine Association
39 Presbury Road
Cheltenham
Glos GL52 2PT
Helpline: 01242 226770

Council for Complementary and Alternative Medicine
206-8 Latimer Road
London W10 6RE

Association of Physical and Natural Therapists
68A The Avenue
Worcester Park
Surrey KT4 7HJ

MANIPULATIVE AND ENERGETIC THERAPIES

A variety of techniques are used to improve the structure and functioning of
the body and balancing of vital energy.

ACUPRESSURE

British Acupuncture Council (BAC)
Park House
206-208 Latimer Road
London W10 6RE

The Hale Clinic
7 Park Crescent
London W1N 3HE

BIOENERGETICS
A variety of body-work methods is used to help an individual become aware of bodily tensions and how to relieve them. Many also involve verbal exploration of emotional conflict.

Gerda Boyesen Centre
Acacia House
Centre Avenue
London W3 7JX

BODY WORK/MASSAGE
Therapeutic massage is the pleasurable manipulation of muscles, joints and soft tissue to alleviate pain or promote well-being. Holistic practitioners can help release emotional tension stored in the body, which may be associated with symptoms of illness. There are many styles of massage – experiment until you find the one that suits you best.

The British Massage Therapy Council
Greenbank House
65A Adelphi Street
Preston PR1 7BH

Academy of Aromatherapy and Massage
50 Cow Wynd
Falkirk
Stirlingshire FK1 1PU

London College of Massage
5 Newman Passage
London W1P 3PF

Massage Training Institute
24 Highbury Road
London N5 2DQ

BODY-WORK TECHNIQUES

There is a range of approaches based on the use of touch and manipulation that is applied to heal the body and alleviate tension.

ALEXANDER TECHNIQUE

Realignment of the body to correct postural imbalances.

Society of Teachers of the Alexander Technique

20 London House
266 Fulham Road
London SW10 9EL

Alexander Teaching Network

PO Box 53
Kendal
Cumbria LA9 4UP

APPLIED KINESIOLOGY

Muscle testing is used to identify structural, chemical and mental imbalance. Uses nutrition, manipulation, diet and exercise to restore well-being.

Association of Systematic Kinesiology

39 Browns Road
Surbiton
Surrey KT5 8ST

FELDENKRAIS

Use of movement to enhance body and mind coordination.

Feldenkrais Guild UK

PO Box 370
London N10 3XA

HELLERWORK

Holistic approach to restoring and maintaining health through deep-tissue body work, movement re-education and therapeutic dialogue.

European Hellerwork Association

c/o Roger Golten
The MacIntyre Gallery
29 Crawford Street
London W1H 1PL

POLARITY THERAPY

Bodywork, involving a light touch that focuses on the spine, is used to balance the body's energy flow. Diet and exercise may also be used as adjuncts.

UK Polarity Therapy Association

Monomark House
27 Old Gloucester Street
London WC1N 3XX

Zero Balancing Association UK
36 Richmond Road
Cambridge CB4 3PU

International Society of Polarity Therapists
Shelaymah National Healing Centre
42 Braydon Road
London N16 6QB

Polarity Therapy Educational Trust
Ashburton Hotel
79 East Street
Ashburton
Devon TQ13 7AL

REFLEXOLOGY

Reflex points on the feet (or hands) have been thought to correspond to every part of the body, a plausible notion since their sensory nerves make many contacts through the spinal cords. Precise pressure is applied to activate natural internal healing. Usually an extremely pleasurable and relaxing experience.

Irish Reflexologists Institute
4 Ruskin Park
Lisburn
Co. Antrim
Northern Ireland

The International Federation of Reflexologists
78 Edridge Road
Croydon
Surrey CR0 1EF

The British Reflexology Association
Monks Orchard
Whitbourne
Worcester WR6 5RB

The Reflexologists Society
39 Prestbury Road
Cheltenham
Gloucester GL52 2PT

The Association of Reflexologists
Flat 6
Sillwood Mansions
Sillwood Place
Brighton BN1 2LH

The Holistic Association of Reflexologists
The Holistic Healing Centre
92 Sheering Road
Old Harlow
Essex CM17 0JW

International Institute of Reflexology UK
15 Hartfield Close
Tonbridge
Kent TN10 4JP

The Scottish Institute of Reflexology
17 Cairnwell Avenue
Mastrick
Aberdeen AB2 5SH

REIKI
Gentle healing touch is used to restore the body's energy. Also a system for self-healing.

The Raphael Clinic
211 Sumatra Road
West Hampstead
London NW6

Council for Complementary and Alternative Medicine
206-8 Latimer Road
London W10 6RE

ROLFING
Manipulation of the body's myofascial system to achieve correct alignment, balance and poise.

Loan Tran
Neals Yard Therapy Rooms
Neals Yard
Covent Garden
London WC2

TRAGERWORK
Combination of body work and rhythmic exercises to alleviate the body's tensions and increase self-awareness.

Independent Professional Therapists International
8 Oldsall Road
Retford
Notts DN22 7PL

Independent Register of Manipulative Therapists Ltd
32 Lodge Drive
London N13 5JZ

CHIROPRACTIC

This long-established medical system uses spinal and joint adjustments to alleviate pain and stimulate the body's natural defence mechanisms. It has been scientifically proven to help back pain and headaches. Some practitioners have a gentle style with no 'cracking'; most emphasize mind and spirit as well as body, and offer nutritional diagnosis and advice.

The British Chiropractic Association
Equity House
29 Whitley Street
Reading RG2 0EG

British Association for Applied Chiropractic
The Old Post Office
Stratton Audley, Nr Bichester
Oxon OX6 9BA

McTimoney Chiropractic Association
PO Box 126
Oxford OX2 8RH

CRANIOSACRAL THERAPY

Manipulation of the craniosacral system, involving holding the base of the skull and neck to alleviate sensory and motor dysfunctions.

Craniosacral Association
Monomark House
27 Old Gloucester Street
London WC1N 3XX

Craniosacral Therapy Educational Trust
19 Carternowle Road
Sheffield S7 2DW

Craniosacral Therapy Association of Great Britain
8 Warren Road
Colliers Wood
London SW19 2HX

OSTEOPATHY

Uses a holistic approach to combine conventional medicine and manipulation of the musculoskeletal system to restore and maintain wellness.

General Council and Register of Osteopaths
56 London Street
Reading
Berks RG1 4SQ

British Osteopathic Association
8-10 Boston Place
London NW1 6ER

British School of Osteopathy
1-4 Suffolk Street
London SW1Y 4HG

THERAPEUTIC TOUCH
Healing practice of passing the hands lightly over a patient's body to detect energy imbalances, relieve blocks and restore balance.

Association for Therapeutic Healers
Flat 5
54-56 Neal Street
Covent Garden
London WC2

Confederation of Healing Organizations
113 High Street
Berkhamsted
Herts HP4 2DJ

The National Federation of Spiritual Healers (NFSH)
Old Manor Farm Studio
Church Street
Sunbury-on-Thames
Middlesex TW16 6RG

MEDITATION, YOGA

There are various forms of stilling the mind to allow wider awareness and clarity to unfold. Meditative techniques can be used to attain bodily control and altered states of consciousness. Yoga is a system of physical, mental and spiritual development.

TM Training/Stress Management Courses
Contact: Transcendental Meditation
Roydon Hall
East Beckham
Near Tonbridge
Kent TN12 5HN

Transcendental Meditation
Freepost
London SW1P 4YY

Ayurvedic Medical Association UK
The Hale Clinic
7 Park Crescent
London W1N 3HE

Ayurvedic Company of Great Britain
50 Penywern Road
London SW5 9XS

Yoga Therapy Centre
Royal London Homeopathic Hospital
Great Ormond Street
London WC1N 3HR

The Yoga for Health Foundation
Ickwell Bury
Biggleswade
Bedfordshire SG18 9EF

British Wheel of Yoga
1 Hamilton Place
Boston Road
Sleaford
Lincolnshire NG34 7ES

The Iyengar Yoga Institute
233A Randolph Avenue
London W9 1NL

Yoga Biomedical Trust
PO Box 140
Cambridge CB1 1PU

Life Foundation School of Therapeutics
Maristowe House
Dover Street
Bilston
West Midlands WU14 6AL

GENERAL RESOURCE AND REFERRAL LISTINGS

Institute for Complementary Medicine (ICM)
PO Box 194
London SE16 1QZ

British Holistic Medical Association
23 Harley House
Marylebone Road
London NW1 5HE

Alternative Health Information Bureau
12 Upper Station Road
Radlett
Herts WD7 3BX

The Centre for the Study of Alternative Therapies
51 Bedford Place
Southampton
Hants SO1 2DG

Council for Complementary and Alternative Medicine
Park House
106-208 Latimer Road
London W10 6RE

Green Library
9 Rickett Street
London SW6 1RU

Research Council for Complementary Medicine
60 Great Ormond Street
London WC1N 3JF

Natural Health Network
2-4 Hardwicke Road
Reigate
Surrey RH12 9HJ

To obtain additional materials on alternative and complementary medicine practitioners and other professional and resource organizations, contact:

Jackie Wootton, M.Ed.
Alternative Medicine Information
Fax: (001) 301 340-1936
E-mail:jackiew@clark.net
or visit the Web Site: http://www.clark.net/pub/AltMedInfo/

GLOSSARY

Agonist/Antagonist These are terms from pharmacology that refer to two opposing actions associated with the binding of a ligand to its receptor. In the case of a ligand that is an *agonist*, the fit between ligand and receptor is perfect, and binding is followed by transmission of a signal to the cell. With *antagonists*, another situation occurs, which, while considerably rarer, is of tremendous interest from the point of view of drug design and therapeutics. In this case, the ligand fits the receptor well enough to bind to it and to block another ligand (like an agonist), but not well enough to activate the receptor and thereby signal the cell. Typically, the antagonist is an exogenous ligand manufactured synthetically in the laboratory, although there are examples of nature having designed antagonists of her own agonist drugs. By occupying the receptor and preventing an agonist from doing so, the antagonist has the potential to block certain harmful effects. An example is the antagonist naloxone, which, when given to individuals who have overdosed on opiates, is able to nearly instantly reverse the effects of the overdose. Much of modern pharmaceutical research seeks to create antagonists to block hormone action. Tamoxifen is one such hormone antagonist, which was developed to block the action of estrogen in the body in women who have breast cancer.

Amino acids Amino acids are organic compounds that are the building blocks of proteins and the smaller peptides. The name of the acids comes from the stem word *amine*, meaning "derived from ammonia." In terms of structure, each amino acid has at least one carboxyl (COOH) group. Amino acids join together in long chains, the amino group of one amino acid linking with the carboxyl group of another. The linkage is known as a peptide bond, and a chain of amino acids is known as a polypeptide. Proteins are large, naturally occurring polypeptides.

Analog A structural derivative of a drug that often differs from it by a single element and that because of that difference may have desirable properties not present in the parent compound, such as potency, stability, or antagonist activity.

Antibody A large protein molecule secreted by a B lymphocyte. Each antibody from a given cell is unique and specific for one antigen. Collectively, the millions of different antibody-producing cells of the body provide broad ability to recognize and destroy a nearly limitless diversity of antigens. How this occurs, and how it fails in certain illnesses, was one of the big questions of molecular immunology, and is now well understood.

Antigen A substance that, when introduced into the body, is recognized by either B cells, resulting in the stimulation of antibody production, or by T cells, resulting in cellular immunity. Antigens include toxins, bacteria, foreign blood cells, and the cells of transplanted organs.

Artifact A flaw in a research experiment. Typically, an artifact comes to light when differently designed tests of a hypothesis yield conflicting results, suggesting that one of them has incorporated some kind of error.

Assay Every experimental advance requires detecting and quantifying a change in whatever system the researcher is studying. It is necessary, then, to create, often for the first time, the means to make these determinations. This research methodology is called an assay, and since the goal is that it be both accurate and reproducible, meaning that the information is correct and that others will be able to perform it, the assay may be said to be analogous to a recipe.

Atom A unit of matter, the smallest unit of an element, having all the characteristics of that element and consisting of a dense, central, positively charged nucleus surrounded by a system of electrons. Molecules are comprised of atoms.

Autonomic nervous system The two main branches of the autonomic nervous system, emanating from the spinal cord, control involuntary, unconscious actions of smooth and cardiac muscle and glands, and they act in opposition to each other. One branch is known as the sympathetic system, and the nerves controlling it are found in the thoracic and lumbar segments of the spinal cord. The sympathetic system primarily uses the neurotransmitters adrenaline and noradrenaline to mobilize the organism in a "fight or flight" reaction in emergencies. The parasympathetic system, located in the cranial and sacral segments of the spinal cord, uses the transmitter acetylcholine to relax the body.

Axon The usually long extension of a nerve fiber that generally conducts impulses away from the body of the nerve cell.

Brain stem The "lowest" and earliest-evolved of the brain centers, also known as the "reptilian brain." It sits at the base of the skull, well under the cortex, at the top of the spinal cord. It is responsible for such "autonomic" actions as breathing, excretion, and regulation of body temperature.

CCK Cholyocystokinin, a peptide secretion of the pancreas that regulates the release of digestive enzymes and the sensations of satiety.

Cell The smallest structural unit of an organism that is capable of independent functioning, consisting of one or more nuclei, cytoplasm, and various organelles, all surrounded by a semipermeable cell membrane. Cell receptors are located in this membrane, where they are available to bind with various ligands suspended in the extracellular fluid that bathes all cells and serves to transport the various nutrients, waste products, and informational substances.

Central nervous system The nervous system of higher organisms, comprised of the brain and spinal cord.

Chemokine A term that is a hybrid of "chemotactic" and "cytokine," to describe a key biological effect of these peptides, which is to cause chemotaxis of specific immune cells. Neuropeptides such as VIP, enkephalin, or Substance P, to cite a few examples, are also specific chemoattractants for selected immune cells, but are not considered to be chemokines, as the term is applied only to larger peptides containing some 70 to 80 amino acids. This yields a somewhat inaccurate nomenclature, as the term is too restrictive, a not uncommon occurrence in a rapidly expanding research area. In 1996 chemokine receptors were widely recognized for their function as HIV (i.e., virus) receptors, and the chemokine peptides were shown to block HIV replication. Some ten years earlier, neuropeptides related to VIP (*see* Peptide T) were the first peptides described as ligands for HIV receptors. These results opened new possibilities for treatments targeted at blocking viral receptors.

Chemotaxis The ability of cells, including bacteria and other unicellular organisms, to move toward a chemical stimulus. Because cells will move toward (chemotax) higher concentrations of the stimulus, its controlled release enables it to serve as a chemotactic mediator, recruiting cells to specific sites in the body where they are needed, when they are needed.

Cytokine/chemokine (interleukin, lymphokine) As recently as ten years ago, there were numerous small molecules just being identified that mediated intercellular communication among immune cells and the other cells and systems of the body. Each laboratory provided a name for the molecule it was investigating, usually based on whatever function or activity researchers were able to ascribe to it. Eventually, once the "factors" were purified, it was recognized that many of the labs were studying the same molecules. An effort was made to systematize the nomenclature, and, as the identification of these potent biological mediators remains a subject of intense research, this process continues. For example, for a while, the name "interleukin" was used to emphasize the "interleukocyte" nature of the information flow, and a "lymphokine" was the hormonal secretion of a lymphocyte. However, almost as soon as these concepts were established and set forth, it became clear that such communication neither originated solely in lymphocytes nor was confined to lymphocytes. The more general term of "cytokine" was introduced, and "chemokine" emphasizes that some cytokines cause "chemotaxis." *See also* leukocyte.

Dendrite   A branched protoplasmic extension of a nerve cell that conducts impulses from adjacent cells inward toward the cell body. A single nerve may possess many dendrites.

Endogenous   Originating or produced within an organism, a tissue, or a cell. The opposite of exogenous.

Enkephalin   The brain's own morphine (from the Greek "in the head"), a five-amino-acid peptide that binds to the opiate receptor, causing, among other things, analgesia or the euphoria associated with exercise, the "runner's high."

Enzyme   A large peptide, therefore a protein, whose function is to catalyze chemical reactions in biological systems at rates many hundreds to thousands of times faster than would be possible without it. Enzymes can both create larger molecules and break them down into smaller pieces, thereby restructuring the body fabric. But the most interesting enzymes are those that control/regulate the actions of cellular machinery.

Exogenous   Derived or developed outside the body; originating externally or, in medical usage, having a cause external to the body.

Frontal cortex/forebrain   The cortex is the outer layer of gray matter that covers the surface of the cerebral hemisphere. The frontal portion of the cortex, the most recently evolved of the brain structures, sits most forward (behind the forehead) and is present only in primates, such as ourselves. It contains neuronal centers necessary for understanding and producing language, for conceptualization and abstraction, for judgment, and for the capacity of humans to contemplate and exert control over their lives.

Glial cell   Any of the non-neuronal constituent cells of the brain or the peripheral nervous system. Glia are generally considered to support the functions of the neurons. A specialized immune cell derived from the monocyte is the microglial cell, which functions as part of the brain's immune system. The vast majority, some 90 percent, of the brain's cells are glia, not neurons.

Homeostasis   The ability or tendency of an organism or a cell to maintain internal equilibrium by adjusting its physiological processes.

Hormone   A substance, usually a peptide or steroid, produced by one tissue and conveyed by the bloodstream to another to effect a change in physiological activity, such as growth or metabolism. The same problems of nomenclature that limit the applicability of terms like neuropeptide or cytokine apply here.

Insulin A large peptide secretion of the pancreas that acts in a hormonal fashion, that is, by binding to specific receptors on other cells whose primary function is to control blood glucose levels. Insulin and related peptides are also well known for their growth-factor actions, that is, they induce and support the division of numerous cell types.

Leukocyte A white blood cell, a generic term for the lymphocytes, monocytes, and other cells of the immune and host-defense system.

LHRH One of the general class of gonadotropin hormones, this luteinizing hormone-releasing hormone promotes ovulation and egg maturation. When released in the brain, LHRH causes mating behaviors (lordosis) in small animals and probably in humans as well. It is related to the alpha-mating factor, which promotes sexual reproduction in the primitive organisms known as yeasts. This hints at an evolutionary conservation of function (behavior) uniting the simplest and the most complex of organisms.

Ligand From the Latin *ligare*, "that which binds" (same root as religion). Any of a variety of small molecules that specifically bind to a cellular receptor and in so doing convey an informational message to the cell.

Limbic system The limbic system comprises several brain structures associated with memory and emotion. This association was first observed by neurosurgeon Wilder Penfield (1891–1976) during operations on people afflicted by an unusual type of seizure during which they had vivid auditory and visual hallucinations of previously experienced events. Their experiential hallucinations could be reproduced by stimulating the surface of the temporal lobe of the brain. In the past few decades, a number of theories have been advanced as to which part of the brain controls emotion. The hypothalamus, the limbic system, and the amygdala have all been proposed as the centers of emotional expression. Such traditional formulations view only the brain as important in emotional expressivity, and as such are, from the point of view of my own research, too limited. From my perspective, the emotions are what link body and mind into bodymind.

Lymphocyte Cells formed in lymphoid tissue, such as the lymph nodes, spleen, thymus, and bone marrow, constitute between 22 and 28 percent of all white blood cells in the blood of a normal adult human being. They function in the development of immunity and include two specific types, B cells and T cells. B lymphocytes are the source of antibodies, and T lymphocytes are responsible for immunity to tumors, virally infected cells, various hypersensitivity reactions (allergies, poison ivy), or rejection of transplanted organs. A subset of T cells, known as T4 or CD4, are vulnerable to infection by the human immunodeficiency virus (HIV), the cause of AIDS. The lymphocytes as a group are responsible for the ability of the immune system to distinguish what is "self" from what is "foreign." In fact, it is now known that the immune system is programmed shortly after birth to learn what is "self," and everything else then becomes defined as "nonself," or foreign.

Midbrain The mesencephalon. The prefix *meso,* meaning "middle," describes the position on top of the brain stem, below the cortex, or outer covering. The brain stem (medulla, pons, and midbrain) houses the reticular formation, a complex structure that combines many otherwise separate sensory and motor functions. The reticular formation also influences generalized levels of consciousness, including cycles of waking and sleeping. The mesencephalon is involved with more complex functions of sensory-motor processing.

Molecule The smallest particle into which an element or a compound can be divided without changing its chemical and physical properties. A molecule is composed of several, perhaps many, atoms.

Monocyte/macrophage An immune system cell formed from a bone marrow precursor that circulates in the blood for several days before migrating into tissues throughout the body, including the brain, whereupon it matures and differentiates (that is, acquires additional functions and abilities to cause immunity) into a macrophage or microglial cell. Controlled by chemotactic peptide informational molecules, macrophages are among the cells that respond rapidly (in hours or days, rather than weeks) to trauma, injury, and infections. They play prominent roles in wound repair and healing, ingesting and digesting debris (dead cells). But by themselves they do not have the capacity to recognize specific pathogens. This is a key feature that distinguishes them from lymphocytes.

Neuron Any of the impulse-conducting cells that constitute the brain, spinal column, and nerves, consisting of a nucleated cell body with one or more dendrites and a single axon. Also called *nerve cell.* Neurons have usually been associated with the functioning of the brain, but their presence in close association with tissue immune cells is clear evidence that they mediate brain/immune interactions.

Neuropeptide Any of the nearly 100 small peptide informational substances initially described as neuronal secretions. More recent observations that lymphocytes and monocytes both secrete and respond to neuropeptides has, of course, rendered this term somewhat inaccurate, and immunologists favor terms like cytokine or chemokine, but neuroscientists still commonly refer to neuropeptides.

Neurotransmitter A chemical substance, such as acetylcholine or dopamine, that transmits nerve impulses across a synapse.

PAG/Periaqueductal Gray A brain-stem region of neurons and fibers surrounding the aqueduct (fluid-filled space) at the top of the spinal cord in the brain stem. Functionally it serves as a nodal point, enriched in peptide receptors, and processes ascending sensory information arriving to the brain from the extremities. As such it is an early way station by which pain and other perceptual thresholds may be regulated.

Peptide Any of various natural or synthetic compounds containing two or more amino acids linked by the carboxyl group of one amino acid and the amino group of another. By definition, polypeptides are the larger peptides, usually those with in excess of 100 amino acids. But they are smaller than the proteins, which may have 200 or more amino acids as well as other attached molecules, such as sugars or lipids.

Peripheral nervous system The peripheral nervous system is the system of nerves that links the brain and spinal cord—in other words, the central nervous system—to the rest of the human body. The peripheral nerves consist of the cranial nerves (12 pairs), the spinal nerves (31 pairs), and the autonomic nerves (sympathetic and parasympathetic), which are distributed to smooth muscles, cardiac muscle, and glands. The cranial nerves and the spinal nerves, sometimes referred to collectively as the craniospinal nerves, are of three types: sensory (or afferent), motor (or efferent), and mixed (containing both sensory and motor fibers). Sensory nerve fibers carry impulses from sense receptors to the central nervous system. Motor nerve fibers carry impulses from the central nervous system to the muscles and glands. All spinal nerves and most cranial nerves are mixed nerves, containing both motor and sensory fibers. The sympathetic and parasympathetic nerves control involuntary functions—breathing and heartbeat, for example.

Pitocin The synthetic version of the peptide hormone oxytocin. Used pharmacologically to induce labor.

Protein A complex organic macromolecule that is composed of one or more chains of amino acids. Proteins are fundamental components of all living cells and include many substances, such as enzymes, hormones, and antibodies, that are necessary for the proper functioning of an organism.

PNI Psychoneuroimmunology. A term coined in the early eighties to emphasize and promote research that is interdisciplinary in focus and attempts to understand how mental (psychological) function affects immunological activities mediated via traditional neuronal connections. *Neuroimmunomodulation* is another variant term in which psyche is subsumed (implied) within "neuro."

Receptor A molecule, typically a protein or group of proteins, anchored in the outer cell membrane with a site accessible to the outside environment that binds with ligands such as hormones, antigens, drugs, peptides, or neurotransmitters—all those ligands I have been referring to as "informational substances." The receptor is the key player in the communication network of the bodymind, as it is only when the receptor is occupied by the ligand that the information encoded in the informational substances can be received. It is also at the receptor that the earliest informational processing occurs, as the actual signal the receptor transduces to the cell can be modulated by the action of other receptors and their ligands, the physiology of the cell, and even past events and memories of them.

Steroids Fat-soluble (lipid) organic compounds that occur naturally throughout the plant and animal kingdoms and play many important functional roles. Steroids are quite diverse and include molecules like cholesterol, all sex hormones, and the adrenal cortical hormones (corticosteroids). Sex hormones are necessary for many aspects of reproduction and sexual function, while the adrenocortical hormones primarily affect carbohydrate and protein metabolism. The hormonal steroids act via receptors located not on the surface of the cell but deep within, in the nucleus, where they regulate the transcription of various genes. In this respect they differ from the neurotransmitters and peptide informational substances that act rapidly on receptors at the cell surface.

Synapse The junction across which a nerve impulse passes from an axon terminal to a neuron, a muscle cell, or a gland cell.

T4 receptor (CD4) A cell surface molecule that typifies certain T lymphocytes that have "helper" functions (helper cells). When the T4 molecule is activated, it signals the cell to execute its program, which consists of secreting a variety of molecules that then act on other cells to "help" them perform the actual tasks of immunity, killing virally infected cells or tumors, for example.

VIP Vasoactive intestinal peptide, a 28-amino-acid peptide, first identified from intestine extracts. It serves many functions, including acting as a growth hormone for T4 lymphocytes and for certain brain neurons. It also plays a role in digestion, penile erection, and the regulation of blood pressure.

RECOMMENDED
READING

I am used to writing in a certain scientific style that results in a format of fifty citations for each five pages of text and relentlessly documents the origin of each and every factoid. I am grateful to the publisher for allowing me the pleasure and flexibility to write this book in a more popular style that permits me to achieve the best possible communication rather than a dazzling display of the documentation of many ideas requiring hundreds of citations for complete scholarship. Alas, since I find it impossible to strike the middle ground with the occasional footnote by chapter, the book list that follows is less a bibliography (since I have referred over the years to a much longer list of books to obtain the information I write about) and more a recommended reading list of generally nontechnical books that I have found enjoyable, informative, or influential, particularly books I have read or reread recently. There are several of my scientific articles cited.

Achterberg, Jeanne. *Woman as Healer.* Boston: Shambhala, 1992.

Ader, Robert, David Felton, and Nicholas Cohen. *Psychoneuroimmunology.* 2d ed. San Diego: Academic Press, 1991.

Advances: Journal of the Institute for the Advancement of Health. (Institute for the Advancement of Health, 16 E. 53rd St., New York, NY 10022)

Bailey, Covert. *The New Fit or Fat.* Boston: Houghton Mifflin, 1991.

Benson, Herbert, with Marg Stark. *Timeless Healing: The Power and Biology of Belief.* New York: Scribner, 1996.

Bentov, Itzhak. *Stalking the Wild Pendulum: On the Mechanics of Consciousness.* New York: Bantam Books, 1979.

Berry, Linda. *Internal Cleansing: Rid Your Body of Toxins.* Rochlin, Calif.: Prima Publishing, 1997.

Borysenko, Joan. *Minding the Body, Mending the Mind*. Carlsbad, Calif.: Hay House, 1987.

Borysenko, Joan, and Miroslav Borysenko. *The Power of the Mind to Heal*. Carlsbad, Calif.: Hay House, 1994.

Brigham, Dierdre Davis. *Imagery for Getting Well: Clinical Applications of Behavioral Medicine*. New York: W. W. Norton & Company, 1994.

Brown, Melanie, with Veronica Butler, and Nancy Lonsdorf. *A Woman's Best Medicine*. New York: Jeremy P. Tarcher/Putnam, 1993.

Cannon, Walter B. *The Wisdom of the Body*. New York: W. W. Norton, 1931.

Capra, Fritjof. *The Tao of Physics: An Exploration of the Parallels Between Modern Physics and Eastern Mysticism* (revised edition). New York: Bantam, 1984.

———. *The Web of Life*. New York: Doubleday, 1996.

Chopra, Deepak. *Ageless Body, Timeless Mind: The Quantum Alternative to Growing Old*. New York: Harmony, 1993.

———. *Quantum Healing: Exploring the Frontiers of Mind/Body Medicine*. New York: Bantam, 1989.

Clifford, Terry. *Tibetan Buddhist Medicine and Psychiatry: The Diamond Healing*. York Beach, Maine: Samuel Weiser, 1984.

Coleman, Wim, and Pat Perrin. *Marilyn Ferguson's Book of PragMagic*. New York: Simon & Schuster, 1990.

Connelly, Dianne M. *All Sickness Is Home Sickness*. Columbia, Md.: Centre for Traditional Acupuncture, 1986.

Cousins, Norman. *Anatomy of an Illness*. New York: Bantam, 1981.

———. *Head First: The Biology of Hope*. New York: Dutton, 1989.

Dacher, Elliott S. *Whole Healing: A Step-by-Step Program to Reclaim Your Power to Heal*. New York: Dutton, 1996.

Darwin, Charles. *The Expression of the Emotions in Man and Animals*. New York: The Philosophical Library, 1955.

Dawkins, Richard. *The Selfish Gene*. Oxford: Oxford University Press, 1976.

Dienstfrey, Harris. *Where the Mind Meets the Body: Type A, the Relaxation Response, Psychoneuroimmunology, Biofeedback, Neuropeptides, Hypnosis, Imagery—and the Search for the Mind's Effects on Physical Health*. New York: HarperCollins, 1991.

Dossey, Larry. *Healing Words: The Power of Prayer and the Practice of Medicine*. San Francisco: HarperCollins, 1993.

———. *Prayer Is Good Medicine: How to Reap the Healing Benefits of Prayer*. San Francisco: HarperCollins, 1996.

Dychtwald, Ken. *Bodymind*. New York: Putnam, 1977.

Eddy, Mary Baker. *Science and Health with a Key to the Scriptures*. Boston: First Church of Christ, Scientist, 1934.

Ekman, Paul, and Richard J. Davidson, eds. *The Nature of Emotion: Fundamental Questions.* New York: Oxford University Press, 1994.

Epstein, Donald M., with Nathaniel Altma. *The Twelve Stages of Healing: A Network Approach to Wholeness.* Novato/San Rafael, Calif.: Amber-Allen/New World Library, 1994.

Fields, Rick, with Peggy Taylor, Rex Weyler, and Rick Ingrasci. *Chop Wood Carry Water: A Guide to Finding Spiritual Fulfillment in Everyday Life.* Los Angeles: Jeremy P. Tarcher, 1984.

The Gawler Foundation. *Mind, Immunity and Health. Workbook of the 3rd International Conference.* Victoria, Australia, 1997.

———. *The Mind-Body Connection. Workbook of the 2nd International Conference.* Victoria, Australia, 1996.

Gordon, James S. *Manifesto for a New Medicine: Your Guide to Healing Partnerships and the Wise Use of Alternative Therapies.* New York: Addison-Wesley, 1996.

Green, Elmer and Alyce. *Beyond Biofeedback.* New York: Delta, 1977.

Harmon, Willis. *Global Mind Change: The Promise of the Last Years of the Twentieth Century.* San Francisco: Institute of Noetic Sciences, 1988.

Hirshberg, Caryle, and Marc Ian Barasch. *Remarkable Recoveries.* New York: Putnam, 1995.

The Institute of Noetic Sciences with William Poole. *The Heart of Healing.* Atlanta: Turner Publishing, 1993.

Jahnke, Roger. *The Healer Within.* San Francisco: HarperCollins, 1997.

Janov, Arthur. *Why You Get Sick, How You Get Well: The Healing Power of Feelings.* West Hollywood: Dove, 1996.

Johnson, Don Hanlon. *Body, Spirit and Democracy.* Berkeley: North Atlantic Books, 1993.

Judd, Naomi. *Love Can Build a Bridge.* New York: Villard Books, 1993.

Jung, C. G. *Memories, Dreams, Recollections.* New York: Pantheon Books, 1963.

Justice, Blair. *Who Gets Sick: How Beliefs, Moods, and Thoughts Affect Your Health.* New York: Putnam, 1987.

Kabat-Zinn, Jon. *Full Catastrophe Living: The Relaxation and Stress Reduction Program of the University of Massachusetts Medical Center.* New York: Delacorte Press, 1990.

———. *Wherever You Go, There You Are.* New York: Hyperion, 1994.

Kanigel, Robert. *Apprentice to Genius: The Making of a Scientific Dynasty.* Baltimore: The Johns Hopkins University Press, 1986, 1993.

Knaster, Mirka. *Discovering the Body's Wisdom.* New York: Bantam, 1996.

Krippner, Stanley, and Patrick Welch. *The Spiritual Dimensions of Healing.* New York: Irvington, 1992.

Kuhn, T. S. *The Structure of Scientific Revolutions*. Chicago: University of Chicago Press, 1970.

Kunz, Dora, comp. *Spiritual Healing: Doctors Examine Therapeutic Touch and Other Holistic Treatments*. Wheaton, Ill.: Quest, 1995.

LeDoux, Joseph. *The Emotional Brain*. New York: Simon & Schuster, 1996.

Lee, John R., with Virginia Hopkins. *What Your Doctor May Not Tell You About Menopause*. New York: Warner, 1996.

Locke, Steven, and Douglas Collinga. *The Healer Within*. New York: Dutton, 1986.

Love, Susan. *Dr. Susan Love's Book of Hormones: Making Informed Choices About Menopause*. New York: Random House, 1997.

MacLean, Paul D. *The Triune Brain in Evolution: Role in Paleocerebral Functions*. New York: Plenum Publishing Corporation, 1990.

Moss, Richard. *The Second Miracle: Intimacy, Spirituality, and Conscious Relationships*. Berkeley: Celestial Arts, 1995.

Moyers, Bill. *Healing and the Mind*. New York: Doubleday, 1993.

Murray, Michael T. *Natural Alternatives to Over-the-Counter and Prescription Drugs*. New York: William Morrow and Company, 1994.

Northrup, Christiane. *Women's Bodies, Women's Wisdom: Creating Physical and Emotional Health and Healing*. New York: Bantam, 1994.

Oppenheim, Janet. *Shattered Nerves: Doctors, Patients and Depression in Victorian England*. New York: Oxford University Press, 1991.

Ornish, Dean. *Dr. Dean Ornish's Program for Reversing Heart Disease*. New York: Random House, 1990.

———. *Eat More, Weigh Less: Dr. Dean Ornish's Life Choice Program for Losing Weight Safely While Eating Abundantly*. New York: HarperCollins, 1993.

Pelletier, Kenneth. *Sound Mind, Sound Body: A New Model for Lifelong Health*. New York: Simon & Schuster, 1994.

Rossi, Ernest Laurence. *Dreams and the Growth of Personality: Expanding Awareness in Psychotherapy*. New York: Brunnes/Mazel, 1972, 1985.

———. *The Psychobiology of Mind/Body Healing: New Concepts of Therapeutic Hypnosis*. New York: A Norton Professional Book, 1993.

Rossi, Ernest Laurence, ed. *The Collected Papers of Milton H. Erickson on Hypnosis, Vol. II*. New York: Irvington Publishers, 1980, 1989.

Rubik, Beverly. *The Interrelationship Between Mind and Matter*. Philadelphia: The Center for Frontier Sciences, Temple University, 1989.

Siegal, Bernie. *Love, Medicine and Miracles: Lessons Learned About Self-Healing from a Surgeon's Experience with Exceptional Patients*. New York: HarperCollins, 1986.

Simonton, O. Carl, and Reid Hudson. *The Healing Journey*. New York: Bantam, 1994.

Simonton, O. Carl, Stephanie Matthews-Simonton, and James L. Creighton. *Getting Well Again.* New York: Bantam, 1992.

Snyder, Solomon H. *Drugs and the Brain.* New York: Scientific American Library, 1996.

Temoshok, Lydia, and Henry Dreher. *The Type C Connection.* New York: Random House, 1992.

Thondup, Tulku. *The Healing Power of Mind: Simple Meditation Exercises for Health, Well-Being and Enlightenment.* Boston: Shambhala, 1996.

Upledger, John. *Somato-Emotional Release and Beyond.* Palm Beach Gardens, Fl.: UI Publishing, 1990.

Wallace, Robert Keith. *The Physiology of Consciousness.* Iowa: Maharishi International University Press, 1993.

Weil, Andrew. *Spontaneous Healing: How to Discover and Enhance Your Body's Natural Ability to Maintain and Heal Itself.* New York: Knopf, 1995.

Wieland, Theodor, and M. Bodanszky. *The World of Peptides: A Brief History of Peptide Chemistry.* New York: Springer-Verlag, 1963.

Wilson, Edward O. *On Human Nature.* Cambridge, Mass.: Harvard University Press, 1978.

Zukov, Gary. *The Dancing Wu Li Masters: An Overview of the New Physics.* New York: Bantam, 1979.

SELECTED SCIENTIFIC ARTICLES

Pert, Candace B., and Solomon H. Snyder. "Opiate Receptor: Demonstration in Nervous Tissue." *Science* 179 (March 1973).

Pert, Candace B., et al. "Neuropeptides and Their Receptors: A Psychosomatic Network." *Journal of Immunology* 135, no. 2 (August 1985).

Pert, Candace B. "The Wisdom of the Receptors." *Advances* 3, no. 3 (Summer 1986). Also appeared in *Noetic Sciences Review* (Spring 1987), Institute of Noetic Sciences, Sausalito, Calif.

Pert, Candace B., et. al. "Octapeptides deduced from the neuropeptide receptor-like pattern of antigen T4 in brain potently inhibit human immunodeficiency virus receptor binding and T-cell infectivity." *Proceedings of the National Academy of Sciences USA* 83 (1986): 9254–9258.

Ruff, Michael, and Candace B. Pert. "Small cell carcinoma of the lung: macrophage-specific antigens suggest hemopoietic stem cell origin." *Science* 225 (1984): 1034–1036.

Ruff, Michael, and Candace B. Pert. "Neuropeptides Are Chemoattractants for Human Monocytes and Tumor Cells: A Basis for Mind-Body Communication." In *Enkephalins and Endorphins: Stress and the Immune System,* edited by Nicholas P. Plotnikoff, et al. New York: Plenum Publishing Corporation, 1986.

INDEX